Additive Manufacturing for Advanced Applications

The book discusses the latest trends such as 4D printing, wire arc additive manufacturing (WAAM), direct energy deposition, and topological optimization in additive manufacturing (AM), and its compliance with the ASTM/ISO standards. It further explains materials for AM and the development of novel future materials.

The focus of this book is to cover the fundamentals, principles, selection of material and equipment, and applications of AM in a comprehensive manner. It will showcase information about the effective utilization of AM for advanced applications in diverse areas such as biomedical, aerospace, automobile, defence, and reverse engineering. The other main features are:

- Covers a comprehensive discussion on the theoretical aspects of AM such as sintering, diffusion, and photopolymerization.
- Showcases applications of AM in diverse fields including aerospace engineering, automotive engineering, biomedical engineering, and reverse engineering.
- Presents case studies to showcase real-time problems and solutions using AM.
- Includes pedagogical features such as algorithms, exercises, and case studies.

The text is primarily written for senior undergraduate, graduate students, and academic researchers in the fields of manufacturing engineering, industrial engineering, production engineering, mechanical engineering, and aerospace engineering.

Additive Manufacturing for Advanced Applications
Technologies, Challenges and Case Studies

Pawan Sharma
Vishvesh Badheka

CRC Press
Taylor & Francis Group
Boca Raton London New York

CRC Press is an imprint of the
Taylor & Francis Group, an **informa** business

First edition published 2024
by CRC Press
2385 NW Executive Center Drive, Suite 320, Boca Raton FL 33431

and by CRC Press
4 Park Square, Milton Park, Abingdon, Oxon, OX14 4RN

CRC Press is an imprint of Taylor & Francis Group, LLC

© 2024 selection and editorial matter, Pawan Sharma and Vishvesh J. Badheka; individual chapters, the contributors

ISBN: 978-1-032-48094-7 (hbk)
ISBN: 978-1-032-77683-5 (pbk)
ISBN: 978-1-003-48432-5 (ebk)

DOI: 10.1201/9781003484325

Typeset in Times
by MPS Limited, Dehradun

Contents

Editor biographies

Dr. Pawan Sharma completed his B.Tech. degree from Uttar Pradesh Technical University in 2008, securing first position, and got a master's degree from Motilal Nehru National Institute of Technology (MNNIT) Allahabad, with a gold medal in 2013, in production engineering specialization. He earned his Ph.D. in the area of additive manufacturing/3D printing from IIT Delhi in the year 2019.

He has more than eight years of teaching experience and is currently working as Assistant Professor in the Department of Mechanical Engineering, IIT (BHU) Varanasi. Dr. Sharma diversified his research areas in the fields of digital manufacturing, advanced machining, and powder metallurgy and also continued working in the area of additive manufacturing. The key highlight of his research career includes the development of a rapid tooling process using 3D printing and pressureless microwave sintering, the development of an ultrasonic-assisted pressureless sintering process, and the development of topologically ordered iron scaffolds. He has a significant number of research articles in international journals and international/national refereed conferences to his credit. He also has two Indian patents granted in the area of additive manufacturing processes. Currently, he is coordinating the DRDO-DIA Centre of Excellence on Powder Metallurgy at IIT (BHU) and has received research grants from different funding agencies in India.

Dr. Vishvesh Badheka studied metallurgical engineering at the M.S. University of Baroda, India, and earned bachelor's, master's, and a Ph.D. in metallurgical engineering. He qualified for Diploma International Welding Engineer, awarded by International Institute of Welding in December 2011.

Presently, he is working as Professor of Mechanical Engineering at the School of Technology and Dean of Academic Affairs, Pandit Deendayal Energy University (PDEU), Gandhinagar, India. He received funding from various funding agencies including DST-Young Scientist (twice), ISRO, DRDO, and DAE for his research in the area of advanced welding processes, which includes narrow gap welding, using metal cored and flux cored wires, friction stir welding, similar and dissimilar metal, friction stir processing for surface composites and superplasticity, friction welding of pipe for dissimilar metal combinations, and ATIG of reduced activation martensitic steels and 9Cr-1Mo steels. So far he has completed eight sponsored R & D projects and guided 40 M. Tech & 13 Ph.D. research projects to his credit in the area of advanced welding processes. He has published 100+ international journal publications and presented at 60 national and international conferences. He is a member of various professional bodies, including IIW, ISNT, ISTE, IIM, IIF, MRSI, and AWS.

Academic Awards: Dr Vikram Sarabhai Award 2009–10, for "Friction Stir Welding of Aluminum and its alloys", from Gujarat Council on Science and

Technology (GUJCOST), Government of Gujarat; Bronze Medal by the Center for Innovation, Incubation and Entrepreneurship (CIIE) of Indian Institute of Management, Ahmedabad (IIM); several Best Oral Presentation and Best Research Paper awards by the Indian Institute of Welding; National Technical Teacher's Award by AICTE 2022, in recognition of teaching excellence in technical education; Teachers Associateship for Research Excellence (TARE), DST, 2022.

Contributors

Dr. Vishvesh Badheka
Professor
Department of Mechanical Engineering
School of Technology
Pandit Deendayal Energy University
Gandhinagar, Gujarat, India

Neha Choudhary
Ph.D. Research scholar
Additive and Subtractive
    Manufacturing lab
Department of Mechanical and
    Industrial Engineering
IIT Roorkee
Roorkee, India

Debashish Gogoi
Research Scholar
SRM University A.P
Amaravati, India

Vipin Goyal
Ph.D. Research scholar
Department of Mechanical Engineering
Indian Institute of Technology Indore
Indore, India

Dr. Rudranarayan Kandi
Assistant Professor
Department of Mechanical Engineering
National Institute of Technology
    Rourkela
Sundargarh, Odisha, India

Tanyu Donarld Kongnyui
Student
SRM University A.P
Amaravati, India

Ajit Kumar
Post Doc Fellow
Department of Mechanical Engineering
Indian Institute of Science Bangalore
Bangalore, India

Dr. Manjesh Kumar
Assistant Professor
SRM University A.P
Amaravati, India

Dr. Dipesh Kumar Mishra
Department of Mechanical Engineering
Graphic Era (Deemed to be University)
Dehradun, Uttarakhand, India

Dr. J. P. Misra
Assistant Professor
Department of Mechanical Engineering
IIT (BHU)
Varanasi, India

Prof. Pulak Mohan Pandey
Professor
Department of Mechanical Engineering
Indian Institute of Technology Delhi
Hauz Khas, Delhi, India

Aarya Patel
Master Student
Cornell University
Ithaca, United States

Bhumi K. Patel
Deputy Manager
Product Engineering Department
MG Motors
Vadodra, India

Falak Patel
Master Student
Cornell University
Ithaca, United States

Thira Patel
Master Student in Robotics Engineering
Worcester Polytechnic Institute
Worcester, United States

Dr. Pawan Sharma
Assistant Professor
Department of Mechanical Engineering
Indian Institute of Technology (BHU)
Varanasi, India

Dr. Varun Sharma
Assistant Professor
Department of Mechanical and
   Industrial Engineering
IIT Roorkee
Roorkee, India

Aswani Kumar Singh
Ph.D. Research scholar
Additive and Subtractive
   Manufacturing lab
Department of Mechanical and
   Industrial Engineering
IIT Roorkee
Roorkee, India

Dr. Jasvinder Singh
Assistant Professor
Punjab Engineering College
Chandigarh, India

R. Suryanarayanan
Research and Design Engineer
Foxx Life Sciences
Hyderabad, India

Ankit Tripathi
Research Scholar
Department of Mechanical Engineering
Indian Institute of Technology, (BHU)
Varanasi, India

Prof. R. Tyagi
Professor
Department of Mechanical Engineering
IIT (BHU)
Varanasi, India

Dr. Girish C. Verma
Assistant Professor
Department of Mechanical Engineering
Indian Institute of Technology Indore
Indore, India

L. K. Yadav
Research Scholar
Department of Mechanical Engineering
IIT (BHU)
Varanasi, India

1 Materials for Additive Manufacturing
Polymer, Ceramic, Metal, and Smart Materials

Neha Choudhary, Aswani Kumar Singh, and Varun Sharma

1.1 INTRODUCTION

Additive manufacturing (AM) techniques are emerging as a new manufacturing revolution, allowing more freedom in designing and fabricating unique products with complex geometries. Traditional manufacturing (TM) has been completely transformed by AM technology. This technology bridges the gap between idea conceptualization and product development [1]. The 3D modeling of an object through computer-aided design, followed by its slicing into numerous thin layers, creates the digital input file of the object to transfer into AM printers. Furthermore, the controlled deposition of material from AM printers constructs the objects layer by layer. AM has evolved from primarily being a prototyping technique to producing functional parts. AM offers various applications in aerospace, defense, medical, fashion, sports, automobile, electronics, etc. [2]. These applications use different materials in various AM techniques. Materials are essential for extensive knowledge of AM processes. The material must have acceptable service properties for any AM technology to function properly. A diverse range of raw materials is being used, and significant research is also being conducted to develop newer materials for expanding the AM applications.

Based on the raw materials used, AM processes are classified into three major categories: Solid, liquid, and powder-based raw materials. From the early 1980s, various AM techniques were commercially established like Stereolithography (SLA), Fused Deposition Modeling (FDM), Selective Laser Sintering (SLS), Digital Light Processing (DLP), Laminated Object Manufacturing (LOM), Selective Laser Melting (SLM), Direct Metal Laser Sintering (DMLS), Wire Arc Additive Manufacturing (WAAM), etc. [3]. Among these techniques, SLA was the first commercially available AM technique using liquid polymeric resin. A poly jet technique was also developed, combining SLA and 3D printing features to deposit photosensitive polymeric resin (or wax) [4]. Later, in the 1980s, FDM was developed using solid polymer materials in filaments. This shows that during the

DOI: 10.1201/9781003484325-1

1

origin phase of AM, polymers, laminates of paper, and waxes were among the raw materials used in the solid, liquid, and powder phases. With the advancements of AM technologies, other materials such as metals, composites, ceramics, and others have found use in different applications [5]. In the present era, functionally graded and responsive materials are being developed, which comes under the smart material category. As a result, processing via AM now has access to an exceptional material spectrum.

The novelty of the chapter lies in its comprehensive overview of the growth and popularity of AM over recent decades, emphasizing its extensive applicability across diverse fields. It highlights a critical factor in AM, the selection of materials, which significantly influences the performance of the fabricated parts. This chapter introduces the concept of different classes of AM materials based on phases, processes, and applications, presenting a focus on polymers, metals, ceramics, and composites. Furthermore, it acknowledges recent advancements, particularly in the realm of smart materials, offering a glimpse into evolving trends and technologies in the field of AM materials. Last, it hints at the challenges associated with these materials, setting the stage for further exploration and discussion.

1.2   ADDITIVE MANUFACTURING MATERIALS

Materials are essential for a broad understanding of AM techniques. A diverse range of materials are presently being used in various processes for a wide range of applications. The feedstock form or phase should be compatible with the used technology. The feedstock materials, such as polymer, ceramics, metals, and smart materials, are presented in the following sections.

1.2.1   State of AM Materials

The state of feedstock during any process is a vital criterion for understanding the suitability and compatibility of the process and material together. There are two states of material, i.e., solid and liquid, where solid contains powders, filaments, and sheets [6]. A group of AM techniques, such as FDM, WAAM, and LOM uses filaments and sheets, respectively, of metals, polymers, etc. Another group of processes like SLA, DLP, and some extrusion-based uses liquid resins and inks to print objects with different applications. The powder-based AM materials (polymers, metals, and composites) were sintered or melted during SLS, SLM, DMLS, etc. Generally, the processing of polymers is easy in comparison to metals and ceramics due to their lower processing temperatures. In addition to this, liquid materials provide better printing and finishing. Therefore, it is necessary to understand the type of material to be used in different AM processes.

1.2.2   Polymer

Polymers are the most commonly used materials for AM due to their easy availability and lower cost. Polymers can be processed by approximately all fusion-based AM processes with careful selection of method and polymer.

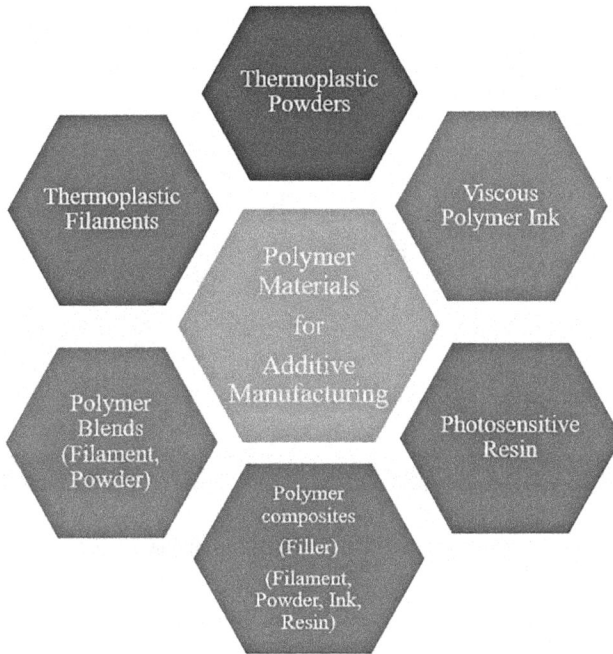

FIGURE 1.1 Polymers for polymeric AM processes.

However, vat photopolymerization, material extrusion, direct ink writing, and material jetting are major AM techniques that frequently use polymer materials in different phases. Polymers have lower processing temperatures, improved chemical stability, and desirable flowability in molten and semi-molten states [7]. Some key applications of polymers in AM include the fabrication of plastics, polymeric-graded materials, and polymer matrix composites. Polymeric materials can be processed in any physical state (liquid, powder, sheets, or filaments). The primary types of polymers used in AM are thermoplastic, thermosetting, or UV-curable polymers and their composites [6], as shown in Figure 1.1.

1.2.2.1 Thermoplastic Polymers

Thermoplastic polymers soften at some high temperature and solidify on cooling or at room temperature. Polylactic acid (PLA), Acrylonitrile Butadiene Styrene (ABS), Polyamide (PA), and nylon are commonly available thermoplastic polymers. Thermoplastic polymers are generally used in material extrusion and powder bed fusion (PBF)-based AM processes due to their adjustable nature during the heating-cooling cycle.

1.2.2.1.1 Thermoplastic Polymers Employed in Material Extrusion

FDM is the most common material extrusion-based AM process that uses solid filament material of 1.75 mm or 2.8 mm diameter. The material flows through the nozzle upon heating and then solidifies. Thermoplastic polymers that are amorphous

PRICE & PERFORMANCE

PRODUCTION VOLUME

PSU
PPS
PEI
PES
PEEK PVDF PEKK PPSU
PAI

PET PBT PCL PC
PA6 PA12 PA11 PVA
POM
PLA PLGA TPU ASA
TPE

SAN PVC
PEVA PS
PP ABS HIPS PMMA
UHMWPE

Semi-crystalline   Amorphous

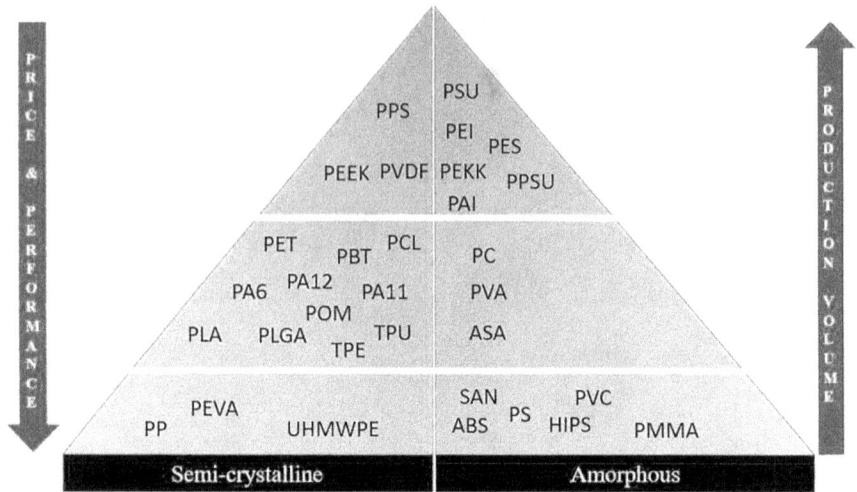

FIGURE 1.2 FDM filament material (commercially available and lab research).

in nature are suitable for the extrusion process because of their low processing temperatures (glass transition and melting temperature) and thermal expansion coefficient. The viscosity of amorphous thermoplastic polymers variates gradually changes with temperature [8]. By modulating pressure and nozzle's temperature, the material's extrusion can be precisely controlled.

In contrast, the fusion of crystalline polymers results in a distinct phase transition when heated. It is challenging to control a homogeneous extrusion process in this situation. A wide range of polymers such as Polycarbonate (PC), Polycaprolactone (PCL), PA, High-Density Polyethylene (HDPE), PLA, ABS, and biopolymers, and their blends of two polymers have been employed in the extrusion-based AM technique for different applications (as shown in Figure 1.2). ABS and PLA are the most commonly used thermoplastic polymers.

1.2.2.1.1.1 Acrylonitrile Butadiene Styrene ABS is popular primarily due to its accessibility and advantageous mechanical characteristics. ABS combines Acrylonitrile, Butadiene, and Styrene monomers produced by polymerizing acrylonitrile and styrene in the presence of polybutadiene, making its production simple and affordable. The variation of these monomers, including 15–35% acrylonitrile, 40–60% styrene, and 5–30% butadiene, are used in various combinations [9]. ABS is an amorphous solid with a 105°C glass transition temperature but no true melting point. It is typically used between 20 and 80°C. ABS is a recyclable material but produces voids caused by the development of volatile substances.

1.2.2.1.1.2 Polylactic Acid PLA is a biodegradable and biocompatible material derived from natural resources such as sugarcane, corn starch, and tapioca roots. It comprises two main monomers, cyclic di-ester, lactide, and lactide acid, and is made by ring-opening polymerization of lactide. When compared to traditional polymers, PLA has superior mechanical properties. PLA is a semi-crystalline and amorphous solid with a melting temperature of 180°C and a glass transition temperature of 55°C.

1.2.2.1.1.3 Polycarbonate PC is also a thermoplastic polymer with the largest consumption at present. It has high impact strength, low shrinkage, non-toxic, and good flame-retardant properties. PC is produced using polycondensation of bisphenol A and phosgene. The glass transition temperature of PC is above 145°C and starts flowing above 155°C. It has difficulty adhering to the build plate, so the platform temperature is kept a bit high, which is in the range of 80–120°C. It is applied in optical glass production, bulletproof windows, motorcycle helmets, etc.

1.2.2.1.1.4 Polyethylene Terephthalate Glycol Polyethylene Terephthalate Glycol (PETG) thermoplastic polymer has good strength and toughness. It is developed by combining polyethylene terephthalate with glycol, improving its wear and corrosion resistance. PETG extrusion temperature is 250–260°C, has a glass transition temperature of 85°C, and melts at 260°C. PETG has good impact strength, resistance to shock, and flexibility. It is suitable for fabricating prototypes, jigs, fixtures, and food containers.

1.2.2.1.1.5 Polycaprolactone PCL is a semi-crystalline polymer being used in extrusion through setup modifications. It is produced by cationic or anionic ring-opening polymerization of ε-caprolactone in the presence of a suitable catalyst such as dibutyltin dilaurate. It has a low melting point of about 60°C. PCL has biomedical applications such as fabricating scaffolds of different geometries and porosities. A small porous PCL block has been used to create biomimetic meniscus scaffolds that mimic type I radial and circumferential structures.

The overhanging design of the parts requires support material along with the build material, which must be simple to remove once the part is complete. Polyvinyl alcohol (PVA) and High Impact Polystyrene (HIPS) are the two most commonly used support structures [9]. These materials are easily available and have low cost. PVA and HIPS have good solubility in water and d-limonene, respectively. Other than this, Stratasys created a number of support materials, including SR-20TM for ABS and PC-ABS, SR-30TM for ABS, SR-100TM for PC, and SR-110TM for nylon [10]. Different types of polymers and their properties are listed in Table 1.1.

1.2.2.1.1.6 Polymer Blends and Composites In extrusion-based processes, the composite feedstock was developed prior to deposition. Blending is the process of combining two or more polymers to obtain unique characteristics. The commercially available UltemTM 9085 is an example of a blended polymer. It is a mixture of polyetherimide (PEI) resin and PC where both polymers have different glass transition temperatures, i.e., PEI has T_g of 216°C and PC has 147°C T_g [11]. The obtained blend has a lower T_g = 186°C and has aerospace application. In addition to having acceptable mechanical properties like high strength and stiffness, UltemTM 9085 exhibits improved ductility, flowability, and flame-retardant properties. It also has good chemical resistance to a variety

TABLE 1.1
FDM polymer filament properties

Property/Material	Density (g/cm³)	Glass Transition Temp. (°C)	Tensile strength (MPa)	Tensile Modulus (GPa)	Elongation at break (%)	Flexural strength (MPa)	Flexural modulus (GPa)	Izod impact strength (J/m)
PLA	1.264	63	48	3.039	2.5	84	2.93	27
ABS+	1.17–1.18	47–53	55–60	2.6–3.0	25–40	65–75	1.7–2.2	90–115
PP	0.89–0.92	140–150	26–41.4	0.95–1.776	100–650	55.2	0.83–1.73	21.4–267
PC	1.2	140	35	2.1	1.8	75	1.88	25
ULTEM	1.27	180	58	1.98	1.9	60	1.93	2.98
TPU 95 A	1.28–1.66	−50–0	28–96	0.621–5.50	10.0–86.0	19–95.1	0.520–4.5	214
PEI	0.050–1.9	220	1.0–650	.00280–56	0.500–110	17.9–770	0.00896–46	2.98–200
PEEK	1.26–1.61	145	50.3–265	2.14–7.58	1.50–110	86.2–380	2.48–24.0	21.4–107
ABS/PC	1.11	105.33	36.5	1.99	4.7	61.9	1.86	241

of solvents. Similarly, a PC blend with ABS also has high impact strength, toughness, heat resistance, and optical transparency.

Besides polymer blends, nano, micro, and fibrous materials have also been incorporated into the polymer matrix. Additives like surfactants and plasticizers are also frequently used in feedstocks, where plasticizers enhance rheology and surfactants alter the dispersion character of the developed composite. In order to produce an extrudate with a low enough viscosity to deposit and provide strength throughout the part build time, reinforcements to polymer feedstock must be of suitable composition. For instance, incorporating conductive particles like carbon-based fillers and metal particles like iron and copper in thermoplastic polymers for extruding electrically conductive filaments is one strategy for fabricated 3D-printed electrical devices [11]. The non-traditional bio-fabrication of PLA polymer with the ceramics using the FDM system has been performed to investigate the effect of different bio-ceramics on mechanical and tribological performance. The group also compared the PLA/alumina and PLA/alumina hydroxyapatite-coated scaffolds for mechanical and biological properties [12]. Fiber-reinforced polymers, generally carbon or glass fibers, differ in mechanical strength depending on fiber orientation, size, and matrix-fiber interface design. The fibers of length <50 μm are whiskers, >50 μm are short, and continuous is the full length of the component. The ceramic polymeric composite usually contains ~40% ceramics part in the FDM process [13]. Adding particles to pure polymer caused nozzle clogging; therefore, ideal nozzle sizes and the right amount of wt.% of particles provide improved adhesion depending on the filament material and print parameters.

1.2.2.1.2 Thermoplastic Polymers for Powder Bed Fusion
AM's PBF process uses an IR laser or UV heat source to fuse polymers, mostly semi-crystalline powder polymers. The powder fuses together due to the heat above their glass transition and melting temperature for amorphous and semi-crystalline polymers, respectively. The polymer softens as the temperature rises and forms a neck at the point when two particles come in contact. Furthermore, the neck grows due to surface tension, and two particles gradually merge into a single sphere. The CO_2 lasers are the most common heat source used in PBF techniques [14]. There are various polymers, including semi-crystalline thermoplastic, such as PA, polyaryletherke-tone (PAEK), polyethylene (PE), polybutylene terephthalate (PBT); and amorphous thermoplastics such as PC, polystyrene (PS), and elasto-mers. Additionally, some biocompatible polymers such as PCL, PLA, polyglycolide (PGA), etc. [15] can also be processed by the PBF technique, as shown in Figure 1.3.

1.2.2.1.2.1 Polyamide The most popular PBF polymer includes PA 11 and PA 12. Generally, PAs have a glass transition temperature of around 40°C. PA 11 and PA12's melting temperatures range between 183–190°C and 176–188°C, respectively. However, compared to PA11 and PA12, PA 6 has a higher melting temperature of 223°C [16]. PA 11 has a narrower sintering window than PA12, resulting in higher distortion due to significant thermal defects upon solidification. It experiences re-crystallization and degradation during pre-heating, resulting in a

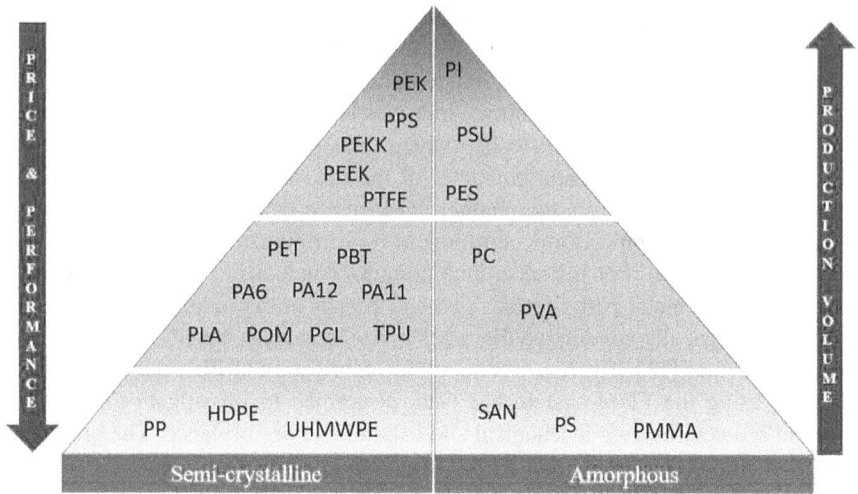

FIGURE 1.3   Polymer powders for SLS (commercially available and lab research).

highly ordered crystalline phase. PAs are liable to thermal cycles, residual stresses induced, and microstructural changes. The condensation of carboxylic groups and active amine in PAs over a number of thermal cycles results in molecular chain growth, which raises melting point and melt viscosity and degrades powder properties [17].

1.2.2.1.2.2 Polyaryletherke-Tone   PAEKs are another family of superior polymers being used in the SLS process. PAEKs include PEK, PEEK, and PEKK polymer materials. PEK and PEEK are semi-crystalline in nature, providing good thermal and chemical resistance. Various works investigated these polymers through their flowability, viscosity, and mechanical and thermal properties. The thermal investigation showed that PEK exhibits T_g and T_m of 164°C and 372°C, respectively, and PEEK has lower critical points, i.e., T_g and T_m of 143°C and 343°C. T_g of PEEK varies from 170–310°C, providing a wider sintering window than PEK [7]. PEKK has good biocompatibility, stiffness, and density. It does not react much and shows excellent chemical resistance.

1.2.2.1.2.3 Polycaprolactone   PCL is a non-toxic, biodegradable polymer. It is a semi-crystalline polymer with a melting range of 58–63°C and a glass transition temperature of 60°C is produced by the ring-open polymerization of -caprolactone in PCL. PCL comprises five non-polar methylene groups and one polar easter group, which can degrade in the presence of a hydrolytically unstable aliphatic-ester linkage. The mechanical properties of PCL can be tailored by changing its composition. It has applications in bone and cartilage repair.

1.2.2.1.2.4 Polypropylene   PP has chemical resistance properties with acids, bases, mineral oils, and alcohols. It has a corrosion resistance application.

Additionally, it is biocompatible, and its components can be used in the fabrication of permanent surgical implants.

1.2.2.1.2.5 Poly Alpha-Hydroxy Acid Poly alpha-hydroxy acid (PLAGAs) is derived from glycolic and lactic acid; high molecular weight PE (HMWPE) also has good biocompatibility and medical application.

PS and PC are the amorphous thermoplastic polymers being used as feedstock in indirect laser sintering. PS is an economic polymer with a high melting point temperature above 270°C. It has a low dielectric constant, gas permeability, mechanical ductility, and good chemical resistance. PC holds high impact strength and thermal and dimensional stability. It has poor ductility and elasticity, but its composite showed improved toughness and elastic properties useful for industrial purposes [18].

Elastomers such as thermoplastic polyurethane (TPU) can also be printed using powder-based fusion AM technique. TPU consists of a soft segment of polyether and a hard segment of a short-chain diol to an isocyanate. Various researchers studied the laser sintering of TPUs and investigated the flow of powder, and volume change upon melting and joining. The un-sintered TPU can be reused, as the powders do not age after one use [7].

1.2.2.1.2.6 Polymer Composites Employed in Powder Bed Fusion The polymer blends or reinforcements have been used to develop the composite to be used in the PBF AM technique. Reinforcements into the polymer matrix can be categorized as bio-ceramics, glass fillers, carbon-based fillers, and metallic fillers. Incorporating bio-ceramics such as HA in PCL [19,20], CAP in PLLA, and PEEK [21] improved the bioactivity of the fabricated scaffolds. The nano-sized alumina particles have been used as fillers to reinforce PS to prepare a scaffold through SLS. The scaffold's impact was improved by 50%, and its tensile strength was enhanced by 300% [22]. On the other hand, a carbon-based filler, such as CNT, graphite, or graphene, was printed, as shown in Figure 1.4, improving polymeric's mobility and light-absorbing ability [23]. So, reinforcements are generally added to improve their mechanical, electrical, and thermal properties, biocompatibility, and other properties of the printed parts.

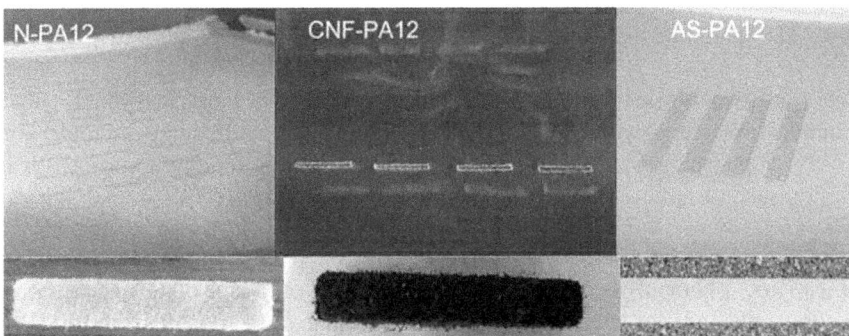

FIGURE 1.4 Powder bed of SLS (top) and printed parts (bottom) through SLS [23].

1.2.2.2 Thermosetting Polymers

Thermoset polymers exchange electrons to form new chemical bonds during curing and become irreversibly hard. These polymers include viscous inks (typically called resins, which are photocurable or thermally curable) and polymer solutions. Photopolymers are mainly composed of monomers, oligomers, and photoinitiators. A variety of other additives, such as dyes, inhibitors, toughening agents, etc., can also be added that help to tune the properties of the resin [24]. The polymer solutions are formed by dissolving polymers into a solvent to get a polymer liquid. AM techniques such as SLA and DIW use these resins and inks to process through an external irradiance source like UV light or heat. Various resins such as gelatine methacrylate, epoxy, polyelectrolyte solutions, alginate, and collagen are used as AM materials. Three types of polymerizations are used in the resin-based AM process, i.e., free radical photopolymerization, cationic photopolymerization, and hybrid photopolymerization [25].

1.2.2.2.1 Free-radical Photopolymerization

The free radical photopolymerization reaction includes three phases: chain initiation, growth, and termination. The reaction initiates with the photoinitiator's absorption of UV light of a particular wavelength and changes its state. The structure's covalent bonds undergo singlet or triplet states, resulting in a primary active free radical. The primary free radical then combines with the monomer to form the monomer free radical. The reaction is then followed by propagation or transference, which means that the generation of new free radicals via reaction with the second monomer or oligomer molecule continues [26]. The activity of the new radicals is not diminished, and they continue to combine with other reactive molecules to form chain radicals with increased activity. Finally, one of the three following mechanisms may terminate the polymerization [25]:

a. Combination of two chains to make a single lengthy chain. The molecular weight of the reproducing species makes it simple to keep track of this termination mode.
b. Radical disproportionation, where polymers are created, one with an unsaturated terminal and another with a saturated one, by abstracting hydrogen from one chain to another
c. The fusion of a photoinitiator and an active chain end.

There are two types of photoinitiators, i.e., Type 1 and Type 2. Type 1 photo-initiators are unimolecular free radical generators such as hydroxy acetophenone (HAP) or phosphineoxide (TPO). Type 2 photoinitiators require a co-initiator to trigger, such as amines or alcohol.

1.2.2.2.2 Cationic Photopolymerization

In cationic photopolymerization, when UV light of a specific wavelength is exposed, the initiator creates a cationic center that starts polymerization. Then, the initiator molecules go through a series of decomposition reactions to create super protonic acid or Lewis acid, which starts the photopolymerization [14].

The cationic component absorbs UV light, while the anionic component produces strong acid. Then, the polymerization mechanism's propagation, termination, and chain transfer steps follow a similar pattern as mentioned earlier. There are various cationic photoinitiators like diazonium salt, diaryliodonium salt, triarylsulfonium salt, alkylsulfonium salt, iron arene salt, sulfonyloxyketone, and triarylsiloxysiloxane [25].

1.2.2.2.3   Hybrid Photopolymerization

In hybrid photopolymerization, cationic photoinitiators are used with free radical photoinitiators. When exposed to UV light, the system can simultaneously produce cations and free radicals, leading to the polymerization of cationic and free radical monomers. The cationic initiators do not absorb high wavelengths of UV light, so some sensitizers are typically added. On the other hand, free radical initiators sensitize the cationic initiator by electron transfer [14].

Initially, acrylates and vinyl ether monomers were used for fabricating the parts using photopolymerization. These resins exhibited shrinkage of about 20%, which induced residual stresses and warpage of the printed parts. Furthermore, in the 1990s, epoxy resins came into existence that shrink less and have higher dimensional stability than acrylates. Epoxy resins undergo cationic photopolymerization, unlike acrylate, which undergoes free radical polymerization. These two monomers do not react with each other but form an interpenetrating polymer network when mixed. Acrylate polymerizes more broadly in the presence of epoxy, leading to higher molecular weights. (Meth)acrylates and poly(meth)acrylates or their mixture are the members of the acrylate family. Alkyl (meth)acrylates (SR 313 A and 31313) monofunctional monomers and polyfunctional monomers (SR 238, SR 350, SR 454, and SR 8335) are commercially available monomers by Sartomer Co.'s and Ageflex FM6 [14].

Other than acrylate, polyglycidyl, non-glycidyl, and epoxy phenol novolac compounds are examples of epoxy compounds. Almost always, functional groups with ring-opening reaction capabilities are employed. Oxirane-(epoxide), oxetane, tetrahydrofuran, and lactone rings are a few examples [14]. Various researchers have mentioned different resins based on the applications using SLA to fabricate 3D structures. In this context, poly (trimethylene carbonate) (PTMC) with a three-armed methacrylate PTMC monomer has been used to fabricate scaffolds for tissue engineering applications. Similarly, poly (propylene fumarate) (PPF)/diethyl fumarate was used for manufacturing biodegradable and biocompatible kidney scaffolds using a micro SLA system [27]. Another author has also used PPF as a biomaterial, DEF as the solvent, and BAPO as a photoinitiator to produce a scaffold after optimizing the process parameters such as laser exposure, penetration depth, and resin composition.

1.2.2.2.4   Composite Polymer Resin

Obtaining the desirable properties when using single thermosetting resins is very difficult. In order to mitigate this, reinforcement such as ceramics, metals (nano or micron size), and fibers has been employed in the resins. SLA has been used to process bioactive glass scaffolds [28], multi-polymeric microstructural arrays

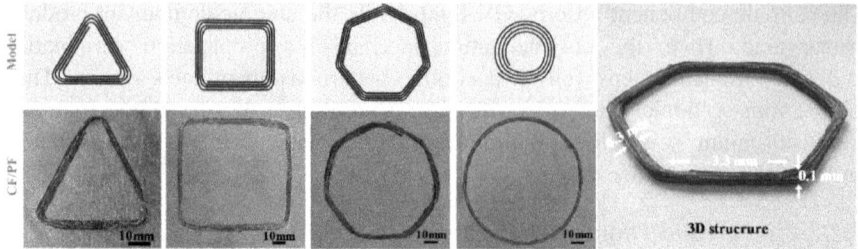

FIGURE 1.5   Carbon fiber composite printed through SLA [32].

Alessandra Vitale, and graphene oxide reinforcement into resins [29]. In order to make acrylate resin conductive, diamond particles of micron size of about 30 wt.% have been added by simply suspension. It was found through the IR camera that the composite heat sink's temperature has been increased by approx. 8°C [30]. In another work, bio-ceramics such as hydroxyapatite, calcium phosphate, and graphite have been mixed with Poly (ethylene glycol) diacrylate (PEGDA) resin and characterized for mechanical, rheological and sedimentation behavior for medical application.

Similarly, PEGDA/vitreous carbon composite with different wt.% of carbon (1–5 wt.%) has been developed and checked for SLA printing. Besides micron or nano-sized reinforcement, fibers are incorporated to enhance the mechanical strength of resins by controlling the fiber directions [31]. The continuous carbon fiber-reinforced phenolic resin composite has been developed (as shown in Figure 1.5) and printed at different pre-curing temperatures to obtain optimal mechanical strength. It was obtained that 260°C is the optimum pre-curing temperature, and the degree of curing was 80.7% after printing [32].

1.2.3   CERAMICS

Ceramics are non-metallic, inorganic materials that are typically brittle. They possess distinct physical, chemical, and mechanical properties such as high heat, wear, corrosion resistance, hardness and stiffness, and a low friction coefficient [14]. These noteworthy properties are due to their atomic arrangements and bonding, i.e., covalent and ionic bonds. Ceramics, due to their remarkable properties, has wide application in different fields like aerospace, energy, medical, military, chemical, and so on [5]. The ceramics are generally grouped into two categories: traditional and advanced, based on their nature and use. Traditional ceramics consist of clay, cement, whiteware, refractories, and glasses. The advance ceramics are broadly non-oxides and oxides [26]. The advance ceramics are generally synthetic, unlike naturally produced conventional ceramics. This section focuses mainly on advance ceramics.

1.2.3.1   Non-Oxide

Non-oxide ceramics are generally carbides, nitrides, borides, and silicates. These are thermally stable with high melting points and exhibit good chemical inertia. Due to the development of new technologies, the synthesis and fabrication of materials enhanced the application of high-performance ceramics in different sectors like engineering, electronics, medical, nuclear, and aerospace [26]. Different forms of non-oxide ceramics include powders, fibers, and monocrystals. Carbide powders like Tungstan Carbide (WC) and tantalum carbide (TaC) can be produced at high temperature by mixing (solid-solid reaction). A variety of gas-gas reactions can also be used to produce carbides such as B_4C, SiC, and others. Nitrides are divided into two groups: interstitial (TiN, ZrN, HfN, TaN) and macromolecular covalent (Si_3N_4, AlN, and BN nitrides) [33]. The oldest method of producing nitrides consists of nitriding the element with nitrogen or ammonia. Additionally, a gas-gas reaction consists of a high-temperature reaction between gas precursors, and a liquid–liquid reaction consists of a liquid medium reaction from mineral precursors by pyrolysis, which leads to amorphous nitrides [34].

1.2.3.2   Oxides

Oxide ceramics are generally inorganic compounds of metallic or metalloid elements like Al, Zr, Mg, Ti, and Si with oxygen. Oxides are the highest oxidation state of metals, so they are stable in their micro and macro structures. They have excellent mechanical, wear, and corrosion properties [26]. Also, oxide ceramics have good biocompatibility for prostheses and other biomedical applications. In the late 1970s, alumina (Al_2O_3) attracted researchers' interest due to its remarkable strength and biocompatibility. Furthermore, zirconia (ZrO_2) came into the picture due to its relatively good fracture toughness. Other advanced ceramics such as titanium oxide (TiO_2), silica (SiO_2), magnesia (MgO), zinc oxide, and Mn oxide also have a broad range of applications [34].

1.2.3.2.1   Alumina

Alumina is produced by Bayer process from bauxite ore. Initially, the refractory industries consumed the most alumina, followed by abrasive, electronics, chemistry, ceramics, and medical industries. Alumina crystallizes to form sapphire monocrystals in corundum structures [35]. The corundum structure has a rhombohedric lattice crystal structure. The crystal has compact hexagonal stacking of O^{2-} anions, in which 2/3rd of octahedral interstices is occupied by Al^{3+} cations [34]. This indicates that alumina has an ionic bond for 2/3rd and a covalent bond for remaining 1/3rd. As a result, alumina is one of the most closely bonded compounds with high hardness, high young modulus, high compression strength, wear resistance, good tribological and refractivity properties with high melting point (2050°C) and boiling temperatures (3500°C). Alumina has a formation enthalpy of 1600 kJ/mole and a density of 4 g/cm^{-3} [34].

Alumina has wide applications as bio-ceramics in orthopedics due to its good chemical and tribological properties. It has been used in the fabrication of heads of femoral prostheses and acetabulum. Alumina has also been proposed as a bioactive ceramic that can be used for bioactive coating. Hip prosthesis stems made entirely of alumina have also been developed due to inadequate mechanical properties that differ greatly from bone tissue. Alumina has numerous other applications, such as inner ear ossicles, ocular prostheses, electrical insulation for pacemakers, cardiac pumps, and catheter orifices [36].

1.2.3.2.2   Zirconia

Zirconia ceramics, also known as zirconium dioxide, are naturally found in the form of baddeleyite and can also be prepared by high heat and chemical treatment from zirconium silicate sands. Zirconia has a high melting temperature of ~2880°C, which solidifies in a cubic phase and transforms to a tetragonal phase (T < ≈ 2370°C) and finally becomes monoclinic (T < ≈ 1170°C) on treatment [34]. During the transition from the tetragonal to monoclinic phase, zirconia accompanied high dimensional changes resulting in material fragmentations. This shows that the pure form of zirconia can only be in powder form. In order to produce zirconia part, it must be combined with stabilizers such as CaO, MgO, or Y_2O_3, which helps zirconia to stabilize at different temperatures [33].

Zirconia stabilized with Y_2O_3 has wide medical applications. The tetragonal phase of yttria zirconia is retained at room temperature due to its small grain size, i.e., 0.5 microns and homogeneous distribution of yttria [35]. This also enhances the mechanical properties of Y-TZP. The transformation from a tetragonal to a monoclinic phase increases the material's toughness. This has the application in the heads of osteoarticular prostheses. Zirconia has good failure strength, bending strength, and resistance to fatigue than alumina [36].

1.2.3.2.3   Titanium Oxide

Titanium oxide (TiO_2) is one of the oxides of titanium, which crystallizes in three structural forms: brookite, anatase, and rutile. Rutile is the most important mineral ore that has 95% of TiO_2. TiO_2 is characterized by high photocatalytic activities such as Sulfate or chloride process [37]. The produced TiO_2 can be used in different applications such as pigments, ceramics, etc. It has antimicrobial properties, good biocompatibility and osteoconductivity, and potential application in bone regeneration and implant fabrication.

1.2.3.3   Bio-ceramics: Bioactive and Biodegradable

Bio-ceramics are the special materials used in tissue engineering applications to repair, restore, improve, or replace damaged tissues inside the human body. Porcelain and plaster of paris (POP) were the initial materials used in the 18th and 19th centuries to treat dental disorders. Due to technological improvement, the usage of bio-ceramics is increased in the 20th century [38]. The success of bio-ceramics depends upon their functionality and compatibility. Therefore, bio-ceramics are classified into three subclasses i.e., bioinert, bioactive, and biodegradable [39]. Bioinert ceramics have good biocompatibility and physicochemical properties with hard bone tissues. When bioinert ceramics are implanted inside the human body, they do not react and are rejected by the tissues. They have remarkable fracture resistance, corrosion, and wear resistance. Alumina and zirconia are the conventional bioinert ceramics with the application in orthopedic and dental areas as described in the previous section.

Bioactive ceramics are the class of ceramics that chemically integrate with bone and soft tissues and provide strength for a long period of time in the regeneration, repair, and restoration of tissues. These ceramics are generally used for repairing orthopedic, dental, craniofacial, and chronic osteomyelitis. These are glass ceramics, bio-glass, hydroxyapatite, etc. [40]. Besides this, biodegradable ceramics can react and break down rapidly when in contact with human body fluid. These ceramic implantations do not require a second surgery to remove implants from the body. These materials become part of the hard tissue inside the body, and the normal metabolic pathways treat the chemicals produced through them in the human body. Calcium phosphate ceramics such as tricalcium phosphate, hydroxyapatite, calcium oxide, calcium carbonate, and gypsum are considered as biodegradable bio-ceramics [34].

1.2.3.3.1 Bio-glass

L. Hench first described the term bio-glass in 1969. Bio-glass develops chemically stable bonds and attachments with bone tissues, forming an apatite layer in body fluid [38]. Thus, bioactivity depends upon the apatite formation that shows bio-glass compatibility with tissues. Different types of glass and vitroceramics belong to the family of $Na_2O–CaO–SiO_2–P_2O_5$, which mainly have the constituents of CaO and P_2O_5. Ca and P are the main elements of bone's mineral phase, which is why bio-glass can adhere to bone. The SiO_2 content should be between 42–52% to show the chemical bonding with bone tissue. Other compounds like MgO or CaF_2, and Na_2O by K_2O are also the components of bioactive glass. In order to avoid bioactivity inhibition, the glass fabrication process was altered by adding small amounts of Al_2O_3 and B_2O_3, which may substitute SiO_2. Bio-glasses are brittle, therefore, cannot be implanted in areas with high mechanical stresses [33]. In order to compensate for this shortcoming, bio-glasses are subjected to heat treatment to form a vitroceramic or even used as a deposit on a metallic substrate. The heat treatment offers good mechanical strengths against impact. Vitoceramic A/W is a biphasic material with excellent mechanical properties composed of a wollastonite phase (CaO SiO_2), an apatitic phase ($Ca10(PO_4)_6(OH, F)_2$), and a residual vitreous phase $MgO–CaO–SiO_2$, is clinically used, and are being used particularly in vertebral reconstructive surgery [34].

1.2.3.3.2 Calcium Phosphates

Calcium phosphate is ceramic that has been preferred as bone substitute for orthopedic and maxillofacial surgeries for 20 years. It is an inorganic phase of bone minerals that aids in bone calcification and resorption through improved biological affinity and activity. Calcium phosphate has the same chemical composition and structure as bone and is found in hydroxyapatite, alpha, and beta-tricalcium phosphate and their composites [4].

Synthetic hydroxyapatite ($Ca10(PO_4)_6(OH)_2$) is a mineral form of apatite that is similar to the natural mineral form found in bones. It is a crystalline alloplastic hexagonal material. It has a nominal atomic ratio of Ca/P of 1:67 [38]. Despite its high constituent similarity to bone, hydroxyapatite has very low mechanical properties compared to natural bone. It is brittle and has low Young's modulus and fracture toughness.

Similarly, beta TCP has an atomic ratio of Ca/P of 1.5 and is also biocompatible and biodegradable. It is also capable of developing chemical bonds with bone and has proved to be a substitute for bone. The biphasic compound HA-beta TCP was also developed, having the physicochemical properties of each other. They can be prepared for controlled resorption and substitution [33].

1.2.3.4 3D Printing of Ceramics

There are different traditional methods to form ceramics, such as casting, injection molding, dry pressing, and isostatic pressing, with or without binder, followed by different post-processing for structure densification. These traditional methods have a high cost, time consumption, and complexity in fabrication of intricate shapes. The machining of ceramics is also very difficult due to their brittleness and hardness. Therefore, the introduction of AM in the fabrication of ceramics provided new possibilities for addressing the previously mentioned limitations. Fabrication of ceramic using AM first came into the picture by Marcus and Sachs in the 1990s. Based on the pre-processing of materials, ceramics fabrication through AM are classified into three forms of feedstock as powder-based, slurry-based, and solid-based feedstock. Here, powder ceramics are directly printed by laser-based techniques such as SLS, SLM; and slurry and solid-based feedstock comes into indirect printing, where ceramics are used to mixed with different binders and additives and printed through SLA, DLP, IJP, DIW and LOM, FDM. Although the processing of ceramics is considerably difficult in comparison to polymers and metals, most of the AM technologies can be adopted to shape the ceramics as discussed later:

1.2.3.4.1 Powder-based Feedstock

The AM techniques that use loose powder of ceramics as feedstock are known as PBF techniques. They generally use a focused heat source to selectively fuse the powder over the bed of machine. Most common techniques are SLS, SLM, and 3DP, which are used to shape the ceramics. The sintering can be performed in two ways: first, the direct sintering of powders and second, indirect sintering, where the powders are mixed with binders that act as temporary phases which will be removed upon heat treatment. In indirect sintering generally, the powders of approx. ~90 μm size was used to mix with low melting binder phase, which may be inorganic or polymeric. The local sintering of the ceramic mix melts the binder and provides a solid matrix upon re-densification, ensuring the formation of a green body. Furthermore, the green body was used to transfer into the furnace for heat treatment [34]. Various ceramics like alumina, zirconia, silicon carbide, ZrB_2, SiO_2, and glass ceramics have been sintered by indirect sintering using AM PBF techniques.

Lakshminarayan first performed ceramic part printing using SLS in the 1990s. The alumina powders have been sintered using 0.43 volume % of ammonium phosphate inorganic binder to fabricate a green body. The binder was heat treated at 850°C and converted into aluminum phosphate matrix, resulting in a composite of

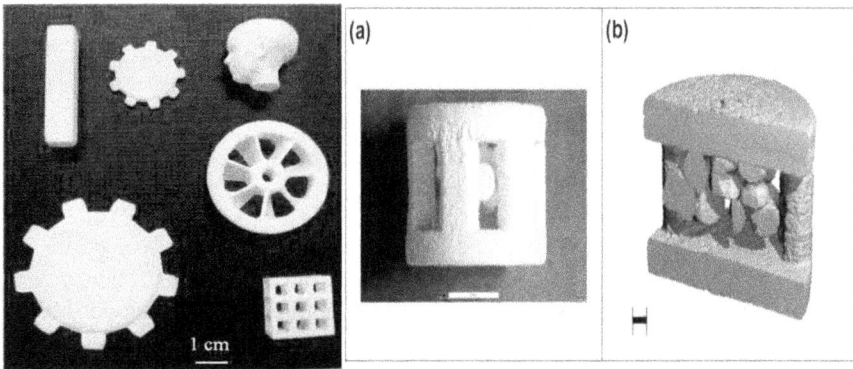

FIGURE 1.6 SLS printed Alumina-PP printed parts [42], 3DP printed part (a, b) [47].

alumina and aluminum phosphate [26]. Similarly, alumina mixed with metaboric acid (HBO_2), an inorganic binder, resulted in alumina-boron oxide composite after post-heat treatment and provided good mechanical strengths. Alumina-PA composite fabricated using SLS assisted by quasi-isostatic pressing at high temperature provides 94% of theoretical density [41]. In another work, alumina with 30vol.% PP composite has been printed (refer Figure 1.6) with infiltration and warm isostatic pressing post-processing that enhanced the density by 80% [42]. The ball-milled mixture of the stearic acid binder with alumina and ZrB_2 enhanced the theoretical density and flexural strength of laser-sintered bars.

Furthermore, researchers have used PVA epoxy-coated alumina powders to manufacture gears with laser sintering followed by cold isostatic pressing and hot isostatic pressing in order to obtain a higher density of about 96%. In another work, a group of researchers used selective laser curing to develop the parts of an organosilicon pre-ceramic polymer powder mixed with SiC ceramic particles [43]. The fabricated parts were high density with no observable porosity and very high flexural strength but with a visible rough surface.

SLM of ceramics is difficult in comparison to metals and composites. In the year 2007, ZrO_2 was fabricated using SLM but resulted in defects like pores, and cracks. Furthermore, alumina, yttria-stabilized zirconia ceramic have been fabricated using SLM with lower density i.e., 85% and 56%, respectively [43]. In order to prevent the lower density and cracking of parts, slurry-based mixture of ceramics has been prepared, and a modified method known as laser-engineered net shaping (LENS) has been developed. In LENS technique, ceramic powders are deposited coaxially at laser spot area. Dense alumina has been fabricated using LENS with relative density of 94% [44]. A dental restoration has been printed using SLM with the mixture of alumina and zirconia powder. The cracks development through thermal gradient has been prevented using pre-heating system through laser. A dense part with a good flexural strength of about 500 MPa has been developed [45].

The 3DP ceramic printing method was first reported in 1990 with alumina, silicon carbide particles, and silica as binders. 3DP method has extended its application in the biomedical field in the fabrication of porous structure. A $CaSiO_3$ ceramic has been used to fabricate a part of 64% porosity, and an in-vitro test has also been performed, which showed no toxicity of the material [46]. Other bio-ceramics like HA calcium phosphate have also been used to fabricate scaffolds for bone tissue engineering applications using 3DP technique (refer Figure 1.6 (a,b)) [47].

1.2.3.4.2   Slurry-based Feedstock

The slurry-based feedstock includes a liquid or semi-liquid system dispersed with ceramic particles to yield a colloidal solution in the form of ink or paste. The viscosity of the solution depends upon the loading percentage. The ceramic powder can be used of a submicron size that will help to improve the packing density and reduce the power consumption of the laser. The developed slurry or suspension should possess appropriate viscosity, stability, and homogeneity. A good suspension retains its satisfactory viscosity for a reasonable time and during the printing process. The slurry-based feedstock can be 3D printed by photopolymerization, extrusion, inkjet printing, etc. The 3D-printed green part must be processed further with treatments such as sintering to remove organics to achieve high density [26].

The polymerization kinetics of a suspension with SiO_2, Al_2O_3, ZrO_2, and SiC ceramics up to 40 wt.% has been shown that smaller particle size of SiO_2, Al_2O_3 has better light scattering effect than ZrO_2 and SiC particles for curing using the SLA printing technique [26]. Similarly, using SLA, zirconia toughened alumina of 0.35 micron grain size has been fabricated. The sintered density was 99.5%, with great hardness, flexural strength, and toughness [43]. SLA can also be used for the printing of SiC and Si_3N_4 ceramics with photocurable resins. Complex SiC structure were manufactured from liquid allyl hydrido polycarbosilane mixed with methacrylate monomer and photoinitiator [26]. The fabricated structures have only microporosity that does not hinder the mechanical strength much (as shown in Figure 1.7).

FIGURE 1.7   SiC parts produced by SLA [26].

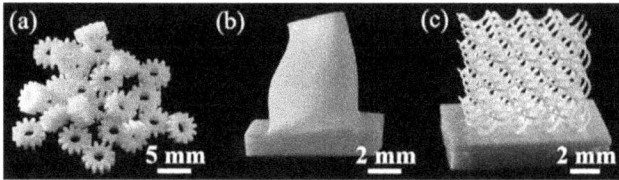

FIGURE 1.8   DLP fabricated alumina ceramic parts a) gear wheels; (b) turbine blade; (c) cellular cube 3D structures [43].

Researchers have widely explored ceramic fabrication using DLP for years. In 2015, lithography-based ceramic printing technology has been commercialized. Various work has been performed using the Lithography-based Ceramics Manufacturing (LCM) technique. The complex structures of alumina (as shown in Figure 1.8) [43] and bio-glass have been printed with a relative density of above 90%, and the mechanical strengths were also found to be comparable with conventional processes. Similarly, zirconia and alumina with high loading of 99% and zirconia and β-tricalcium phosphate (β-TCP) with about 50% volume loading were also printed successfully [26]. The two-photon polymerization (TPP) technique was first used to print 3D SiCN-loaded structures of submicron size. The 40 wt.% of silica particles of 10 nm was loaded for microstructure fabrication. Furthermore, the works have been reported with SiC, SiOC, and Zr-Si ceramic loading in polymers using the TPP technique. In the field of nanostructures, a high-quality ceramic nano hollow tube lattice using SiO_2, TiN, and alumina have been successfully printed using TPP [48].

Another 3D printing technique i.e., inkjet printing, uses slurry-based feedstock material to print small structures. It was first described by Blazdell and his team in 1995 using ZrO_2 and TiO_2 ceramic ink. The group has worked with different weight % of ceramics from 5 vol.% to 14 vol.% and studied the viscosity and surface tension properties of ink in the printing process (refer Figure 1.9 (a,b,c, and d)) [43]. Furthermore, various researchers have developed ceramic inks with alumina, barium strontium titanate, yttria-stabilized zirconia, etc. The growing interest in ceramic ink fabrication printed the micropillar array of PZT (2.5 vol.%) and TiO_2 (15 vol.%) with an application of transducers in medical imaging. Direct ink writing technique has wide application in the medical field. It is well suited for the fabrication of porous ceramic structures with periodic features. Therefore, it has been used for the fabrication of tissue engineering scaffolds of hydroxyapatite, calcium phosphate, and Al_2O_3–ZrO_2 gel paste [26]. In addition, high-quality ceramic printing with functionally graded material has also been printed with $CaCO_3$, Al_2O_3, and ZrO_2 (refer Figure 1.9 (e.f, and g)) [49].

1.2.3.4.3   Solid Feedstock
Solid feedstock to fabricate ceramic objects includes ceramic-loaded filaments, ceramic sheets, green tapes, or papers. Ceramic LOM fabricates the desired geometry by stacking, cutting, and bonding the ceramic green tapes or pre-ceramic papers. The outline of the desired shapes is first cut by CO_2 laser, and then a heated roller is applied onto the surface of the layer. Due to the applied pressure

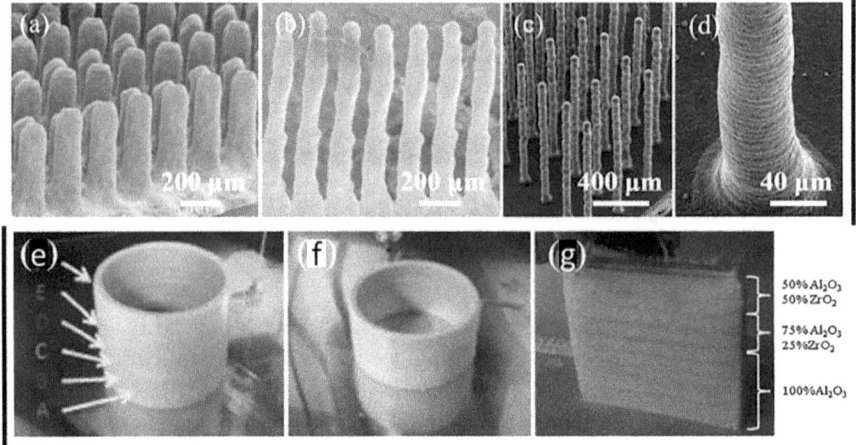

FIGURE 1.9 Micro-arrays of pillars: (a,b) PZT pillars,(c,d) TiO$_2$ pillars [43], Functionally graded part fabricated through DIW (e) and (f) Pink- and green-colored CaCO$_3$, (g) test bar fabricated with 100% Al$_2$O$_3$ to 50% Al$_2$O$_3$ + 50% ZrO$_2$ [49].

through the roller it binds with the adjacent layer. LOM has been used to fabricate components with Al$_2$O$_3$, Si$_3$N$_4$, Si/SiC, ZrO$_2$/Al$_2$O$_3$, TiC/Ni, Ti$_3$SiC$_2$ composites, HA bioimplants [43,50]. LiO$_2$–ZrO$_2$–SiO$_2$–Al$_2$O$_3$ (LZSA)glass–ceramic parts have been produced using LOM. LZSA tape was prepared by tape casting with PVA binder [51,52] (refer Figure 1.10). The delamination and bending strength have

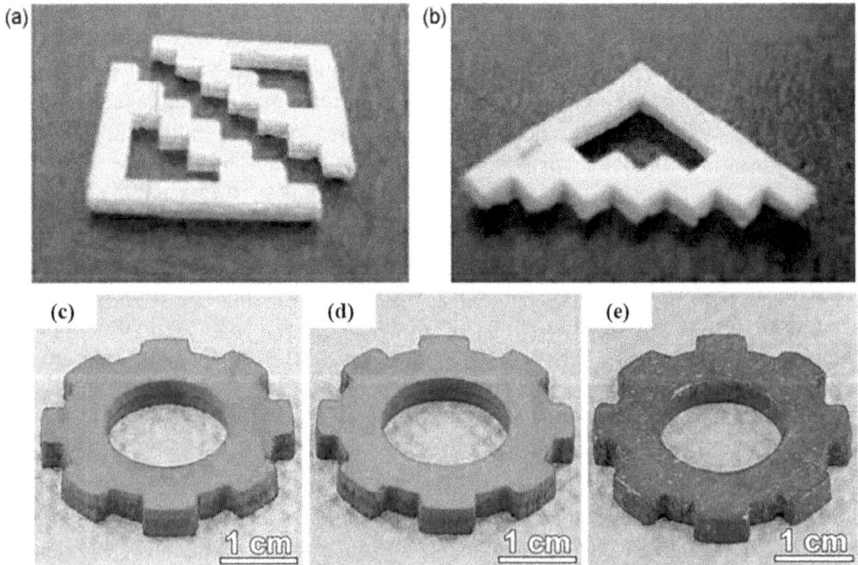

FIGURE 1.10 LZSA stair-like geometry a) green part b) sintered part [52]; 3D Gear of Ti$_3$SiC$_2$ c) green part d) sintered part d) part after silicon infiltration [50] produced by LOM.

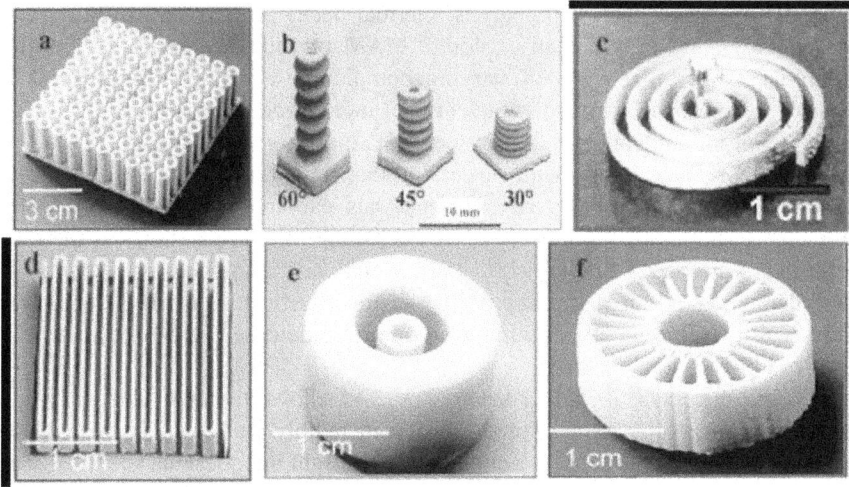

FIGURE 1.11 Transducers made by FDM: (a) tube array; (b) bellows; (c) spiral; (d) curved; (e) telescoping; (f) radial actuators [53].

been examined at different stacking directions and weight percentages of ceramics. However, high surface roughness was observed due to the staircase effect. Other works have also been conducted with different ceramics to improve the process, including build time and process parameters. A variant of LOM with a curved surface and stack approach has also been developed to fabricate curved ceramic components and other ceramic parts [26].

During FDM of ceramics, highly loaded ceramic filament feedstock is used to extrude out using a standard FDM printer. The binder of the feedstock includes a base polymer, elastomer, plasticizer, and wax. The printed green parts are made to sinter to debind the polymeric part. The ceramic loading percentage ranges between 45 and 65 vol.%. FDM has fabricated the 3D ceramic part from alumina, SiO_2, Si_3N_4, and barium titanate [43]. A lead–zirconate–titanate (PZT) and lead–magnesium niobate (PMN) composite filament has been used to fabricate electronic transducers of different shapes using FDM ceramic printing technique as shown in Figure 1.11 [53]. Similarly, multi-nozzles have been used in another work to print multilayer PZT sensors. Ceramic printing using FDM has applications in biomedical scaffolds, electronic bandgap structures, etc. [43].

1.2.4 METALS

Metal materials play a pivotal role in AM, revolutionizing the production of complex and customized components by layering metal powders, precisely fusing them together using advanced techniques, and enabling unprecedented design flexibility and engineering possibilities. This section discusses different types of metal materials used for metal additive manufacturing (MAM) and sheds light on their specific properties and applications. The selection criteria for metal materials, their mechanical properties, thermal behavior, and compatibility with AM processes have also been explored. MAM provides exceptional flexibility in producing metallic parts of varying complexities, ranging from simple to intricate designs. In contrast to conventional metal processing techniques, MAM offers numerous advantages. Over the past 25 years, the field of MAM has experienced remarkable growth, surpassing the limited progress achieved during its initial introduction. This advancement has captured the attention of both researchers and industry professionals, establishing MAM techniques as the focal point in the modern industrial world [54]. Metal parts manufactured through AM have demonstrated an increasing ability to meet critical industries' demanding and specific requirements, including automotive, defense, aerospace, construction, and electronics [55–58].

Metallic materials, including steels, Ni-based alloys, titanium alloys, specific grades of lightweight metal alloys, etc., are incredibly well-matched with MAM systems [59,60]. The following section delivers insight into these metallic materials in sequence.

1.2.4.1   Ferrous Alloys

Ferrous alloys (precipitation hardened, duplex, austenitic, martensitic, etc.) are extensively used in the MAM process and processed with laser-based AM techniques such as PBF and directed energy deposition (DED). The grades of austenitic stainless steels such as SS316, SS304, 304 L, and 316 L AISI types are generally used [61–63]. As compared to conventional manufacturing techniques, MAM generates fine-grained steel components due to rapid solidification and non-equilibrium conditions [64]. In order to attain the desirable properties of additively manufactured steel parts, post-processing processes are commonly applied.

1.2.4.2   Titanium Alloys

Titanium alloys are broadly used as MAM materials for research due to their outstanding properties, such as fatigue resistance, high strength-to-weight ratio, good fracture, better corrosion resistance, and formability. Their excellent properties make them a better option to be utilized in various sectors, including biomedical, aerospace, and automobiles [65,66]. Several attempts have been made to fabricate parts using MAM techniques with these alloys. Among these alloys, Ti–6Al–4 V is widely taken as MAM materials. This can be attributed to the compatibility of this material with several biomedical applications [67]. Titanium has two phases in its pure form, commonly referred to as α and β, out of which the preceding phase is vital with less ductility, whereas the latter is more ductile. Alloys with both ($\alpha+\beta$) phases possess high strength and formability. In order to achieve the desirable properties of Titanium alloys, these phases can be adjusted carefully. The higher β phase content is required for the bone mimics, and part density should be matched with neighboring material. In Ti–6Al–4 V alloy, Al and V stabilize α-phase and β-phase, respectively. The quenched β-phase results in the formation of a less stable α-phase. Therefore, the proper selection of printing parameters is required to attain better properties like corrosion resistance, strength, density, and ductility.

1.2.4.3   Lightweight Metal Alloys

Aluminum (Al) alloys find extensive application in various engineering sectors due to their favorable strength-to-weight ratio and corrosion resistance. However, AM of Al alloys remains limited due to challenges such as poor weldability and low laser absorption [68]. Another contributing factor is the susceptibility of Al alloys to melting during the fusion-based AM process, resulting in the potential solubility of hydrogen. Subsequent solidification of the molten pool leads to the entrapment of hydrogen and the formation of pores, which significantly compromises the mechanical properties of the final part. In order to address these issues, it is crucial to shield the process zone using additional shielding gas [69]. Apart from DED and PBF, other indirect AM processes, such as arc welding-based AM techniques [70] and emerging solid-state AM techniques like friction-based AM techniques [71], have also been developed to overcome these limitations.

1.2.4.4   Magnesium (Mg) Alloys

Magnesium alloys exhibit great potential as degradable biomaterials, offering a stiffness comparable to bone and minimizing stress-shielding effects [72]. The applications of Mg alloys are rapidly expanding across various medical fields, including orthopedics, urology, cardiology, respiratory, and more. Compared to TM methods, the ease of design provided by AM techniques has attracted significant interest in the AM of Mg alloys. AM can fabricate biodegradable implants, making it a highly desirable approach. Different AM techniques, such as PBF, laser AM, wire-arc AM, and friction stir-based AM, are employed to develop biodegradable materials based on Mg alloys. Each of these processes has its own unique mechanics and utilizes different forms of raw materials. Some AM techniques, like PBF (SLM and electron beam melting), face challenges related to oxidation and evaporation of Mg during processing. However, this issue can be overcome by printing Mg alloys in an inert atmosphere while optimizing the process parameters. In such cases, indirect AM processes are crucial in developing biodegradable Mg alloys.

1.2.4.5   Nickel Alloys

Ni-superalloys are widely utilized in the aerospace industry, particularly for constructing components that operate under high-temperature conditions. These materials exhibit exceptional mechanical strength, creep resistance, and resistance to oxidation and corrosion at elevated temperatures. Aerospace engine components, requiring sustained mechanical strength over extended periods under elevated temperatures, often rely on Ni-superalloys. Ni-based superalloys such as Inconel 625 [73] and 718 [74] have been used for these applications. Given the complex geometries of these components, MAM emerges as an appealing manufacturing technology. Metallurgical investigations of MAM-produced Ni-superalloys typically revealed melt pool characteristics spanning the entire volume of the fabricated part. These features consist of fine columnar growth resulting from microstructural growth perpendicular to the melt pool boundary, resulting from rapid cooling and heat dissipation in that region.

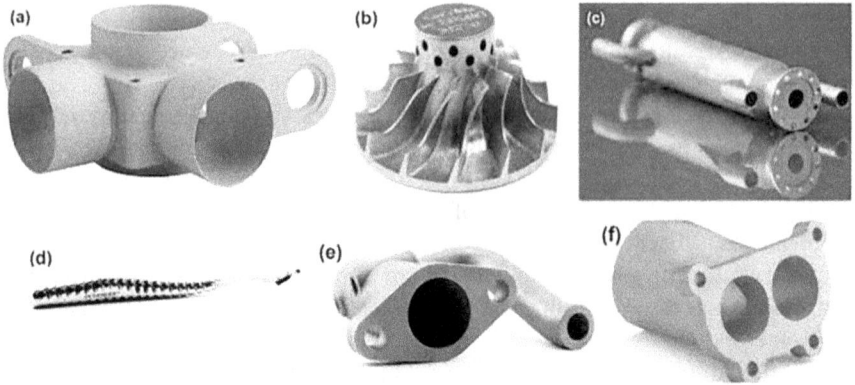

FIGURE 1.12 Metallic materials printed using MAM techniques (a) AlSi10Mg, (b) Cobalt Chrome CoCrMo, (c) Copper C18150, (d) Ni Alloy 718, (e) SS316 L, and (f) Ti6Al4V [75].

Consequently, MAM-produced components often exhibit a pronounced texturing effect along the build direction, potentially leading to mechanical property anisotropy. In Ni-superalloys, preferential growth typically occurs along the (100) plane. The fine grain structure and higher dislocation density contribute to the elevated mechanical strength observed in MAM-produced samples compared to wrought annealed and cast samples. Figure 1.12 depicts the parts printed using MAM with different materials.

1.2.5 SMART MATERIALS

Smart materials are sensitive to a particular stimulus and tend to change their state once exposed. They are classified into various categories based on their capabilities: shape-changing, self-actuating, self-sensing, self-diagnostic, and self-healing. Each material exhibits a different behavior or response when exposed to a stimulus. Among the development of smart materials, metamaterials have gained significant attention in the last decade. It is formed by combining the Greek word *meta,* meaning 'beyond', and the Latin word *materia,* denoting 'material'. It is an artificially designed composite structure with periodic or non-periodic arrangements of patterns that offer unique properties. It possesses negative indexes, including compressibility, swelling, thermal expansion, stiffness, poisson ratio, moisture expansion, etc. The metamaterials have found groundbreaking novel applications in niche areas of science and technology like optics, aerospace, medical, marine, textile, electronics, etc. Figure 1.13 shows the classification of smart materials.

The metamaterials are broadly classified as electromagnetic, acoustic, and mechanical metamaterials [77]. The mechanical metamaterials are conceptualized with a negative Poisson ratio, known as auxetic mechanical metamaterials, which expand transversely when stretched axially. The four main classes of auxetic metamaterials are re-entrant, chiral, rotating-rigid, and origami structures. The emerging concept of designing chiral auxetic metamaterial has shown promising applications in soft electronics, biomedicine, tissue engineering, etc. The chiral metamaterials are the structures that are not superimposable on their mirror image

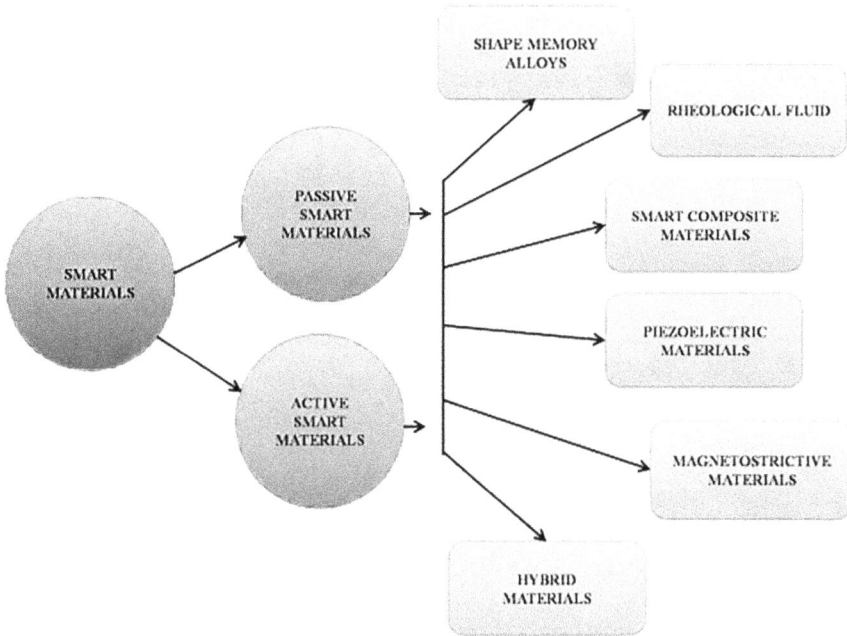

FIGURE 1.13 Classifications of smart materials [76].

and anti-chiral are the opposite to it. Chirality is widely spread in nature, such as shell of a snail, DNA, protein, and artificially exists in structures such as screws, propellers, etc. The chiral unit consists of a central cylinder surrounded with tangentially attached ligaments, constructed either left-handed or right-handed (refer Figure 1.14). On the application of mechanical load, the cylinder rotates, and the ligaments bend. In this way, the folding and unfolding of the structure result in a negative poison ratio based on its design arrangements.

In order to create a chiral structure, the constraints of rotational symmetry should be obeyed, i.e., the number of ligaments attached to a node should be equal to order n of rotational symmetry. When the constraints are relaxed, metachiral strutures may be created otherwise, trichiral, tetrachiral, hexachiral, antitrichiral, and antitetrachiral structures may be obtained [78]. These structures' units or periodic arrays are very complex to design through conventional methods. Such difficult design concepts can be achieved by 3D printing technology. Various 3D printing technologies like DLM, Multi Photon Lithography (MPL), WAAM, SLM/S, SLA, and FDM can fabricate complex lattice and hierarchical structures at micro or nano levels. Hence, the uncommon properties of auxetic metamaterials and 3D printing techniques have significant potential in biomedical applications.

Auxetic metamaterials proved its application as stents, annuloplasty rings, implants, and scaffolds for tissue engineering. Esophageal cancer develops the tumor and blocks the intake through the esophagus. The auxetic stents expand at the blocked area and provide passage for the intake. Researchers have used different materials to fabricate stents like polyurethane and polypropylene [79]. Besides this,

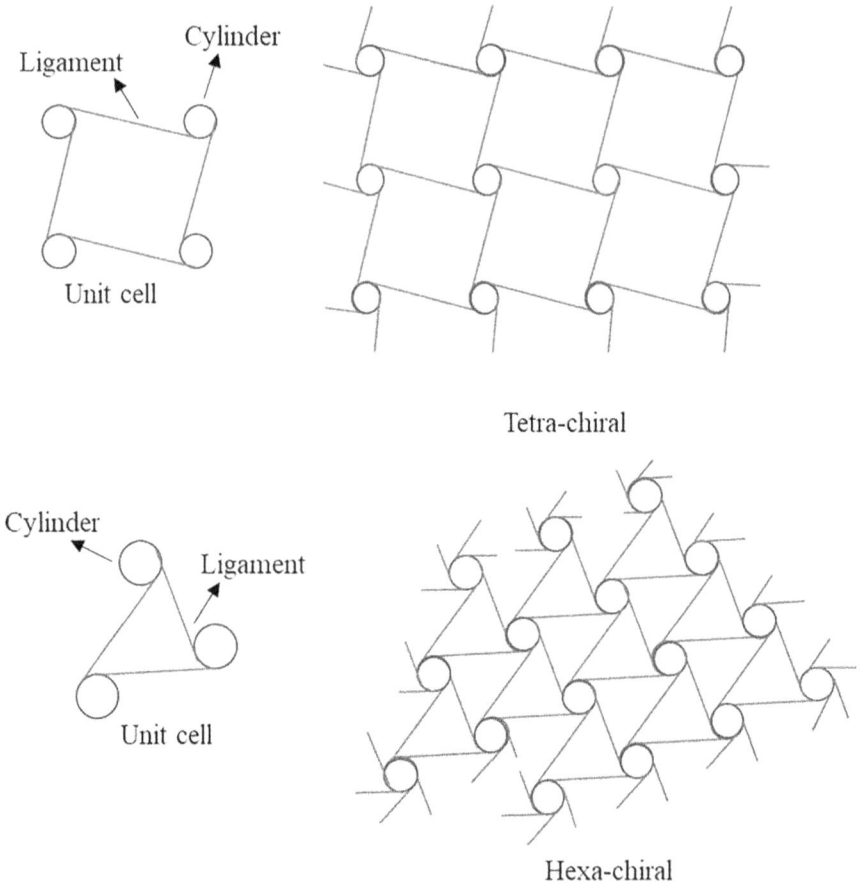

FIGURE 1.14 Chiral auxetic structures.

an advanced 3D auxetic scaffold facilitates biological activity, showing the next-generation scaffolds for tissue engineering applications [80]. Along with this, unlike conventionally designed implants, rationally designed metabiomaterial implants improve implant-bone contact in off-axial load and life. Other than this, researchers have developed the auxetic structures and performed the simulation [81] and topology optimization study [82] for biomedical applications.

Shape memory alloys (SMAs) are a popular smart material that possesses two phases: martensite phase (low temperature) and austenite phase (high temperature). These two phases allow SMAs to alter and return to their original shape when exposed to high temperatures. Another popular class of smart materials is shape memory polymers (SMPs). SMPs can remember a permanent shape and transform it to a temporary shape when exposed to a number of external stimuli such as temperature, light, electric conductivity, magnetic field, pH, moisture, near-infrared, and others [83,84]. Currently, the most used 3D printing techniques include SLA, FDM, SLM, direct inkjet writing, laser-assisted bioprinting, etc. 4D printing has an

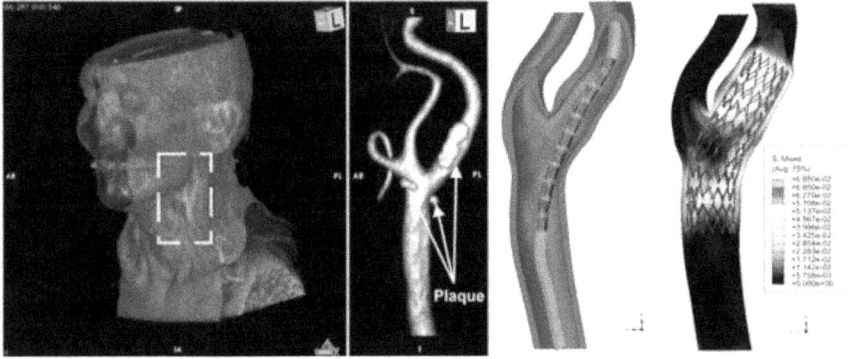

FIGURE 1.15   Applications of SMAs in stent and orthopedic [76].

emerging trend and great potential in biomedical applications, including smart stents, medical tools, drug delivery systems, tissue engineering, etc. A stent is a biocompatible metal or plastic mesh coil that is used to keep open the blocked passageway of an artery or a vein or treat atherosclerosis. Atherosclerosis is a severe and common form of cardiovascular disease [85]. Accumulation of cells, collagen, calcium, lipids, and inflammatory infiltrates can cause artery occlusion, resulting in restriction in blood flow, which can be treated by inserting a stent. The applications of SMAs are shown in Figure 1.15.

Based on the previously mentioned findings, the advantages, disadvantages, and applications of materials used in the AM process are tabulated in Table 1.2.

TABLE 1.2

Advantages, disadvantages, and applications of materials for AM

Material	Advantages	Disadvantages	Applications
Polymers	• Economic • Easy processing • Fast prototyping • Lightweight • Flexible	• Limited availability • Anisotropic • Mechanically weak • Limited heat resistance	• Aerospace • Toys • Drug delivery • Orthopedics • Sports • Architectures
Ceramics	• Control of porosity • High temperature resistance • Chemical resistance	• Limited printable ceramics • Brittle • Poor surface finish • Inaccuracy • Complexity in printing	• Orthopedics • Dentistry • Automotive • Chemical industries • Electronic insulators
Composites	• Tailorable properties • High strength to weight ratio	• Limited availability of composites • Complexity in printing	• Medical • Aerospace • Automotive • sports

(Continued)

TABLE 1.2 *(Continued)*
Advantages, disadvantages, and applications of materials for AM

Material	Advantages	Disadvantages	Applications
Metals	• Good strength • Heat resistance • Versatility	• Expensive • Complex post-processing	• Implants • Dentistry • Aerospace • Automotive • Tooling and industrial parts
Smart Materials	• Tailorable properties • Innovative and adaptable	• Limited availability • Expensive	• Biomedical devices (e.g., stents). • Aerospace components with self-healing properties. • Wearable technology and sensors.

1.3 CONCLUSIONS

In conclusion, AM has witnessed significant growth and widespread popularity over the past few decades. Its versatile applications have expanded across almost every field, making it a crucial technology in various industries. However, the periodic evaluation of AM techniques reveals inherent differences among them, with one of the most critical distinctions lying in the selection of materials.

The choice of appropriate materials holds paramount importance in determining the performance and quality of the fabricated parts. Material selection involves producing the feedstock in a suitable form for utilization in different AM processes, ultimately resulting in the creation of desirable geometries. AM encompasses different classes of materials categorized based on their phase, the type of process employed, and the intended application.

This chapter primarily focuses on gathering pertinent information about the different types of AM materials, including polymers, metals, ceramics, and their composites. It explores the unique properties and challenges associated with each material category. Moreover, the chapter also delves into the exploration of new advancements in AM materials, including the development and application of smart materials, which have opened up exciting possibilities for the future of AM.

Overall, this chapter serves as a comprehensive resource that provides insights into the diverse range of materials employed in AM processes. By shedding light on these materials' characteristics, limitations, and advancements, it aims to equip readers with a deeper understanding of the materials utilized in AM and the challenges that arise in their implementation.

REFERENCES

[1] X. Tian *et al.*, "Roadmap for Additive Manufacturing: Toward Intellectualization and Industrialization," *Chinese J. Mech. Eng. Addit. Manuf. Front.*, vol. 1, no. 1, p. 100014, 2022, doi: 10.1016/j.cjmeam.2022.100014.

[2] N. Choudhary, V. Sharma, and P. Kumar, "Polylactic acid-based composite using fused filament fabrication: Process optimization and biomedical application," *Polym. Compos.*, vol. 44, no. 1, pp. 69–88, Jan. 2023, doi: 10.1002/pc.27027.

[3] R. Durga Prasad Reddy and V. Sharma, "Additive manufacturing in drug delivery applications: A review," *Int. J. Pharm.*, vol. 589, p. 119820, Nov. 2020, doi: 10.1016/j.ijpharm.2020.119820.

[4] M. Kumar, S. Ghosh, V. Kumar, V. Sharma, and P. Roy, "Tribo-mechanical and biological characterization of PEGDA/bioceramics composites fabricated using stereolithography," *J. Manuf. Process.*, vol. 77, no. January, pp. 301–312, 2022, doi: 10.1016/j.jmapro.2022.03.024.

[5] M. Bhuvanesh Kumar and P. Sathiya, "Methods and materials for additive manufacturing: A critical review on advancements and challenges," *Thin-Walled Struct.*, vol. 159, p. 107228, Feb. 2021, doi: 10.1016/j.tws.2020.107228.

[6] M. Srivastava, S. Rathee, V. Patel, A. Kumar, and P. G. Koppad, "A review of various materials for additive manufacturing: Recent trends and processing issues," *J. Mater. Res. Technol.*, vol. 21, pp. 2612–2641, 2022, doi: 10.1016/j.jmrt.2022.10.015.

[7] S. Yuan, F. Shen, C. K. Chua, and K. Zhou, "Polymeric composites for powder-based additive manufacturing: Materials and applications," *Prog. Polym. Sci.*, vol. 91, pp. 141–168, 2019, doi: 10.1016/j.progpolymsci.2018.11.001.

[8] R. Durga Prasad Reddy and V. Sharma, "Investigations of hybrid infill pattern in additive manufactured tablets: A novel approach towards tunable drug release," *J. Biomed. Mater. Res. Part B Appl. Biomater.*, vol. 111, no. 11, pp. 1869–1882, Nov. 2023, doi: 10.1002/jbm.b.35290.

[9] Y. Shi *et al.*, *Polymer material for additive manufacturing–filament materials*. 2021.

[10] Stratasys, "FDM Support Materials," *Stratasys*. https://support.stratasys.com/en/materials/fdm/fdm-support-materials.

[11] L. J. Tan, W. Zhu, and K. Zhou, "Recent progress on polymer materials for additive manufacturing," *Adv. Funct. Mater.*, vol. 30, no. 43, pp. 1–54, 2020, doi: 10.1002/adfm.202003062.

[12] N. Choudhary, C. Ghosh, V. Sharma, P. Roy, and P. Kumar, "Investigations on effect of pore architectures of additively manufactured novel hydroxyapatite coated PLA/Al2O3 composite scaffold for bone tissue engineering," *Rapid Prototyp. J.*, vol. 29, no. 5, pp. 1061–1079, May 2023, doi: 10.1108/RPJ-09-2022-0303.

[13] N. Choudhary, V. Sharma, and P. Kumar, "Reinforcement of polylactic acid with bioceramics (alumina and YSZ composites) and their thermomechanical and physical properties for biomedical application," *J. Vinyl Addit. Technol.*, vol. 27, no. 3, pp. 612–625, Aug. 2021, doi: 10.1002/vnl.21837.

[14] D. Bourell *et al.*, "CIRP Annals – Manufacturing Technology Materials for additive manufacturing," *CIRP Ann. Manuf. Technol.*, vol. 66, no. 2, pp. 659–681, 2017, [Online]. Available: 10.1016/j.cirp.2017.05.009.

[15] C. M. González-Henríquez, M. A. Sarabia-Vallejos, and J. Rodriguez-Hernandez, "Polymers for additive manufacturing and 4D-printing: Materials, methodologies, and biomedical applications," *Prog. Polym. Sci.*, vol. 94. pp. 57–116, 2019, doi: 10.1016/j.progpolymsci.2019.03.001.

[16] I. Gibson and D. Shi, "Material properties and fabrication parameters in selective laser sintering process," *Rapid Prototyp. J.*, vol. 3, no. 4, pp. 129–136, 1997, doi: 10.1108/13552549710191836.

[17] C. Ramesh, "Crystalline Transitions in Nylon 12," *Macromolecules*, vol. 32, no. 17, pp. 5704–5706, Aug. 1999, doi: 10.1021/ma990494o.

[18] H. C. H. Ho, I. Gibson, and W. L. Cheung, "Effects of energy density on morphology and properties of selective laser sintered polycarbonate," *J. Mater. Process. Technol.*, vol. 89–90, pp. 204–210, 1999, doi: 10.1016/S0924-0136(99)00007-2.

[19] P. Feng, K. Wang, Y. Shuai, S. Peng, Y. Hu, and C. Shuai, "Hydroxyapatite nanoparticles in situ grown on carbon nanotube as a reinforcement for poly (ε-caprolactone) bone scaffold," *Mater. Today Adv.*, vol. 15, 2022, doi: 10.1016/j.mtadv.2022.100272.

[20] S. Eosoly, N. E. Vrana, S. Lohfeld, M. Hindie, and L. Looney, "Interaction of cell culture with composition effects on the mechanical properties of polycaprolactone-hydroxyapatite scaffolds fabricated via selective laser sintering (SLS)," *Mater. Sci. Eng. C*, vol. 32, no. 8, pp. 2250–2257, 2012, doi: 10.1016/j.msec.2012.06.011.

[21] H. Schappo, K. Giry, G. Salmoria, C. Damia, and D. Hotza, "Polymer/calcium phosphate biocomposites manufactured by selective laser sintering: an overview," *Prog. Addit. Manuf.*, vol. 8, no. 2, pp. 285–301, Apr. 2023, doi: 10.1007/s40964-022-00332-4.

[22] H. Zheng, J. Zhang, S. Lu, G. Wang, and Z. Xu, "Effect of core-shell composite particles on the sintering behavior and properties of nano-Al2O3/polystyrene composite prepared by SLS," *Mater. Lett.*, vol. 60, no. 9–10, pp. 1219–1223, 2006, doi: 10.1016/j.matlet.2005.11.003.

[23] R. D. Goodridge *et al.*, "Processing of a Polyamide-12/carbon nanofibre composite by laser sintering," *Polym. Test.*, vol. 30, no. 1, pp. 94–100, 2011, doi: 10.1016/j.polymertesting.2010.10.011.

[24] I. Gibson, D. W. Rosen, and B. Stucker, *Additive Manufacturing Technologies*. Boston, MA: Springer US, 2010.

[25] Y. Shi *et al.*, "Polymer materials for additive manufacturing: liquid materials," in *Materials for Additive Manufacturing*. Academic Press, 2021, pp. 361–400.

[26] Y. Lakhdar, C. Tuck, J. Binner, A. Terry, and R. Goodridge, "Additive manufacturing of advanced ceramic materials," *Prog. Mater. Sci.*, vol. 116, no. September 2020, p. 100736, Feb. 2021, doi: 10.1016/j.pmatsci.2020.100736.

[27] J. W. Choi, R. Wicker, S. H. Lee, K. H. Choi, C. S. Ha, and I. Chung, "Fabrication of 3D biocompatible/biodegradable micro-scaffolds using dynamic mask projection microstereolithography," *J. Mater. Process. Technol.*, vol. 209, no. 15–16, pp. 5494–5503, 2009, doi: 10.1016/j.jmatprotec.2009.05.004.

[28] L. Elomaa, A. Kokkari, T. Närhi, and J. V. Seppälä, "Porous 3D modeled scaffolds of bioactive glass and photocrosslinkable poly(ε-caprolactone) by stereolithography," *Compos. Sci. Technol.*, vol. 74, pp. 99–106, Jan. 2013, doi: 10.1016/j.compscitech.2012.10.014.

[29] K. Markandan and C. Q. Lai, "Enhanced mechanical properties of 3D printed graphene-polymer composite lattices at very low graphene concentrations," *Compos. Part A Appl. Sci. Manuf.*, vol. 129, no. November 2019, 2020, doi: 10.1016/j.compositesa.2019.105726.

[30] U. Kalsoom, A. Peristyy, P. N. Nesterenko, and B. Paull, "A 3D printable diamond polymer composite: A novel material for fabrication of low cost thermally conducting devices," *RSC Adv.*, vol. 6, no. 44, pp. 38140–38147, 2016, doi: 10.1039/c6ra05261d.

[31] J. J. Martin, B. E. Fiore, and R. M. Erb, "Designing bioinspired composite reinforcement architectures via 3D magnetic printing," *Nat. Commun.*, vol. 6, pp. 1–7, 2015, doi: 10.1038/ncomms9641.

[32] W. Dong *et al.*, "Fabrication of a continuous carbon fiber-reinforced phenolic resin composites via in situ-curing 3D printing technology," *Compos. Commun.*, vol. 38, no. January, 2023, doi: 10.1016/j.coco.2023.101497.

[33] Y. Shi *et al.*, *Ceramic materials for additive manufacturing*. 2021.

[34] P. Boch and J.-C. Nièpce, *Ceramics Materials. Process, Properties and Applications*, vol. 98. John Wiley & Sons, 2007.

[35] G. Maccauro, P. Rossi, L. Raffaelli, and P. Francesco, "Alumina and Zirconia Ceramic for Orthopaedic and Dental Devices," in *Biomaterials Applications for Nanomedicine*, no. July, InTech, 2011.

[36] C. Piconi, G. Maccauro, F. Muratori, and E. B. D. E. L. Prever, "Alumina and zirconia ceramics in joint," *J. Appl. Biomater. Biomech.*, vol. 1, pp. 19–32, 2003, doi: 10.1177%2F228080000300100103.

[37] E. Romanczuk-Ruszuk, B. Sztorch, D. Pakuła, E. Gabriel, K. Nowak, and R. E. Przekop, "3D printing ceramics—Materials for direct extrusion process," *Ceramics*, vol. 6, no. 1, pp. 364–385, 2023, doi: 10.3390/ceramics6010022.

[38] P. Kumar, B. S. Dehiya, and A. Sindhu, "Bioceramics for hard tissue engineering applications: A review," *Int. J. Appl. Eng. Res.*, vol. 13, no. 5, pp. 2744–2752, Feb. 2018, doi: 10.3844/ajbbsp.2006.49.56.

[39] K. Shanmugam and R. Sahadevan, "Bioceramics—An introductory overview," in *Fundamental Biomaterials: Ceramics*, Elsevier, 2018, pp. 1–46.

[40] L. C. Hwa, S. Rajoo, A. M. Noor, N. Ahmad, and M. B. Uday, "Recent advances in 3D printing of porous ceramics: A review," *Curr. Opin. Solid State Mater. Sci.*, vol. 21, no. 6, pp. 323–347, 2017, doi: 10.1016/j.cossms.2017.08.002.

[41] J. Deckers, K. Shahzad, J. Vleugels, and J. P. Kruth, "Isostatic pressing assisted indirect selective laser sintering of alumina components," *Rapid Prototyp. J.*, vol. 18, no. 5, pp. 409–419, 2012, doi: 10.1108/13552541211250409.

[42] K. Shahzad, J. Deckers, J. P. Kruth, and J. Vleugels, "Additive manufacturing of alumina parts by indirect selective laser sintering and post processing," *J. Mater. Process. Technol.*, vol. 213, no. 9, pp. 1484–1494, 2013, doi: 10.1016/j.jmatprotec.2013.03.014.

[43] Z. Chen *et al.*, "3D printing of ceramics: A review," *J. Eur. Ceram. Soc.*, vol. 39, no. 4, pp. 661–687, Apr. 2019, doi: 10.1016/j.jeurceramsoc.2018.11.013.

[44] V. K. Balla, S. Bose, and A. Bandyopadhyay, "Processing of bulk alumina ceramics using laser engineered net shaping," *Int. J. Appl. Ceram. Technol.*, vol. 5, no. 3, pp. 234–242, 2008, doi: 10.1111/j.1744-7402.2008.02202.x.

[45] H. Yves-Christian, W. Jan, M. Wilhelm, W. Konrad, and P. Reinhart, "Net shaped high performance oxide ceramic parts by Selective Laser Melting," *Phys. Procedia*, vol. 5, no. PART 2, pp. 587–594, 2010, doi: 10.1016/j.phpro.2010.08.086.

[46] A. Zocca *et al.*, "3D-printed silicate porous bioceramics using a non-sacrificial preceramic polymer binder," *Biofabrication*, vol. 7, no. 2, 2015, doi: 10.1088/1758-5090/7/2/025008.

[47] A. Butscher, M. Bohner, N. Doebelin, S. Hofmann, and R. Müller, "New depowdering-friendly designs for three-dimensional printing of calcium phosphate bone substitutes," *Acta Biomater.*, vol. 9, no. 11, pp. 9149–9158, 2013, doi: 10.1016/j.actbio.2013.07.019.

[48] T. A. Pham, D. P. Kim, T. W. Lim, S. H. Park, D. Y. Yang, and K. S. Lee, "Three-dimensional SiCN ceramic microstructures via nano-stereolithography of inorganic polymer photoresists," *Adv. Funct. Mater.*, vol. 16, no. 9, pp. 1235–1241, 2006, doi: 10.1002/adfm.200600009.

[49] M. C. Leu, B. K. Deuser, L. Tang, R. G. Landers, G. E. Hilmas, and J. L. Watts, "Freeze-form extrusion fabrication of functionally graded materials," *CIRP Ann. - Manuf. Technol.*, vol. 61, no. 1, pp. 223–226, 2012, doi: 10.1016/j.cirp.2012.03.050.

[50] M. Krinitcyn *et al.*, "Laminated object manufacturing of in-situ synthesized MAX-phase composites," *Ceram. Int.*, vol. 43, no. 12, pp. 9241–9245, 2017, doi: 10.1016/j.ceramint.2017.04.079.

[51] C. Gomes *et al.*, "Laminated object manufacturing of LZSA glass-ceramics," *Rapid Prototyp. J.*, vol. 17, no. 6, pp. 424–428, 2011, doi: 10.1108/13552541111184152.

[52] C. M. Gomes, A. P. N. Oliveira, D. Hotza, N. Travitzky, and P. Greil, "LZSA glass-ceramic laminates: Fabrication and mechanical properties," *J. Mater. Process. Technol.*, vol. 206, no. 1–3, pp. 194–201, 2008, doi: 10.1016/j.jmatprotec.2007.12.011.

[53] A. Safari, "Processing of advanced electroceramic components by fused deposition technique," *Ferroelectrics*, vol. 263, no. 1, pp. 45–54, 2001, doi: 10.1080/00150190108225177.

[54] A. Bandyopadhyay, K. D. Traxel, M. Lang, M. Juhasz, N. Eliaz, and S. Bose, "Alloy design via additive manufacturing: Advantages, challenges, applications and perspectives," *Mater. Today*, vol. 52, pp. 207–224, Jan. 2022, doi: 10.1016/J.MATTOD.2021.11.026.

[55] V. Petrovic, J. Vicente Haro Gonzalez, O. Jordá Ferrando, J. Delgado Gordillo, J. Ramón Blasco Puchades, and L. Portolés Griñan, "Additive layered manufacturing: sectors of industrial application shown through case studies," *Int. J. Prod. Res.*, vol. 49, no. 4, pp. 1061–1079, Feb. 2011, doi: 10.1080/00207540903479786.

[56] Y. W. D. Tay, B. Panda, S. C. Paul, N. A. Noor Mohamed, M. J. Tan, and K. F. Leong, "3D printing trends in building and construction industry: a review," *https://doi.org/10.1080/17452759.2017.1326724*, vol. 12, no. 3, pp. 261–276, Jul. 2017, doi: 10.1080/17452759.2017.1326724.

[57] F. Segonds, "Design by additive manufacturing: an application in aeronautics and defence," vol. 13, no. 4, pp. 237–245, Oct. 2018, doi: 10.1080/17452759.2018.1498660.

[58] N. Saengchairat, T. Tran, and C. K. Chua, "A review: additive manufacturing for active electronic components," vol. 12, no. 1, pp. 31–46, Jan. 2016, doi: 10.1080/17452759.2016.1253181.

[59] S. Gorsse, C. Hutchinson, M. Gouné, and R. Banerjee, "Additive manufacturing of metals: a brief review of the characteristic microstructures and properties of steels, Ti-6Al-4V and high-entropy alloys," http://www.tandfonline.com/action/journalInformation?show=aimsScope&journalCode=tsta20#.VmBmuzZFCUk, vol. 18, no. 1, pp. 584–610, Dec. 2017, doi: 10.1080/14686996.2017.1361305.

[60] J. H. K. Tan, S. L. Sing, and W. Y. Yeong, "Microstructure modelling for metallic additive manufacturing: a review," vol. 15, no. 1, pp. 87–105, Jan. 2019, doi: 10.1080/17452759.2019.1677345.

[61] T. DebRoy *et al.*, "Additive manufacturing of metallic components – Process, structure and properties," *Prog. Mater. Sci.*, vol. 92, pp. 112–224, Mar. 2018, doi: 10.1016/J.PMATSCI.2017.10.001.

[62] P. Vallabhajosyula and D. L. Bourell, "Modeling and production of fully ferrous components by indirect selective laser sintering," *Rapid Prototyp. J.*, vol. 17, no. 4, pp. 262–268, 2011, doi: 10.1108/13552541111138388.

[63] Y. Han *et al.*, "Microstructure and properties of a novel wear- And corrosion-resistant stainless steel fabricated by laser melting deposition," *J. Mater. Res.*, vol. 35, no. 15, pp. 2006–2015, Aug. 2020, doi: 10.1557/JMR.2020.70/METRICS.

[64] A. Bandyopadhyay, Y. Zhang, and S. Bose, "Recent developments in metal additive manufacturing," *Curr. Opin. Chem. Eng.*, vol. 28, pp. 96–104, Jun. 2020, doi: 10.1016/J.COCHE.2020.03.001.

[65] L. C. Zhang and Y. Liu, "Additive manufacturing of titanium alloys for biomedical applications," *Addit. Manuf. Emerg. Mater.*, pp. 179–196, Jan. 2018, doi: 10.1007/978-3-319-91713-9_5/COVER.

[66] T. Majumdar, N. Eisenstein, J. E. Frith, S. C. Cox, and N. Birbilis, "Additive manufacturing of titanium alloys for orthopedic applications: A materials science Viewpoint," *Adv. Eng. Mater.*, vol. 20, no. 9, p. 1800172, Sep. 2018, doi: 10.1002/ADEM.201800172.

[67] J. Mezzetta *et al.*, "Microstructure-properties relationships of Ti-6Al-4V parts fabricated by selective laser melting," *Int. J. Precis. Eng. Manuf. - Green Technol.*, vol. 5, no. 5, pp. 605–612file:///C:/Users/NEHA/Downloads/s40194-019-, 2018, doi: 10.1007/s40684-018-0062-1.

[68] Y. Ding, J. A. Muñiz-Lerma, M. Trask, S. Chou, A. Walker, and M. Brochu, "Microstructure and mechanical property considerations in additive manufacturing of aluminum alloys," *MRS Bull.*, vol. 41, no. 10, pp. 745–751, Oct. 2016, doi: 10.1557/MRS.2016.214/METRICS.

[69] A. Langebeck, A. Bohlen, H. Freisse, and F. Vollertsen, "Additive manufacturing with the lightweight material aluminium alloy EN AW-7075," *Weld. World*, vol. 64, no. 3, pp. 429–436, 2020, doi: 10.1007/s40194-019-00831-z.

[70] K. S. Derekar, "A review of wire arc additive manufacturing and advances in wire arc additive manufacturing of aluminium," vol. 34, no. 8, pp. 895–916, May 2018, doi: 10.1080/02670836.2018.1455012.

[71] J. John Samuel Dilip, H. Kalid Rafi, and G. D. Janaki Ram, "A new additive manufacturing process based on friction deposition," *Trans. Indian Inst. Met.*, vol. 64, no. 1–2, pp. 27–30, Feb. 2011, doi: 10.1007/S12666-011-0005-9/METRICS.

[72] R. Karunakaran, S. Ortgies, A. Tamayol, F. Bobaru, and M. P. Sealy, "Additive manufacturing of magnesium alloys," *Bioact. Mater.*, vol. 5, no. 1, pp. 44–54, Mar. 2020, doi: 10.1016/J.BIOACTMAT.2019.12.004.

[73] I. Yadroitsev, L. Thivillon, P. Bertrand, and I. Smurov, "Strategy of manufacturing components with designed internal structure by selective laser melting of metallic powder," *Appl. Surf. Sci.*, vol. 254, no. 4, pp. 980–983, Dec. 2007, doi: 10.1016/J.APSUSC.2007.08.046.

[74] C. Körner, H. Helmer, A. Bauereiß, and R. F. Singer, "Tailoring the grain structure of IN718 during selective electron beam melting," *MATEC Web Conf.*, vol. 14, p. 08001, 2014, doi: 10.1051/MATECCONF/20141408001.

[75] M. Bhuvanesh Kumar and P. Sathiya, "Methods and materials for additive manufacturing: A critical review on advancements and challenges," *Thin-Walled Struct.*, vol. 159, no. October 2020, 2021, doi: 10.1016/j.tws.2020.107228.

[76] U. Shukla and K. Garg, "Journey of smart material from composite to shape memory alloy (SMA), characterization and their applications-A review," *Smart Mater. Med.*, vol. 4, pp. 227–242, Jan. 2023, doi: 10.1016/J.SMAIM.2022.10.002.

[77] K. Parveen, "Metamaterials: Types, applications, development, and future scope," *Int. J. Adv. Res. Ideas Innov. Technol.*, vol. 4, no. 3, pp. 2325–2327, 2018.

[78] H. M. A. Kolken and A. A. Zadpoor, "Auxetic mechanical metamaterials," *RSC Adv.*, vol. 7, no. 9, pp. 5111–5129, 2017, doi: 10.1039/C6RA27333E.

[79] M. N. Ali, J. J. C. Busfield, and I. U. Rehman, "Auxetic oesophageal stents: Structure and mechanical properties," *J. Mater. Sci. Mater. Med.*, vol. 25, no. 2, pp. 527–553, 2014, doi: 10.1007/s10856-013-5067-2.

[80] G. Flamourakis *et al.*, "Laser-made 3D auxetic metamaterial scaffolds for tissue engineering applications," *Macromol. Mater. Eng.*, vol. 305, no. 7, p. 2000238, Jul. 2020, doi:10.1002/mame.202000238.

[81] R. Gatt *et al.*, "Hierarchical auxetic mechanical metamaterials," *Sci. Rep. 2015 51*, vol. 5, no. 1, pp. 1–6, Feb. 2015, doi: 10.1038/srep08395.

[82] H. Xue, Z. Luo, T. Brown, and S. Beier, "Design of self-expanding auxetic stents using topology optimization," *Front. Bioeng. Biotechnol.*, vol. 8, p. 736, Jul. 2020, doi: 10.3389/fbioe.2020.00736.

[83] Z. Fang *et al.*, "Modular 4D printing via interfacial welding of digital light-controllable dynamic covalent polymer networks," *Matter*, vol. 2, no. 5, pp. 1187–1197, May 2020, doi: 10.1016/J.MATT.2020.01.014.

[84] D. Kokkinis, M. Schaffner, and A. R. Studart, "Multimaterial magnetically assisted 3D printing of composite materials," *Nat. Commun. 2015 61*, vol. 6, no. 1, pp. 1–10, Oct. 2015, doi: 10.1038/ncomms9643.

[85] I. Pericevic, C. Lally, D. Toner, and D. J. Kelly, "The influence of plaque composition on underlying arterial wall stress during stent expansion: The case for lesion-specific stents," *Med. Eng. Phys.*, vol. 31, no. 4, pp. 428–433, May 2009, doi: 10.1016/J.MEDENGPHY.2008.11.005.

2 Metal-based Additive Manufacturing Technologies

Vipin Goyal, Ajit Kumar, Pawan Sharma, and Girish C. Verma

2.1 INTRODUCTION

Additive manufacturing (AM) is an advanced manufacturing technique that facilitates the fabrication of objects by layer upon layer [1]. There is remarkable growth in the manufacturing area to cater to the stringent requirements of present industries. Some of these requirements involve high product quality, complex features, defect-free products, etc., which are almost impossible to achieve with conventional manufacturing techniques. AM provides several advantages over the conventional technique, such as digital planning, freedom of part geometry design, easy customization, minimum wastage of material, elimination of production steps, low pollutant emission, and very complicated parts as a single unit [2]. In addition, with AM technique, we can tailor material composition and develop FGMs for specific uses [3,4]. AM techniques can be used in numerous fields, including aerospace, defense, automotive, construction, biomedical, and other industries [5]. Wohlers et al. [6] published a report on the fast growth of AM processes and predicted that in the next 5 years, AM-based industries are going to generate business of more than USD 22.9 billion worldwide. Figure 2.1 shows the revenue of the end market in the year 2018 in different industries' adoption of AM.

AM technology is able to fabricate a wide range of materials such as polymers, ceramics, and metals. Among these materials, metallic materials have garnered significant interest from both industry professionals and scholars. Many materials such as Ti alloys (Ti6Al4V), SS (SS304, SS3016L, tool steel), aluminum alloys (AlSi10Mg, AlSi12), Al–Cu–Mg), Ni-based superalloys, and Co–Cr-based alloys are used on the different AM techniques such as powder bed fusion (PBF) and direct energy deposition (DED) [7,8].

Many authors have reviewed the effects of processing parameters and materials (Ti, Al, Ni-based alloys, etc.) on the microstructure and mechanical properties [9,10]. It can be concluded that components fabricated through different AM techniques or deposited with different process parameters exhibit unique microstructure and mechanical properties.

DOI: 10.1201/9781003484325-2

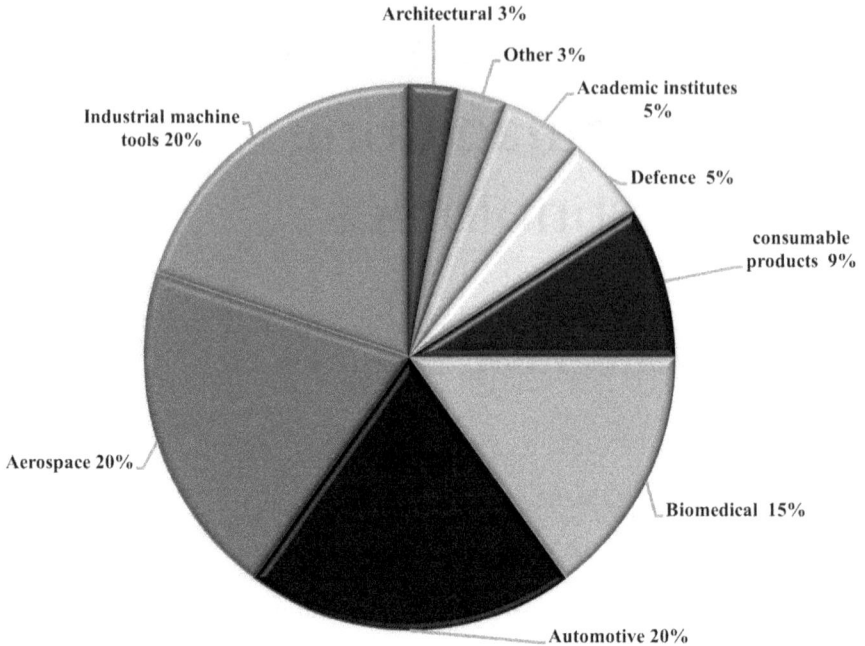

FIGURE 2.1 Market of metal additive manufacturing.

The aim of this chapter is to provide a detailed summary of metals-based AM techniques, highlighting their benefits, applicability in many industry areas, current challenges, and limitations.

2.2 AM TECHNOLOGIES FOR METALS

The AM techniques are classified mainly into seven categories by the American Society for Testing and Materials (ASTM) 52900–2015. Material extrusion (ME), PBF, direct energy deposing (DED), material jetting (MJ), binder jetting (BJ), and sheet lamination (SL) [11]. Among these AM techniques, metal-based AM techniques have been developed for manufacturing typical and complex metallic parts. These techniques can be categorized based on the feedstock material as shown in Figure 2.2.

It consists of three main fabrication techniques such as power bed, powder feed, and wire feed. All these techniques differ in their manner of spreading or layering the material as well as the way to fuse or bond the feedstock.

According to the ISO/ASTM standard 52900, the metal powder can be effectively used in PBF technique. It offers better reproducibility and dimensional accuracy as compared to conventional manufacturing techniques. PBF techniques are classified into three major techniques, namely, electron beam melting (EBM), selective laser melting (SLM)/selective laser sintering (SLS), and direct metal laser sintering (DMLS). All these processes are discussed in detail here [12].

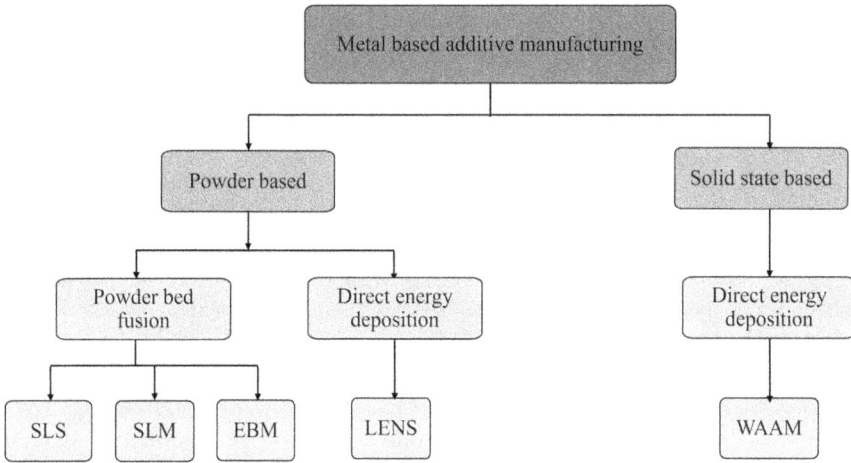

FIGURE 2.2 Classification of metal-based AM techniques.

2.2.1 POWDER BED FUSION

PBF is the most common AM technique that is employed to fabricate the metal parts with high dimensional accuracy and precision. Through this technique, numerous materials can be used, making it most useful for industries. Due to this more than 60% of metal parts are fabricated using the PBF process. The PBF process involves the high-powered laser or electron beam power that focuses on the metal power with a layer thickness of 20–100 μm and selectively fuses metal powder particles and fabricates the desired component. In SLM, the metal powder melted completely in an inert atmosphere of argon. During the process, high internal and thermal stresses are developed, which may be eliminated through post-processing techniques such as heat treatment, HIP, etc. The PBF process is shown in Figure 2.3 [13].

The process parameters such as 1) laser-related parameters (laser power, spot size, pulse duration, pulse frequency), (2) scan-related parameters (scan speed, scan spacing, scan rotation, scan pattern), (3) powder-related parameters (particle size and shape, distribution, layer thickness), and (4) auxiliary process parameters

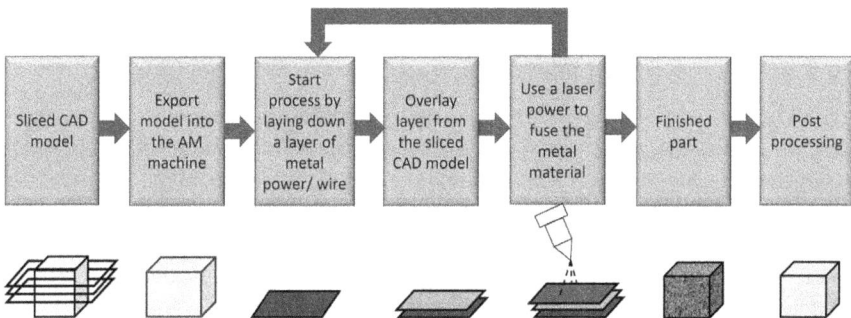

FIGURE 2.3 Schematic diagram of PBF process.

(bed temperature, gas flow, gas type) play a key role on the microstructure and mechanical properties [14].

SLM is extensively used in automotive and aerospace sectors to produce complicated parts including tooling, fuel injection components, and turbine blades with intricate channels that allow sufficient cooling and heating and prolonged service life. In the biomedical industry, medical tools, prosthetic hip and knee implants, and dental crowns are manufactured through SLM [15].

However, in DMLS, the powder is sintered in place of complete melting which makes the products porous as compared to SLM fabricated parts. Higher product accuracy and porous morphology are the key advantages of the DMLS process, which are highly required in the biomedical domain [16].

In EBM a high-intensity electron beam (upto 60cc/h), fuses the metal powder. The process is conducted inside a highly evacuated environment facilitating energy efficiency of over 95%, which is more than five times higher than the PBF process. EBM-manufactured parts are fabricated at an elevated temperature and subsequently annealed, resulting into lower internal stress [15].

Many parameters such as power melting temperature, energy source, powder size and shape, flowability, and layer thickness influence the properties of final components [17]. The microstructure and mechanical properties of the fabricated parts are highly affected by the manufacturing techniques. As compared to the conventional techniques, PBF offers many advantages such as high metal density, minimum internal defects, low manufacturing time, and minimum of human error. L-PBF-processed material has fine microstructure and the absence of intermetallic particles. Moreover, it exhibits better corrosion resistance as compared to traditionally manufactured components [18].

2.2.2 DIRECT ENERGY DEPOSITION (DED)

The metallic powder or wire is commonly used to fabricate large parts because of the high deposition rate. The powder or wire form is supplied via the nozzle and melted by the thermal energy (from electron beam or laser or arc) to be deposited into a product [19]. DED is less popular as compared to PBF, due to lower accuracy, and surface finish that requires post-processing. However, wire-based DED is 2–10 times faster than the PBF technique and it can also be used to produce large components [15]. The DED-based techniques, wire arc AM (WAAM), and laser-engineered net shaped (LENS) are explained later in detail.

2.2.2.1 Wire Arc Additive Manufacturing

WAAM uses a solid metallic wire (Ti alloy, nickel, copper, steel, magnesium, chromium) as the feedstock material, and an electric arc is used as a heating source. In this process, the metal wire melted and deposited layer upon layer. Conventional welding techniques, such as gas metal/tungsten/plasma arc welding, can also be employed in the WAAM process [20]. During the deposition, recrystallization may occur which leads to a fine or coarse grain structure dependent on the deposition temperature. Compared to the laser-based PBF process, WAAM requires low equipment costs and can deposit the material at high rates. This makes this process cost-effective and suitable for many applications such as construction, automobile, aerospace, and biomedical applications [21]. Moreover, feedstock material (wire) for WAAM is easy to manufacture through wire drawing, and it is easy to handle (transport and storage).

The WAAM systems include simulation software, data communication lines, a power supply, a wire feed system, and an execution actuator. The dimensional accuracy and surface finish is the big issue in the WAAM technique, and it is highly dependent on the process parameters including mode of welding, feed rate, deposition rate, current, and voltage as an energy input [22].

2.2.2.2 Laser-Engineered Net Shaped

This technology was first invented by Stanford University and U.S. Sandia National Laboratories and commercialized by Optomec [23]. In this technique, the metallic powder is supplied by an inert gas and melted using a high-intensity laser beam. This technique is used to fabricate the net-shaped products and repair the highly complex shapes and components. The heat-affected zone is very small in the LENS process, which results in low distortion after deposition. The summary of application, advantages, and limitations of the metal-based AM techniques are summarized in Table 2.1 [24] (Table 2.2).

TABLE 2.1

Different Metallic-based Additive Manufacturing Processes, Advantages, Limitations, and Applications

AM Category	AM Process	Advantages	Limitations	Application	Material Used
PBF	SLS SLM EBM	No support structure needed. no post-processing wide range of material used	Size limitations, High surface roughness, Expensive process	Biomedical aerospace, automobile, (turbine blades, heat exchanger turbine airfoils, nozzle, bracket)	Al, Cu, Ni, Co–Cr, SS, Ti6Al4V, Ni-based alloy, Al alloy, SS 316 L
DED	LENS WAAM	Excellent mechanical strength,	Post-processing, supporting structures needed Not environment friendly	Biomedical, aerospace, automobile	Al, Cu, Ni, Co–Cr, SS, Ti6Al4V, super alloy

TABLE 2.2

Mechanical Properties of AM-Processed Materials [25, 26]

AM Process	Material	Young Modulus (GPa)	Yield Strength	Ultimate tensile Strength	Elongation
PBF	SS316L	190–220	410–450	567–640	8–36
	Ti6Al4V	110–130	896–1034	965–1103	6–13.8
	Co–Cr alloy	210–240	640–1150	655–1896	4–6
	Co–Cr–Mo alloy	205–250	900–1200	1150–1450	>6

DED parts often require post-processing operations such as heat treatment and machining to obtain the desired dimensional accuracy, which are discussed later on.

2.3 ADDITIVELY MANUFACTURED METALS

Additively manufactured metals such as stainless steel, aluminum, titanium alloy, Co–Cr alloy, Ni-based alloys (Inconel), and Mg, as well as precious metals including gold and silver, are extensively used in various applications including automotive, aerospace, and defense [27]. Additively manufactured metals such as Ti-6Al-4V, SS 316 are preferred for the fabrication of patient-specific orthopedics implants. Some of the commonly used materials fabricated through AM are elucidated in Figure 2.4.

2.3.1 AM-Processed Stainless Steel

Various types of steel such as Austenitic stainless steels (304 L, SS316L), maraging steel, and tool steel are fabricated through different AM techniques. The mechanical properties of steel are highly dependent on process parameters, cooling rate, and elemental composition. The AM processes metallic parts are widely used in different typical applications such as automobile, aerospace, and biomedical [13].

2.3.2 AM-Processed Ti6Al4V Alloys

High material costs make it difficult to fabricate through conventional or WAAM techniques, therefore laser or electron-based AM is the best technique. Laser-processed materials have rapid cooling, where most of the β phase transformed α martensite, which is attributed to the unique microstructure, the higher hardness, and better mechanical strength. Ti–Nb–Zr, Ti–Nb–Zr–Sn, Ti–Al–Mo–V–Cr, Ti–Nb–Zr, and Ti-6Al–7Nb alloys are the commercially available powder and widely used in many biomedical applications, processed through PBF technologies [28].

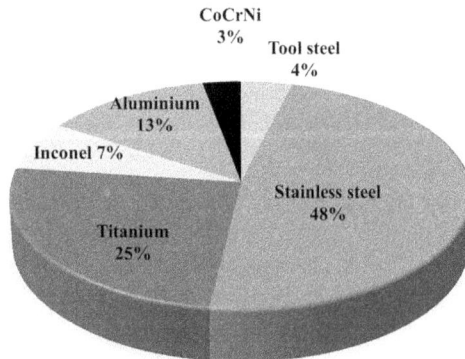

FIGURE 2.4 Commonly used metals in AM.

2.3.3   Iron and Iron-based Alloys

The iron can be used in 3D printing techniques to create porous orthopedic and dental implants that may reduce the stress shielding effect and promote cell proliferation, osseointegration, and bone growth. The biodegradation and cell growth rate of iron is quite low however it may be improved using alloying iron such as Mn, Pd. Moreover, the alloying elements may reduce the susceptibility of magnetic, which makes it the best suitable material for magnetic resonance imaging (MRI) compatibility [29].

2.3.4   Chromium and Cobalt Alloys

Chromium (Cr) is one of the major alloying elements often found in stainless steel enabling passivation and resist the formation of corrosion [30]. DED-processed Co–Cr–Mo is used in satellites [31]. Around 12% of the Co is used as the metal binder in WC-Co cemented carbides that can be processed by binder jetting due to their superior wetting, adhesion, and mechanical properties [32]. Moreover, Co exhibits excellent genotoxic and cancerogenic properties.

2.3.5   Zirconium (Zr) and Alloys

In prostheses and implants, Zr is often used as a prominent alloying element rather than an individual material. Zr and its alloy alloys have emerged as a promising class of biomaterials. Zr, particularly in the state of powder form possesses high reactivity and strong chemical affinity toward O, N, and H [15]. The Zr powders must be chemically pure because impurities such as H, O, C, N, and S may induce mechanical properties.

2.3.6   Magnesium (Mg), Zinc (Zn), and Alloys

Mg alloy is an attractive material for automotive, consumer electronics aerospace, and biomedical applications due to its high strength and low density. Moreover, Mg is biodegradable and has a young modulus (45 GPa) that is very close to the human bone. There is growing interest in biomedical applications (orthopedic and vascular surgery) due to the development of porosity in the orthopedic component, which is not possible with conventional technology [33].

Zn is an important mineral element in the body and is used as biomaterial. Zn is generally used as a porous material in the Mg alloy in AM. Zn can be added in the Mg as a minor quantity. A very low level of porosity can be developed when the Zn concentration in the alloy is very low (<1Wt%). If the Zn concentration is more than 6%, not only a large pore develops, but a severe crack formation may also occur [34].

PBF fabricated Al-based Mg alloy showed high strength and fine grain and equiaxed structure by superheating or inoculation and made the intermetallic phase. The intermetallic material can be used to develop the porosity, which may enhance cell viability and cell growth, which is a primary requirement for biomedical applications [35].

2.3.7 Nickel (Ni) and Ni-based Alloys

Ni-based alloys have been developed through the PBF process for numerous applications. L-PBF-processed alloy Ni-based alloys were found to be highly corrosive and fatigue resistant than the wrought counterpart material due to the higher surface hardness. Moreover, Karimi et al. [36] found the microstructure and corrosion behavior was highly influenced by the build direction. The scanning direction had the maximum corrosion resistance in 3.5% of NaCl solution as compared to the scanning direction.

2.3.8 Aluminum Alloys

Cracking is a common issue in the Al fabricated components using the conventional technique. However, the laser-based AM technique has a higher working temperature and lower cooling rate, which may effectively reduce the numerous defects such as cracking and surface finish by using high energy density [37].

The GE aviation company manufactured the nozzle in which more than 20 parts were consolidated and reduced weight by more than 25%. Overall, the company reduced a total number of 825 parts to consolidate into 12 parts. These results showed the drastic weight reduction in the aircraft and increased the fuel efficiency by more than 20% [38].

2.3.9 Other Metals

Other than steel, titanium alloy and metal high-performance alloys, such as high-entropy alloy, Inconel, Ni–Co–Cr, Fe–Co–Cr–Ni, Al–Co–Cr–Fe–Ni, and Co–Cr–Fe–Mn–Ni alloys, have also been investigated by AM for high-temperature applications. All these materials give better mechanical strength, microstructural stability, and high corrosion resistance at high temperatures [39]. In Ti-5AL-5MO-3Cr, the addition of Mo and Cr strongly influences the microstructure and mechanical properties of AM-processed components. It shows better fracture elongation (>10%) and relatively low tensile strength (<800 MPa) with a negligible work hardening capacity. Various materials processed through different AM techniques, their advantages, limitations, and applications are shown in Table 2.3.

TABLE 2.3

Various Materials Are Processed through Different AM Techniques, Their Advantages, Limitations, and Applications

Material	AM Process	Advantages	Applications	Refs.
SS 316 L	PBF	Excellent mechanical properties	Aerospace, automobile, biomedical, engineering industries	[13]
Ti6Al4V alloy	PBF DED WAAM	High strength, lightweight, biocompatible corrosion resistance	Automotive, aerospace, and healthcare industries.	[28]

TABLE 2.3 *(Continued)*
Various Materials Are Processed through Different AM Techniques, Their Advantages, Limitations, and Applications

Material	AM Process	Advantages	Applications	Refs.
Iron	PBF DED	Excellent mechanical properties, nontoxic, biodegradable	Artificial cardiovascular stents bone tissue engineering, autologous bone grafts	[29]
Co- Cr	DED	Excellent weldability, hot corrosion thermal fatigue, and wear resistance	Satellite, tool materials turbine blades for jet aircraft engines, biomedical implants	[16,40]
Zr Nb Ta	PBF DED	Biocompatible and mechanical properties	Biomedical, aerospace	[15]
Mg and alloy	PBF, WAAM	Lowest density, high strength, biodegradable, low young modulus (45 GPa)	Automotive, aerospace, biomedical (orthopedic and vascular surgery)	[33] [34]
Ni and alloys	PBF	High-temperature resistance	Aerospace applications	[41]
Al and alloy	PBF DED	High thermal conductivity that reduce the thermally induced stresses and increases the higher processing speeds	Automobile (heat exchanger), aerospace, and biomedical.	[37]
Cu and alloy	PBF WAAM	High thermal conductivity, high corrosion resistance	Aerospace, automotive, and electric fields	

2.4 DEFECTS

Different types of defects occur in different AM processes like low accuracy in WAAM, porosity in PBF, etc. In PBF technique, defects generally occur due to the following reasons: 1) powder-related defects, and 2) processing-related defects [13]

  i. **Powder-related defects:** AM-processed parts are affected by powder characteristics such as powder size, shape, and distribution. The spherical powder affects the surface finish, density, and flowability. The particle size influences the powder layer thickness and surface finish.
 ii. **Processing-related defects:** The processing parameters including laser power, laser scan speed, layer thickness, and hatching spacing create defects such as pores, keyholes, voids, lack of fusion, and porosity. Moreover, thermal heat generated during the process may cause to cracking, distortion, etc. The type of defects and their effects are summarized in Table 2.4 [13].

TABLE 2.4
The Classification and Source of Defects

Powder-related defects	Powder characterization
	Surface contamination and oxidation
	Trapped gas
	Porosity
Processing-related defects	Powder spreading anomalies
	Lack of fusion porosity
	Key hole porosity
	Formation of columnar grains
	Loss of alloying elements
	Spattering
	Residual stresses, cracking and delamination
	Surface finish and roughness
	Geometrical and dimensional defects

2.4.1 Porosity

The formation of porosity occurs due to powder characteristics (powder size, shape) and processing parameters such as low laser power, high layer thickness, and high velocity. The pores (>50 μm) generally occur on and beneath the surface. The porosity can be controlled by adjusting the processing parameters and appropriate selection of powder characteristics [42].

2.4.2 Residual Stress, Cracking, and Delamination

The residual stresses in AM-built components are due to the repeated heating and cooling cycles and processing parameters that lead to excessive cracking and delamination. [43]. In addition, the cracking and delamination susceptibility are also significantly influenced by the microstructural and metallurgical variables.

2.4.3 Surface Roughness

An important challenge in PBF fabricated components is surface finish leading to reduced service of parts [44]. Powder shape, size, and spreading and processing parameters such as build direction, laser power, scan speed, and layer thickness significantly affect the surface roughness. The surface roughness results into the premature failure of the fabricated parts due to the stress concentrations at the surface. Many researchers have reported that the surface roughness can be strongly controlled by processing parameters such as laser power, velocity, and beam profile [45].

2.5 POST-PROCESSING OPERATIONS

Post-processing operations such as powder removal, stress relief, support structure removal, and machining are the primary needs for ensuring dimensional accuracy and surface finishing for AM-fabricated metallic parts. These operations may include operations [46]. Many techniques such as heat or laser treatment, shot peening, hot isostatic pressing (HIP), coating, and machining are generally employed to enhance the surface finish, wear resistance, porosity, and microstructure of AM metallic components [47–49]. Cold rolling may also be employed for improving microstructure and decreasing the porosity of DED-made titanium and other metallic parts.

2.5.1 Heat Treatment

Heat treatment is also used to decrease residual and thermal stress that may result in the desirable grain structure and higher mechanical strength of the end-use metal AM parts. However, heat treatment may cause several internal (keyholes, gas porous) and external defects (surface roughness) that can be removed through HIP machining or surface treatment [47].

2.5.2 Hot Isostatic Pressing (HIP)

The HIP process is generally used on additively manufactured parts to reduce porosity and stress. HIP is performed at high temperature and pressure under an inert atmosphere on metallic components. High temperature and pressure cause to coarsen microstructural, resulting in decreased strength and undesirable microstructure [50]. HIP is a well-suited process for a wide range of materials, including Ti, SS, Al, and superalloys.

2.5.3 Machining and Surface Treatment

Machining and surface treatment are necessary to increase the dimensional accuracy, surface finishing, and elimination of the partially melted particles from the components [51]. Surfaces of the fabricated components may also be chemically treated or coated to enhance both physical and functional properties including reactivity, biocompatibility, corrosion, and wear resistance.

2.6 APPLICATIONS OF METAL-BASED AM TECHNIQUES

Metal AM has novel prospects for enhancing production capabilities, such as low lead times, development of new materials, low material wastages, and manufacturing of typical components, which are crucial factors that encourage industries to employ metal AM technologies. The application of different AM techniques in different sectors is discussed here.

2.6.1 Automotive

In recent years, the automotive industry has been the leading industry to adopt the AM technique, with a notable growth rate of 5%. The Formula Student Germany 2012 used EOS DMLS technology to fabricate the automotive components and resulted in a significant reduction of weight by 35% as compared to existing cast

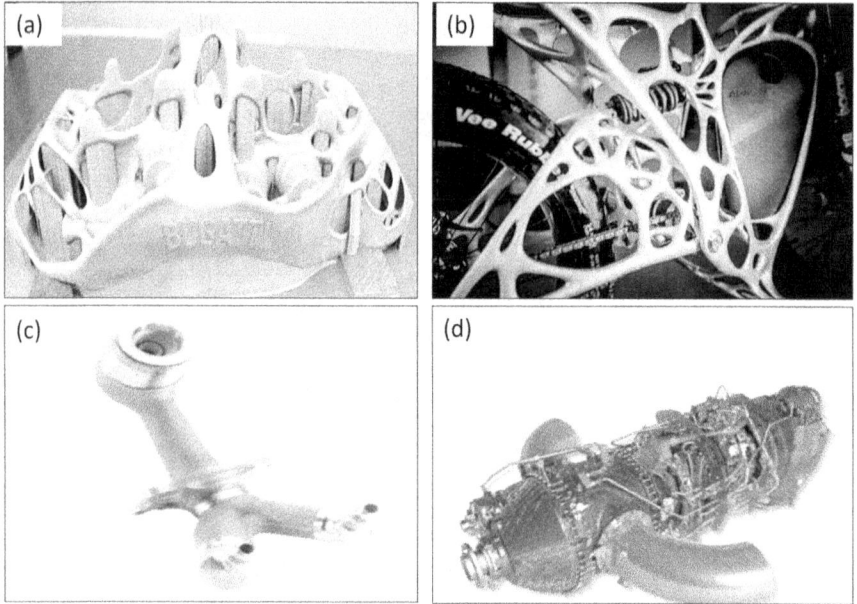

FIGURE 2.5 AM components fabricated for Aerospace: (a) Bugatti brake caliper, (b) lightweight component fabricated by scalmalloy Al alloy, (c) GE leap nozzle, (d) GE turboprop engine.

parts. German-based automotive company BMW, with HP and EOS, is able to manufacture 100 window guide rails in 24 hours and several components for engines, including cylinder heads and pumps, using different AM techniques [24].

Bugatti, a renowned car manufacturer, manufactured one of the biggest brake calipers in the world (2.9 kg). The feat was accomplished by the use of Ti6Al4V through an SLM technique. The weight of the bracket caliper was reduced by more than 40% from the original Aluminum bracket (Figure 2.5a) [52]. The Honda Motor German Airbus subsidiary AP Works fabricated the body shell parts and lightweight structure using the Al, Mg, and scalmalloy (Figure 2.5b) [24]. Scalmalloy is a notably corrosive resistant, lightweight material, therefore it is a prevalent choice for many high-performance applications [6].

2.6.2 Aerospace Application

Currently, PBF and DED are popular technologies that are extensively used in various sectors such as aerospace and defense to the fabrication of many components including aircraft and missile systems [16]. Aerospace components manufacturers, including Rolls-Royce, Airbus, Bombardier, Boeing, GE, and Air Force Research Lab-US implement SLM technique in civil and military aircraft and space vehicle applications [16,53].

General Electric (GE) aviation fabricated the nozzle for the LEAP engine (Figure 2.5c) using the laser-based Arcam metal printer that is 25% lighter in weight than the made through conventional casting with an efficiency of more than 15%. Moreover, the turbo pump engine (Figure 2.5d) assembled the 12 parts in one unit and enhanced the efficiency by more than 20%.

LENS technologies are furthermore used in repairing turbine airfoils and engine combustion chambers. The finding indicated that the repaired airfoil matched the geometry of the original geometry with a mean accuracy of 0.03 mm. Additionally, enhanced the carbon footprint by 45% and overall energy upto 36%.

2.6.3 BIOMEDICAL APPLICATION

The AM is well-suited for medical and dental applications to fabricate custom-ized parts for end users parts. Many companies such as Lima Corporate, Italy, and Nexxt Spine-USA have developed many Ti alloy-based porous orthopedic and spline implants using the PBF process. Many authors have concluded that AM can develop the porosity, pore geometry, surface roughness, and relative density inside the metallic parts that promote the new bone formation and improve the osseointegration process. Laser and electron-based AM techniques are used in major five areas medical model, physical model, surgical guides, and surgical aids. The physical model that is used to design and develop medical implants as well as to give a better solution to surgeons before performing any surgery.

AM also played a key role during the COVID-19 pandemic to fabricate essential personal protective equipment (PPE). This included the fabrication of various items such as protective safety goggles, masks, contact-free door knobs, manikins, respirators, face shields, and safety glasses, as well as ventilators and their components, for the safety and well-being of healthcare personnel [54]. Many companies and research institutions have started to use 3D printing (3DP) technology for the manufacturing of safety eyewear and small quarantine enclosures intended for healthcare settings [55]. Figure 2.6a–f illustrates many components of the 3DP pandemic, including preventive and diagnostics measures

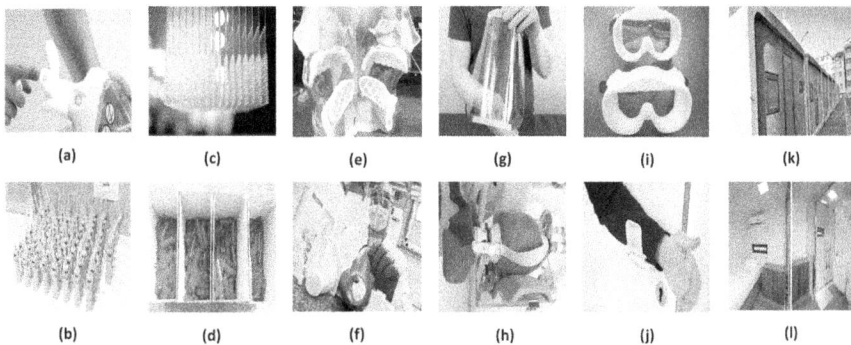

(a)   (c)   (e)   (g)   (i)   (k)

(b)   (d)   (f)   (h)   (j)   (l)

FIGURE 2.6 3D printing equipment's safety kit: (a) valves for ventilators, (b) valves for respiratory devices, (c) nasopharyngeal swabs, (d) syringes, (e) medical manikins, (f) silicon masks, (g) face shields, (h) respiratory equipment, (i) safety goggles, (j) door handles, (k) isolation wards, (l) quarantine enclosures.

like as nasopharyngeal swabs face syringes, shields, ventilator valves, testing equipment, medical gloves, test tubes, and other connections [5,15].

The new development of the design, process, and material reduces the stress shielding effect, low young modulus that is the primary requirement and increases the life of the implants.

2.6.4 ELECTRONICS AND COMMUNICATION

In recent years, metal AM has been gaining attention with the manufacturing of electronic and communication components, such as resistors, circuit boards, antennas, waveguides, and couplers, by using functional materials, including metal, semiconductor, and polymer. A laser's PBF technology-made RF metallic micro-antenna was manufactured by Optisys LLC for aerospace and defense applications, as shown in Figure 2.7a. In comparison with the conventionally manufactured

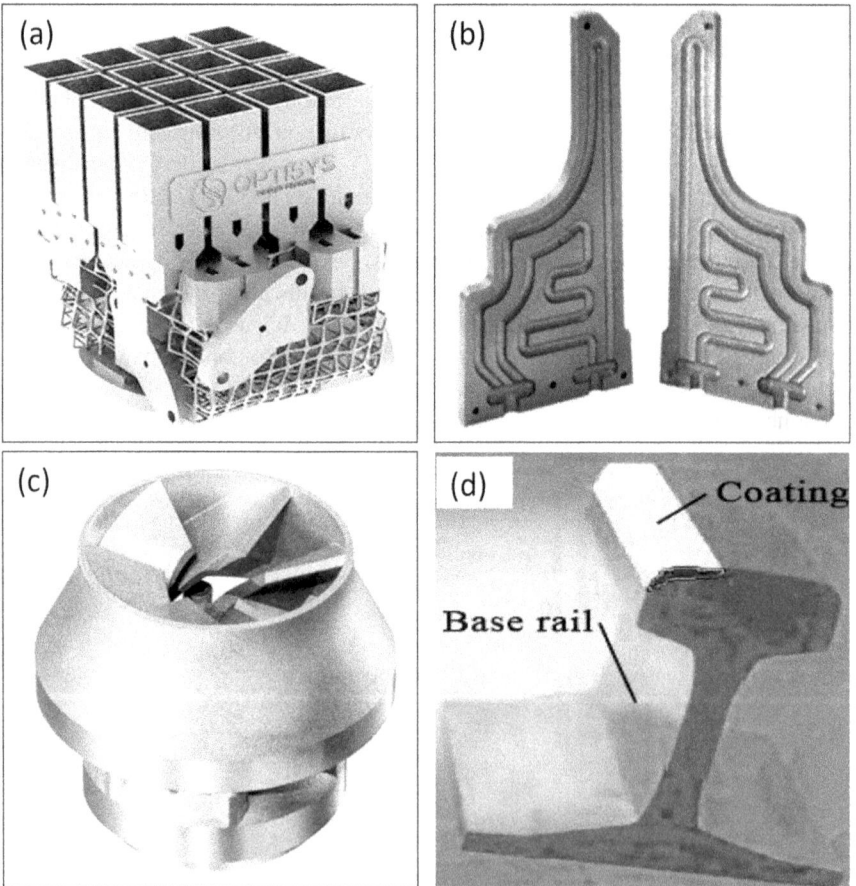

FIGURE 2.7 AM components (a:) lightweight AM antenna, (b) tool die set equipped with cooling channels, (c) Siemens-made burner head for oil and gas industry, (d) two-material rail.

antenna, the AM offers many advantages, including a substantial weight reduction of 95%, decreased manufacturing cost by 20%, and a significant reduction in lead time from 11 months to two months. Verploegh et al. [56] designed a W-band component using maraging steel through DMLS techniques. The component consists of a straight waveguide of 10 cm in length and a 20 dB coupler that ensures structural integrity and satisfactory conductivity.

2.6.5 TOOLS AND MOLDS

The use of metal AM is progressively gaining attention from manufacturers specializing in mold and tool production, enabling the fabrication of tooling components like mold inserts, jigs, fixtures, and gauges. Moreover, AM techniques have been used to fabricate different types of cooling channels in tools and mold (Figure 2.7b), which would be a challenging task using traditional manufacturing methods [24]. This additively manufactured mold offers rapid and uniform cooling, less deformation, residual stress, and casting flaws, as well as a 30% increase in efficiency, to the original mold.

2.6.6 OIL AND GAS

The major advantage of AM technique in the oil and gas sector is its potential to provide cost savings of around USD 30 billion, hence enhancing the efficiency of the supply chain system. The oil and gas sector is expected to invest more than USD 1.4 billion yearly on additively manufactured components during the span of the upcoming ten years [57].

Many industries such as GE, Siemens, and 3 T RPD use metal AM techniques to manufacture very complex and typical parts such as manifold titanium for pipelines, burner swirls for the SGT-750 gas turbine, sealing rings of SST-300 steam turbines, control valve parts, gas turbine blades, fuel nozzles, gas turbine burners, heat shields impellers, and burner swirls that were almost impossible to build with conventional manufacturing systems (Figure 2.7c) [24]. Moreover, several industries such as Barnes and Camisa possess the capability to fabricate lightweight offshore equipment such as separators, water pump impellers, and deck structures through the AM techniques [58].

2.6.7 RAILWAY

The railway sector has begun adopting the AM techniques due to the development of high wear and rolling contact fatigue resistance material that may cause surface failure [59]. Many companies such as Alstom (a French multinational company), MODTRAIN project, and Run2Rail-Europe Dubai's Road Transport, adopt AM techniques for the fabrication of lightweight parts using composite material [24,60]. Webtec, a global rail and transit manufacturer, is producing metal spares in numbers up to 250 for its production lines by 2025. The European Infrastar project employed laser cladding technology to fabricate a reinforced layer on the railhead area (Figure 2.7d). This methodology has the potential to enhance the performance of rolling contact fatigue, decrease the frequency of maintenance intervals, and mitigate the noise generated by the interface between the wheel and rail [59]. Based on the aforementioned information, AM technology not only repairs physical damage but also enhances rail performance and service life.

2.6.8 MINING

Mining projects are often conducted in graphically isolated and high-risk regions. Hence, in instances when any components fail, metal AM might prove to be very suitable, especially for the fabrication of substitute parts as per immediate requirements [61]. Aurora Labs, an Australian metal 3D printer manufacturer, has successfully developed on-demand replacement components in remote areas. Moreover, a collaborative initiative has been established between Downer's Mineral Technologies and the University of Technology Sydney Rapido Center for the manufacturing of mineral separation equipment. By using this approach, mining industries have the potential to achieve a significant reduction of up to 50% in downtime costs and time, while maintaining quality standards [24]. Hence, Sandvik, a Swedish equipment and tool manufacturer, has invested around AUD 30 million (USD 25 million) toward the advancement and proliferation of Metal AM within the mining industry [24].

2.7 CHALLENGES OF METAL-BASED AM TECHNIQUES

Apart from the many advantages and applications, AM technologies present some obstacles to widespread adoption, as well as a need for a comprehensive comprehension of emerging technologies and their potential long-term effects at the first phases of decision making. The following section addresses many significant obstacles associated with the deployment of AM technology.

2.7.1 PRODUCTION VOLUME

Currently, many of the metal-based AM methods are only suitable for high-complexity geometries of customized components and low-volume production [62]. Moreover, metal AM is not suggested for applications that need a great degree of flexibility and low cost per part. Therefore, due to the development of new technology, industries have begun to expand their business for the mass production of metallic components [63].

2.7.2 COSTS

The fabrication of metallic AM components is not simple and cheap. Numerous authors have reported that the total manufacturing cost of components including material, manpower, energy, and machinery must be reduced. The overall cost also includes running costs, buildings, consumables, designing, testing, inspection, and post-processing. The development of a cost analysis model is needed that estimates the AM parts versus a conventionally manufactured product [64].

2.7.3 LIMITED MATERIALS

Although there are various materials available in the market; however, limited materials can be processed through the AM technique. Therefore, research and investigation continue to expand materials with the adoption of current metals such as Gold, Silver, Inconel, Cu, Ni-based superalloys, tool steels, Al alloys, Pt, Palladium, Ta, high-entropy alloys, functionally graded materials (FGMs), and composite structures [65].

2.7.4 STANDARDS COMPLIANCE

The dimensional accuracy, surface finish, material properties, and repeatability for a specific part are the crucial factors in the AM-fabricated parts. Therefore, standards play a key role in the fabrication of end-use metallic components [66]. Currently, AM is progressing into industrial-scale production. Hence, the significance of technical standards, including many perspectives on metal-based, is highlighted. AM has been undertaken mainly by the International Organization for Standardization (ISO), the American Society for Testing and Materials (ASTM), European Committee for Standardization (CEN), Deutsches Institute for Normung (DIN), and British Standards Institution (BSI) to set the technical guidelines and propose the list of standards related to process, testing, methods processing parameters, and tolerance [24].

2.7.5 REPAIRS AND MAINTENANCE

The use of as-printed assemblies presents a significant obstacle, rendering ordinary maintenance or repair tasks unfeasible. In case of failure of any components, it becomes necessary to replace the whole component, which may further increase the wastage and costs throughout the lifespan of the product [67]. In order to address these constraints, it is essential to formulate design strategies, use optimization methods, and establish maintenance processes. When assessing the impacts of design modifications using AM, designers need to possess a comprehensive understanding of the associated expenses related to repair, maintenance, and disposal.

2.8 FUTURE SCOPE

The future scope of AM to the production of metallic parts for enhancing the thermal and mechanical characteristics for various applications such as aerospace, automotive, and biomedical. AM is a cost-effective and time-intensive technique. However, the strength is significantly influenced by processing parameters, hence offering substantial opportunities for future study in this field. Ongoing research and development efforts within the metal AM domain, including innovative materials, design methodologies, and technologies, have effectively addressed the aforementioned obstacles and expanded the scope of industrial application. However, there are certain prospects and opportunities for improvement.

For instance, the comprehensive understanding of the interrelations among AM techniques, processing parameters, and mechanical properties for any metal remains incomplete.

Some theoretical models still need to be developed. These models may include several phenomena like as heat and mass transmission, porosity and residual stress prediction, distortion analysis, and phase change, among others. The issue of sustainability pertaining to metal AM presents a notable research gap that necessitates the exploration of strategies aimed at minimizing waste generation and energy consumption. This may be achieved by the implementation of appropriate metal AM technologies and the subsequent adoption of just-in-time production practices. Furthermore, the quality of produced metal AM parts may be enhanced by the use of innovative materials and/or methods, as well as the implementation of topology optimization approaches. In addition, the use of innovative hybrid manufacturing techniques presents potential avenues for enhancing the surface quality of items as well as facilitating the repair and/or reshaping of pre-existing products. Furthermore, it is important to conduct optimization studies on process parameters and post-processing procedures in order to augment production efficiency in metal AM approaches. In addition, more investigation is necessary to delineate the primary determinants for the adoption of metal AM and to formulate a decision-making framework that may assist various industrial sectors in determining the optimal use of metal AM technology. The establishment of technical standards for metal AM is a continuous endeavor aimed at ensuring the consistent quality of metal prints.

2.9 CONCLUSION

The overview of this book chapter is the metallic-based AM techniques that can be used to produce different metallic components. Based on the study, the following conclusions can be drawn:

1. The metal-based AM technique was found to be excellent for customized and complex parts.
2. PBF-based processes produce accurate and precise products as compared to DED-based processes.
3. The metallic-based AM technique has the potential to fabricate high-performance structural parts for aerospace and biomedical applications.
4. Products fabricated through metal-based AM processes possess better mechanical properties as compared to conventional manufacturing techniques.
5. AM techniques are also used to fabricate components from superalloys which are preferred in high-temperature applications.
6. AM creates an impact in the field of sustainable manufacturing with lower material wastage and eco-friendly processing.
7. Despite several challenges, AM is viewed as a future manufacturing technology for several critical areas like aerospace, biomedical, defense, etc.

REFERENCES

[1] R. Citarella and V. Giannella, "Additive manufacturing in industry," *Appl. Sci.*, vol. 11, no. 2, pp. 1–3, 2021, doi: 10.3390/app11020840.

[2] T. Duda and L. V. Raghavan, "3D Metal Printing Technology," *IFAC-PapersOnLine*, vol. 49, no. 29, pp. 103–110, 2016, doi: 10.1016/j.ifacol. 2016.11.111.

[3] K. V. Wong and A. Hernandez, "A Review of Additive Manufacturing," *ISRN Mech. Eng.*, vol. 2012, pp. 1–10, 2012, doi: 10.5402/2012/208760.

[4] D. Rejeski, F. Zhao, and Y. Huang, "Research needs and recommendations on environmental implications of additive manufacturing," *Addit. Manuf.*, vol. 19, pp. 21–28, 2018, doi: 10.1016/j.addma.2017.10.019.

[5] S. Rajendran *et al.*, "Metal and polymer based composites manufactured using additive manufacturing—A brief review," *Polymers* , vol. 15, no. 11, pp. 1–19, 2023, doi: 10.3390/polym15112564.

[6] J. Wohlers, T. Campbell, I. Diegel, O. Huff, and R. Kowen, "3D Printing and additive manufacturing state of the industry: Annual Worldwide Progress Report," *ISRN Mech. Eng 2012*, pp. 1–10, 2012.

[7] D. G. Ahn, "Direct metal additive manufacturing processes and their sustainable applications for green technology: A review," *Int. J. Precis. Eng. Manuf. Green Technol.*, vol. 3, no. 4, pp. 381–395, 2016, doi: 10.1007/s40684-016-0048-9.

[8] Z. Pan, D. Ding, B. Wu, D. Cuiuri, H. Li, and J. Norrish, "Arc welding processes for additive manufacturing: A review," *Trans. Intell. Weld. Manuf.*, no. 2, pp. 3–24, 2018, doi: 10.1007/978-981-10-5355-9_1.

[9] M. Yakout, M. A. Elbestawi, and S. C. Veldhuis, "A review of metal additive manufacturing technologies," *Solid State Phenom.*, vol. 278 SSP, pp. 1–14, 2018, doi: 10.4028/www.scientific.net/SSP.278.1.

[10] T. DebRoy *et al.*, "Additive manufacturing of metallic components – Process, structure and properties," *Prog. Mater. Sci.*, vol. 92, pp. 112–224, 2018, doi: 10.1016/j.pmatsci.2017.10.001.

[11] S. Kumar and S. Pityana, "Laser-based additive manufacturing of metals," *Adv. Mater. Res.*, vol. 227, pp. 92–95, 2011, doi: 10.4028/www.scientific.net/AMR. 227.92.

[12] S. Bose, D. Ke, H. Sahasrabudhe, and A. Bandyopadhyay, "Progress in materials science additive manufacturing of biomaterials," *Prog. Mater. Sci.*, vol. 93, pp. 45–111, 2018, doi: 10.1016/j.pmatsci.2017.08.003.

[13] A. Mostafaei *et al.*, "Defects and anomalies in powder bed fusion metal additive manufacturing," *Curr. Opin. Solid State Mater. Sci.*, vol. 26, no. 2, 2022, doi: 10.1016/j.cossms.2021.100974.

[14] N. T. Aboulkhair, N. M. Everitt, I. Ashcroft, and C. Tuck, "Reducing porosity in AlSi10Mg parts processed by selective laser melting," *Addit. Manuf.*, vol. 1, pp. 77–86, 2014, doi: 10.1016/j.addma.2014.08.001.

[15] V. V. Popov *et al.*, "Powder bed fusion additive manufacturing using critical raw materials: A review," *Materials*, vol. 14, no. 4, p. 909, Feb. 2021, doi: 10.3390/ma14040909.

[16] M. Revilla-Leon, M. Sadeghpour, and M. Ozcan, "A review of the applications of additive manufacturing technologies used to fabricate metals in implant dentistry," *J. Prosthodont.*, vol. 29, no. 7, pp. 579–593, Aug. 2020, doi: 10.1111/jopr.13212.

[17] L. E. Murr *et al.*, "Metal fabrication by additive manufacturing using laser and electron beam melting technologies," *J. Mater. Sci. Technol.*, vol. 28, no. 1, pp. 1–14, 2012, doi: 10.1016/S1005-0302(12)60016-4.

[18] P. Fathi, M. Mohammadi, X. Duan, and A. M. Nasiri, "A comparative study on corrosion and microstructure of direct metal laser sintered AlSi10Mg_200C and die cast A360.1 aluminum," *J. Mater. Process. Technol.*, vol. 259, no. February, pp. 1–14, 2018, doi: 10.1016/j.jmatprotec.2018.04.013.

[19] M. Schmidt *et al.*, "Laser based additive manufacturing in industry and academia," *CIRP Ann.*, vol. 66, no. 2, pp. 561–583, 2017, doi: 10.1016/j.cirp.2017.05.011.

[20] Y. Xia, H. Teng, X. Zhang, D. Zheng, and G. Quan, "A review of the wire arc additive manufacturing of Ti-6Al-4V alloy: Properties, defects and quality improvement," *Chongqing Daxue Xuebao/Journal Chongqing Univ.*, vol. 45, no. 4, pp. 87–99, 2022, doi: 10.11835/j.issn.1000-582X.2020.265.

[21] J. F. Arinez, Q. Chang, R. X. Gao, C. Xu, and J. Zhang, "Artificial intelligence in advanced manufacturing: Current status and future outlook," *J. Manuf. Sci. Eng. Trans. ASME*, vol. 142, no. 11, pp. 1–16, 2020, doi: 10.1115/1.4047855.

[22] U. Alonso, F. Veiga, A. Suárez, and T. Artaza, "Experimental investigation of the influence of wire arc additive manufacturing on the machinability of titanium parts," *Metals .*, vol. 10, no. 1, p. 24, Dec. 2019, doi: 10.3390/met10010024.

[23] S. Rahmati, *Direct Rapid Tooling*, vol. 10. Elsevier, 2014.

[24] A. Vafadar, F. Guzzomi, A. Rassau, and K. Hayward, "Advances in metal additive manufacturing: A review of common processes, industrial applications, and current challenges," *Appl. Sci.*, vol. 11, no. 3, p. 1213, Jan. 2021, doi: 10.3390/app11031213.

[25] J. J. Lewandowski and M. Seifi, "Metal additive manufacturing: A review of mechanical properties," *Annu. Rev. Mater. Res.*, vol. 46, no. April, pp. 151–186, 2016, doi: 10.1146/annurev-matsci-070115-032024.

[26] V. Goyal, A. Kumar, and G. C. Verma, "Role of additive manufacturing in biomedical application," in *Additive Manufacturing: Advanced Materials and Design Techniques*, Ist., Yashvir Singh, Pulak Mohan Pandey, Nishant K. Singh, Ed. CRC Press, Taylor & Francis Group, LLC, 2023, p. 117.

[27] E. Celik, *Additive Manufacturing of Metals*. De Gruyter, 2020.

[28] Y. Li, C. Yang, H. Zhao, S. Qu, X. Li, and Y. Li, "New developments of ti-based alloys for biomedical applications," *Materials*, vol. 7, no. 3, pp. 1709–1800, 2014, doi: 10.3390/ma7031709.

[29] M. Salama, M. F. Vaz, R. Colaco, C. Santos, and M. Carmezim, "Biodegradable iron and porous iron: Mechanical properties, degradation behaviour, manufacturing routes and biomedical applications," *J. Funct. Biomater.*, vol. 13, no. 2, 2022, doi: 10.3390/jfb13020072.

[30] M. L. Grilli *et al.*, "Solutions for critical raw materials under extreme conditions: A review," *Materials*, vol. 10, no. 3, pp. 1–23, 2017, doi: 10.3390/ma10030285.

[31] M. Moradi, A. Ashoori, and A. Hasani, "Additive manufacturing of stellite 6 superalloy by direct laser metal deposition – Part 1: Effects of laser power and focal plane position," *Opt. Laser Technol.*, vol. 131, no. April, p. 106328, 2020, doi: 10.1016/j.optlastec.2020.106328.

[32] R. K. Enneti, K. C. Prough, T. A. Wolfe, A. Klein, N. Studley, and J. L. Trasorras, "Sintering of WC-12%Co processed by binder jet 3D printing (BJ3DP) technology," *Int. J. Refract. Met. Hard Mater.*, vol. 71, no. July 2017, pp. 28–35, 2018, doi: 10.1016/j.ijrmhm.2017.10.023.

[33] Z. Zeng, M. Salehi, A. Kopp, S. Xu, M. Esmaily, and N. Birbilis, "Recent progress and perspectives in additive manufacturing of magnesium alloys," *J. Magnes. Alloy.*, vol. 10, no. 6, pp. 1511–1541, 2022, doi: 10.1016/j.jma.2022.03.001.

[34] K. Wei *et al.*, "Selective laser melting of Mg-Zn binary alloys: Effects of Zn content on densification behavior, microstructure, and mechanical property," *Mater. Sci. Eng. A*, vol. 756, no. April, pp. 226–236, 2019, doi: 10.1016/j.msea.2019.04.067.

[35] A. Pawlak, M. Rosienkiewicz, and E. Chlebus, "Design of experiments approach in AZ31 powder selective laser melting process optimization," *Arch. Civ. Mech. Eng.*, vol. 17, no. 1, pp. 9–18, 2017, doi: 10.1016/j.acme.2016.07.007.

[36] P. Karimi, E. Sadeghi, J. Ålgårdh, P. Harlin, and J. Andersson, "Effect of build location on microstructural characteristics and corrosion behavior of EB-PBF built Alloy 718," *Int. J. Adv. Manuf. Technol.*, vol. 106, no. 7–8, pp. 3597–3607, 2020, doi: 10.1007/s00170-019-04859-9.

[37] H. Zhang, H. Zhu, T. Qi, Z. Hu, and X. Zeng, "Selective laser melting of high strength Al-Cu-Mg alloys: Processing, microstructure and mechanical properties," *Mater. Sci. Eng. A*, vol. 656, pp. 47–54, 2016, doi: 10.1016/j.msea.2015.12.101.

[38] T. Kellner, "An epiphany of disruption: GE additive chief explains how 3D printing will upend manufacturing," *GE Rep.*, vol. 13, pp. 1–10, 2017.

[39] M. Niinomi, M. Nakai, and J. Hieda, "Development of new metallic alloys for biomedical applications," *Acta Biomater.*, vol. 8, no. 11, pp. 3888–3903, 2012, doi: 10.1016/j.actbio.2012.06.037.

[40] B. Postolnyi, O. Bondar, M. Opielak, P. Rogalski, and J. P. Araújo, "Structural analysis of multilayer metal nitride films CrN/MoN using electron backscatter diffraction (EBSD)," *Adv. Top. Optoelectron. Microelectron. Nanotechnol. VIII*, vol. 10010, p. 100100E, 2016, doi: 10.1117/12.2243279.

[41] T. M. Pollock and S. Tin, "Nickel-based superalloys for advanced turbine engines: Chemistry, microstructure, and properties," *J. Propuls. Power*, vol. 22, no. 2, pp. 361–374, 2006, doi: 10.2514/1.18239.

[42] C. L. A. Leung, S. Marussi, M. Towrie, R. C. Atwood, P. J. Withers, and P. D. Lee, "The effect of powder oxidation on defect formation in laser additive manufacturing," *Acta Mater.*, vol. 166, pp. 294–305, 2019, doi: 10.1016/j.actamat.2018.12.027.

[43] S. Kou, "A criterion for cracking during solidification," *Acta Mater.*, vol. 88, pp. 366–374, 2015, doi: 10.1016/j.actamat.2015.01.034.

[44] A. Townsend, N. Senin, L. Blunt, R. K. Leach, and J. S. Taylor, "Surface texture metrology for metal additive manufacturing: A review," *Precis. Eng.*, vol. 46, pp. 34–47, 2016, doi: 10.1016/j.precisioneng.2016.06.001.

[45] I. Koutiri, E. Pessard, P. Peyre, O. Amlou, and T. De Terris, "Influence of SLM process parameters on the surface finish, porosity rate and fatigue behavior of as-built Inconel 625 parts," *J. Mater. Process. Technol.*, vol. 255, no. December 2017, pp. 536–546, 2018, doi: 10.1016/j.jmatprotec.2017.12.043.

[46] J. J. Lewandowski and M. Seifi, "Metal additive manufacturing: A review of mechanical properties," *Annu. Rev. Mater. Res.*, vol. 46, pp. 151–186, 2016, doi: 10.1146/annurev-matsci-070115-032024.

[47] N. T. Aboulkhair, I. Maskery, C. Tuck, I. Ashcroft, and N. M. Everitt, "Improving the fatigue behaviour of a selectively laser melted aluminium alloy: Influence of heat treatment and surface quality," *Mater. Des.*, vol. 104, pp. 174–182, 2016, doi: 10.1016/j.matdes.2016.05.041.

[48] K. T. Yang, M. K. Kim, D. Kim, and J. Suhr, "Investigation of laser powder bed fusion manufacturing and post-processing for surface quality of as-built 17-4PH stainless steel," *Surf. Coati. Technol.*, vol. 422, no. July, 2021, doi: 10.1016/j.surfcoat.2021.127492.

[49] B. AlMangour and J. M. Yang, "Improving the surface quality and mechanical properties by shot-peening of 17-4 stainless steel fabricated by additive manufacturing," *Mater. Des.*, vol. 110, pp. 914–924, 2016, doi: 10.1016/j.matdes.2016.08.037.

[50] B. Zhang, W. J. Meng, S. Shao, N. Phan, and N. Shamsaei, "Effect of heat treatments on pore morphology and microstructure of laser additive

manufactured parts," *Mater. Des. Process. Commun.*, vol. 1, no. 1, pp. 1–8, 2019, doi: 10.1002/mdp2.29.

[51] E. Lyczkowska-Widlak, P. Lochynski, G. Nawrat, and E. Chlebus, "Comparison of electropolished 316L steel samples manufactured by SLM and traditional technology," *Rapid Prototype. J.*, vol. 25, no. 3, pp. 566–580, 2019, doi: 10.1108/RPJ-03-2018-0060.

[52] T. M. Wischeropp, H. Hoch, F. Beckmann, and C. Emmelmann, "Opportunities for braking technology due to additive manufacturing through the example of a Bugatti Brake Caliper," pp. 181–193, 2019, doi: 10.1007/978-3-662-58024-0_12.

[53] R. Russell *et al.*, *Qualification and Certification of Metal Additive Manufactured Hardware for Aerospace Applications*. 2019.

[54] M. M. Erickson, E. S. Richardson, N. M. Hernandez, D. W. Bobbert, K. Gall, and P. Fearis, "Helmet modification to PPE With 3D printing during the COVID-19 pandemic at Duke University Medical Center: A novel technique," *J. Arthroplasty*, vol. 35, no. 7, pp. S23–S27, 2020, doi: 10.1016/j.arth.2020.04.035.

[55] P. Zhang, Z. Wang, J. Li, X. Li, and L. Cheng, "From materials to devices using fused deposition modeling: A state-of-art review," *Nanotechnol. Rev.*, vol. 9, no. 1, pp. 1594–1609, 2020, doi: 10.1515/ntrev-2020-0101.

[56] S. Verploegh, M. Coffey, E. Grossman, and Z. Popović, "Properties of 50-110-GHz waveguide components fabricated by metal additive manufacturing," *IEEE Trans. Microw. Theory Tech.*, vol. 65, no. 12, pp. 5144–5153, 2017, doi: 10.1109/TMTT. 2017.2771446.

[57] L. Vendra and A. Achanta, "Metal additive manufacturing in the oil and gas industry," in *Solid Freeform Fabrication 2018: Proceedings of the 29th Annual International Solid Freeform Fabrication Symposium – An Additive Manufacturing Conference*, 2018, pp. 454–460.

[58] J. A. Barnes, J. B.; Camisa, "Additive manufacturing for oil and gas-potential of topology optimization for offshore applications," in *In Proceedings of the 29th International Ocean and Polar Engineering Conference, Honolulu, HI, USA*, pp. 16–21.

[59] H. Fu and S. Kaewunruen, "State-of-the-art review on additive manufacturing technology in railway infrastructure systems," *J. Compos. Sci.*, vol. 6, no. 1, 2022, doi: 10.3390/jcs6010007.

[60] A. Killen, L. Fu, S. Coxon, and R. Napper, "Exploring the use of additive manufacturing in providing an alternative approach to the design, manufacture and maintenance of interior rail components," in *Proceedings of ATRF 2018 – Australasian Transport Research Forum*, pp. 1–16, 2018.

[61] C. S. Frandsen, M. M. Nielsen, A. Chaudhuri, J. Jayaram, and K. Govindan, "In search for classification and selection of spare parts suitable for additive manufacturing: a literature review," *Int. J. Prod. Res.*, vol. 58, no. 4, pp. 970–996, 2020, doi: 10.1080/00207543.2019.1605226.

[62] T. Yamazaki, "Development of a hybrid multi-tasking machine tool: Integration of additive manufacturing technology with CNC machining," *Proc. CIRP*, vol. 42, no. Isem Xviii, pp. 81–86, 2016, doi: 10.1016/j.procir.2016.02.193.

[63] U. M. Dilberoglu, B. Gharehpapagh, U. Yaman, and M. Dolen, "The role of additive manufacturing in the era of industry 4.0," *Proc. Manuf.*, vol. 11, no. June, pp. 545–554, 2017, doi: 10.1016/j.promfg.2017.07.148.

[64] A. Vafadar, K. Hayward, and M. Tolouei-Rad, "Sensitivity analysis for justification of utilising special purpose machine tools in the presence of uncertain parameters," *Int. J. Prod. Res.*, vol. 55, no. 13, pp. 3842–3861, 2017, doi: 10.1080/00207543. 2017.1308032.

[65] W. E. King *et al.*, "Laser powder bed fusion additive manufacturing of metals; physics, computational, and materials challenges," *Appl. Phys. Rev.*, vol. 2, no. 4, p. 041304, 2015, doi: 10.1063/1.4937809.

[66] M. D. Monzón, Z. Ortega, A. Martínez, and F. Ortega, "Standardization in additive manufacturing: activities carried out by international organizations and projects," *Int. J. Adv. Manuf. Technol.*, vol. 76, no. 5–8, pp. 1111–1121, 2015, doi: 10.1007/s00170-014-6334-1.

[67] N. Knofius, M. C. van der Heijden, and W. H. M. Zijm, "Consolidating spare parts for asset maintenance with additive manufacturing," *Int. J. Prod. Econ.*, vol. 208, no. November 2018, pp. 269–280, 2019, doi: 10.1016/j.ijpe.2018.11.007.

3 Robotic Systems Used in Additive Manufacturing

Thira Patel and Vishvesh Badheka

3.1 INTRODUCTION

Robotics is an interdisciplinary branch of Computer Science and Engineering that involves the design, construction, operation, and use of robots. The objective is to create machines that can assist humans. Robotics is slowly being utilized for specific tasks in different fields of work [1]. Current applications of robotics include military robots, collaborative robots, construction robots, agricultural robots, medical robots, etc., and many more are in development. In general, robots can be used for many different applications by simply changing the tool heads attached to their end effectors.

Military robots are autonomous robots or remote-controlled mobile robots designed for military applications like transport, search and rescue, and attack. Yadav et al. [2] have described the design and construction of Land Mine Detection Troops Safety Robot (LMDTSR). The autonomous robot scans a specific area and determines the presence of land mines. A PIC microcontroller is used to control the robot autonomously. A cobot, or collaborative robot, is a robot made for direct human–robot interaction within a shared space, or where humans and robots are in close proximity [1]. Gao et al. [3] have developed an intelligent master-slave collaborative robot system for cafeteria service. The developed system can automatically complete the tasks of scooping dishes, taking bowls, and pouring dishes into the bowl based on master-slave collaboration and can thus improve the efficiency and reduce the labor cost of cafeterias. Robots are used for distinct purposes in the field of medicine itself. Robots are used for treatment and diagnostic purposes, as introduced by Duan et al. [4]. They have discussed about a remote robotic ultrasound scanning system so that effective treatment can reach rural areas and small towns as well. Robots in the healthcare sector are being used for a range of applications from developing structural health management systems for health monitoring [5], to the use of intelligent disinfection robots in medical institutions for disinfecting surfaces [6]. Along with these diverse applications, robots are also being used for performing additive manufacturing (AM), which is also called 3D printing.

AM describes a group of processes that produce objects by depositing materials in a layer-by-layer way [7]. It is also called 3D printing. Applications of 3D printing can be found in the construction industry for building structures, in the fashion industry for creating shoes and clothes, in the automobile industry for printing

 DOI: 10.1201/9781003484325-3

mechanical components, and even in the food industry for printing food [8]. As per ASTM, AM is classified into the following seven categories: VAT photopolymerization, sheet lamination, binder jetting, material extrusion, material jetting, powder bed fusion, and direct energy deposition [9]. Over the past few years, a new trend in AM is emerging, called robot-assisted AM, or robot-based AM.

This chapter discusses the concept of robot-assisted AM, the different robots being used in the process, and rather than segregating using the AM techniques, uses the robots as a basis for classification. Examples of robots used in AM are extensively discussed throughout this chapter. We also analyze the general trends observed and identify the challenges as well as future work that can be done in this field.

3.2 ROBOT-ASSISTED ADDITIVE MANUFACTURING: PROCESS AND TOOLS

Robot-assisted Additive Manufacturing is the process in which multi-axis robot systems are used for the process of AM. The conventional AM setup builds parts from planer layers. Thus, curved geometries take longer build times, due to the large number of planar layers required to build them. Also, staircase effect may occur, resulting in poor surface finish. Stacking of planar layers to create parts can lead to undesirable anisotropic material properties, thus decreasing the strength of the manufactured part [10]. Incorporating robots in 3D printing, therefore, offers many advantages. Alsharhan et al. [11] compared the parts created using robot-assisted 3D printing with the same parts printed using planar methods. It could be observed that the part printed using conformal method had higher peak load, and stiffness as compared to the one made using planar method. Also, there was reduction in interlayer failures. A robot with multiple degrees of freedom (generally more than three for this type of AM) has a large workspace. These types of robots can allow for printing in more than one direction, and thus eliminate the need for support structures. Similarly, mobile robots can be said to have an endless workspace. Robot-assisted 3D printing thus allows more complex structures to be created [12].

Among the many papers available for robot-assisted AM, different types of robots with different tool head from different manufacturers have been used. The types of robots used by researchers have been discussed further, along with their specifications. Classification based on robots used seems like a much more effective way of tracking the progress in this field of research.

3.2.1 6 DoF ROBOTIC MANIPULATOR

A Robotic Manipulator is a type of robot that consists of mechanical, electrical, and electronic components and is programmed in such a way that it can perform repetitive work autonomously [13]. It is sometimes called a Robotic Arm or an Industrial Robot. These are commonly used in the industry for welding, painting, assembly, disassembly, pick and place for printed circuit boards, packing and labeling, palletizing, product inspection, and testing, all of which require a high level of endurance, speed, and precision [1]. The Degree of Freedom (DoF) of this type of robot can be defined as the number of independent joints that can perform either translational or rotational motion along or about a unique geometric axis [14]. This section discusses the 6 DoF robotic manipulators that have been used for AM by various researchers. It was observed that the majority of robots being used for Robot-based AM, fall in this category. Table 3.1 states the specifications of these robotic manipulators and other AM-related details.

Yoon et al. [15] have developed an extrusion system for conformal FDM using a single 6 DoF robotic manipulator from Yaskawa Motoman, model GP12. Robotic 3d printing was preferred here as it allows for conformal printing, which highly improves the mechanical properties of the printed final part. The material used for printing is Polylactic Acid (PLA) plastic. In their previous work [16], they used two robotic manipulators, namely, ABB IRB 120 and ABB IRB 2600, to perform the same operation. One of the robots used a high-resolution nozzle to print the surfaces, whereas the other robot used a low-resolution nozzle to fill the inner core of the part. This method reduced the build time, but the substrates had to initially be created by a commercial 3D printer. Thus, to overcome this difficulty, a 3-nozzle extrusion system was proposed in the next paper [15] which had a separate nozzle for building support structures.

Ishak et al. [17–19] used a Motoman SV3X robotic manipulator for performing FDM of PLA plastic. The Fused Deposition modeling method allows creation of 3D lattice structures which have high stiffness properties, based on the size of the struts in the lattice structure. However, in planar FDM, the size of these lattice structures is limited to 2 mm (diameter). Thus, to create lattice structures of even smaller diameters, which will further enhance the stiffness of the part, Ishak et al. [18] have proposed the use of robot-assisted 3D printing, as it is a multi-plane printing process. Several other papers were reviewed, and it was observed that 6 DoF robotic manipulators were used extensively to perform FDM using PLA and Acrylonitrile-Butadiene-Styrene (ABS) [20–27]. The design of shell geometries is very common for architectural and construction-based applications because of their aesthetic appeal as well as structural properties. However, 3D printing of shell-like structures using planar AM methods can only be performed when extensive use of support structures is made. Curved Layer Fused Filament Fabrication (CLFFF) may be better at developing these structures but still has some limitations. Complex geometries and overhangs, as well as insertion of support bars inside the structure becomes a challenge even using CLFFF. Also, the nozzle diameter size increases as material needs to be deposited in bulk. Thus, robot-based 3D printing is not good enough to print these shell structures accurately because for creating these precise shapes a small nozzle will be used in robotic AM. Thus, to solve this problem, [21] have tried to replicate the mechanism from which snails create shells, and thus perform Fused Filament Fabrication (FFF) in a similar way. The build process needs to be continuous. The thermoplastic material is first melted for deposition, and then rapidly cooled, all while deposition takes place in a curved manner using a robot manipulator. A guided heating system is used to attach the next strip of

TABLE 3.1
Specifications of 6 DoF Robotic Manipulators Used in AM [15–28,30–33,35–42]

Robot	Payload	Horizontal reach	Max Speed (wrist angle)	Repeatability	Robot weight	AM Method	Material used	References
Yaskawa Motoman GP12, 6 DoF	12 kg	1440 mm	8.203 rad/sec	±0.02 mm	150 kg	FDM	PLA	[15]
Motoman model SV3X arm	3 kg	677 mm	7.33 rad/sec	±0.03 mm	30 kg	FDM	PLA plastic	[17]
Motoman SV3X; 6 DoF	3 kg	677 mm	7.33 rad/sec	±0.03 mm	30 kg	FDM	PLA plastic (Thermoplastic polymer)	[19]
Motoman SV3X; 6 DoF	3 kg	677 mm	7.33 rad/sec	±0.03 mm	30 kg	SLS and FDM	PLA plastic	[18]
KUKA industrial robot arm; 6 DoF						Shaped Metal Deposition (SMD)	316 Stainless Steel	[31]
KUKA KR5 sixx R850, 6 DoF	5 kg	855 mm	11.51 rad/sec	±0.03 mm	29 kg	FDM, milling sculpting	Acrylonitrile-Butadiene Styrene (ABS) plastic	[27]
KUKA robot arm 6-axis coupled with an extruder head with multiple nozzles						FDM	ABS and PLA	[26]
KUKA-KR30HA	30 kg	2033 mm	5.62 rad/sec	±0.05 mm	665 kg	Laser-based Additive Manufacturing	316 L stainless steel	[28]
EPSON S5-A701S 6 DoF	5 kg	786 mm		±0.02	36 kg	Sheet lamination-based AM	PLA	[35]
Omron Adept Viper s650	5 kg	650 mm	10.46 rad/sec	±0.02	27.987 kg	Stereolithography (SLA)	photopolymers	[30]
Stäubli RX90L, 6 DoF	11 kg	985 mm	10.11 rad/sec	±0.02 mm		FDM	PLA/ABS	[25]

(Continued)

TABLE 3.1 (Continued)
Specifications of 6 DoF Robotic Manipulators Used in AM [15–28,30–33,35–42]

Robot	Payload	Horizontal reach	Max Speed (wrist angle)	Repeatability	Robot weight	AM Method	Material used	References
ABB IRB 6620 6-axis Industrial robotic arm coupled with an extruder head with 2 pumps	150 kg	2200 mm	3.32 rad/sec	±0.03	880 kg	FDM	Concrete	[36]
ABB IRB 2400/10; 6 DoF	10 kg	1550 mm	6.98 rad/sec	±0.03	380 kg	WAAM	Aluminum–silicon alloy AA4047; nickel-based alloy Inconel 625; aluminum–silicon alloy, AA4043	[33]
ABB 2400/L robotic arm	7 kg	1800 mm	5.41 rad/sec	±0.070 mm	380 kg	Anti-gravity AM	very rapidly hardening, two-component thermosetting polymer	[37]
ABB IRB 120 industrial robot arm, 6 DoF	3 kg	580 mm	7.33 rad/sec	±0.01 mm	25 kg	FDM	Blending a masterbatch of flax pellets (brown pellets) with a blend of co-polyesters (white pellets)	[38]
ABB robot arm coupled with extrusion head						FDM	clay and ceramic mixture	[39]
ABB robot arm coupled with extrusion printing head; 6 DoF						FDM	Filament form polymer; ABS	[24]

ABB IRB 120	3 kg	25 kg	580 mm	7.33 rad/sec	±0.01 mm	Freeform Fabrication		[40]
ABB IRB 140	6 kg	98 kg	800 mm	7.85 rad/sec	±0.030 mm			
ABB IRB 120	3 kg	25 kg	580 mm	7.33 rad/sec	±0.01 mm	FDM	PLA	[16]
ABB IRB 2600 manipulator; 6 DoF	20 kg	272 kg	1650 mm	8.73 rad/sec	±0.040 mm			
ABB IRB 1400 robot arm for AM, 6 DoF; ABB 6660 robot for machining and post-processing	5 kg	225 kg	1440 mm	4.89 rad/sec	±0.05 mm	GMAW-based AM		[32]
6 DoF industrial robot						FDM	ABS, SAN, PMMA, PP, PC, PC / ABS, PLA, and PVA all can be used	[20]
6 DoF robot arm coupled with extrusion head						FDM	Thermoplastic polymer	[21]
6 DoF robot arm, equipped with multiple changeable heads and an integrated manufacturing platform						FDM- AM; Milling- SM	PLA(Polylactic acid); the Same setup can also be used for ABS (Acrylonitrile-Butadiene-Styrene) and can further be extended to metals by adjusting the hardware design.	[22]
6 DoF robotic arm						FDM	ABS thermoplastic	[23]

thermoplastic laterally to the previous one, and this time a widened nozzle is used. In this way, consequent layers are attached precisely and without the need for support structures. The fabrication speed as well as freedom of different shape generation is increased in this process. Felsch et al. [20] have described a robotic AM system for printing large-scale structures cost-effectively, whereas Felbrich et al. [21] have focused on architectural applications of robot-based AM. Li et al. [22] and Keating et al. [27] have proposed a robotic platform that can be used for additive (FDM) as well subtractive (Milling) manufacturing. Li et al. [22] realized that the hybrid systems being designed for manufacturing generally used Subtractive Manufacturing (SM) as a post-processing step. Thus, to fill this research gap they proposed a hybrid robotic additive-subtractive system. This has been achieved by developing different tool heads or end effectors for each manufacturing process, that can be attached to the 6 DoF robot. This setup also includes a platform that has a heating bed for AM and a fixturing clamp for SM. Case studies show that this hybrid technique highly decreases the material waste generated in conventional AM or SM, decreases the needs for support structures, and also reduces the machining time. Other advantages of this setup include capability to build freeform parts, as well as increased surface finish, because of the elimination of staircase effect, which is possible because of multi-axis printing using the 6 DoF robot. Moreover, these hybrid systems may affect the life-cycle of the product, as they allow product reuse. This also helps close the supply chain. Materials other than ABS and PLA were also used for robot-assisted FDM. These materials include concrete for large-scale 3D printing, clay and ceramic mixture, and a mixture of flax pellets and co-polyesters.

Some researchers have also used robots for AM techniques other than material extrusion (FDM/FFF). A KUKA-KR30HA robot was used for laser-based AM of 316 L SS by Zheng et al. [28]. Sheet Lamination-based Robotic Additive Manufacturing has also been practiced by Bhatt et al. [29], using PLA as the printing material. They used EPSON S5-A701S robotic manipulator for this experiment. Stevens et al. [30] performed Stereolithography (SLA) of photo-polymers using Omron Adept Viper s650 robotic arm. Development of this system for large-scale production will be beneficial as SLA can be performed with a wide variety of materials. Arc welding-based AM has also been performed with the help of 6 DoF robotic manipulators. Bonaccorso et al. [31] performed Gas Tungsten Arc Welding (GTAW)-based AM technique called Shaped Metal Deposition (SMD) of 316 Stainless Steel (SS) for aerospace applications. A 6 DoF robot from the robot manufacturer KUKA, was used for this process. The main aim of their research was to design an efficient control interface for this Robot-based AM process. For this purpose, they selected four control parameters for the arc welding process: heat source provided, the torch speed, wire feeder rate, and thickness of each layer. Ding et al. [32] have performed GMAW-based AM using the ABB IRB 1400 robotic arm on a completely automated AM system that produces finished parts from the CAD model. Another robot, the ABB IRB 6660 is also incorporated into the system for performing post-processing on the 3D printed part. Evjemo et al. [33] have used ABB IRB 2400/10 robot for Wire Arc Additive Manufacturing (WAAM) of Aluminum-silicon alloy AA4047, nickel-based alloy Inconel 625, and aluminum–silicon alloy, AA4043, for three different experiments. In the first experiment, a thin-walled quadratic box was built, and the path of the robot was programmed manually using the robot's control pad. The build process in this case was continuous to save time and also analyze the effects of this type of build using a robot manipulator. Some unwanted accumulation was observed in the corners of the build. This was because of the inaccuracy of the robot operator as this build was done manually. It was concluded that manual programming may work for simple cases but might lead to a loss of accuracy due to human error. Performing WAAM with aluminum was challenging because of its low melting point. The second experiment was also first performed using Aluminum alloy but the structure being manufactured had intersections where the robot would pass through more than once, and this led to melting and collapsing of some of the structure already welded, due to the walls of the structure that hampered heat dissipation.

Process planning is one of the major challenges of Robot-based Additive Manufacturing with 6 DoF robotic arms. To overcome this, Ding et al. [32] have come up with an automated process planning algorithm for arc welding-based AM. This algorithm allows automated robotic AM by taking only the CAD model as the input. This novel research could be a pioneer for robot-based AM if the algorithm can be somewhat mirrored for other AM processes as well. Due to multi-axis printing, process planning becomes complex and the production time increases. The next section discusses ways to overcome this difficulty (Figures 3.1 and 3.2).

FIGURE 3.1   6 DoF robotic arm schematic [34].

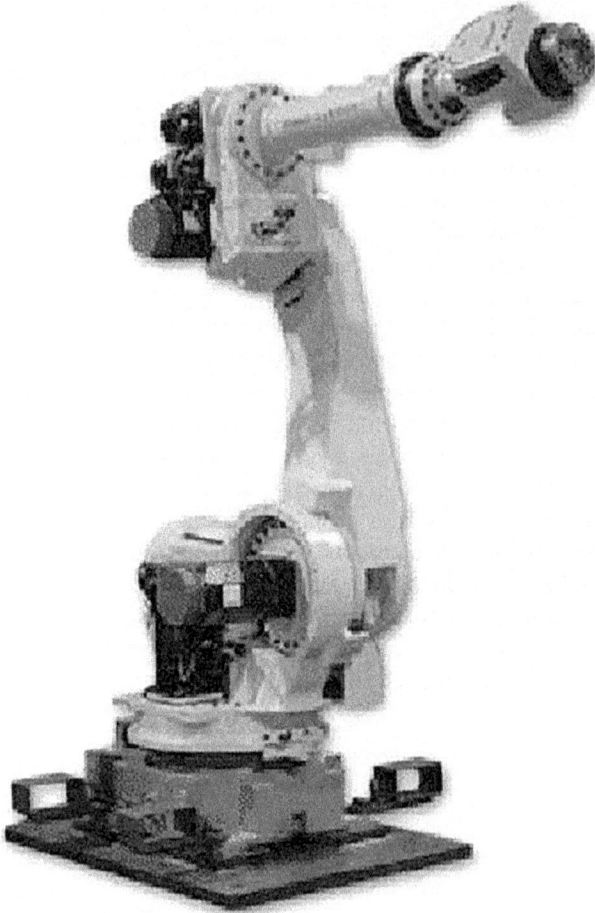

FIGURE 3.2   Image of 6 DoF robotic arm or Industrial Robot Arm [34].

3.2.2   ROBOTIC AM SYSTEM WITH GREATER THAN 6 DoF

All robotic arms do not necessarily have a maximum of 6 DoF. To further expand the accessible workspace of the robot for AM, manipulators with greater than 6 DoF have also been used. Sometimes, 6 DoF robotic arms are coupled with 2-axis positioning systems. This is useful to reorient the build part without having an offset of the AM tool head with respect to the build part. Conventionally, the tool head needs to recover the build part's position every time before deposition of the next layer. This is done to account for the rotation of the worktable during printing. This complicates the process planning and increases production time. However, when a 6 DoF robotic arm is coupled with a 2-axis positioning system, this problem is solved as the tool head moves whenever there is a movement of the build part [43].

Reisch et al. [44] have used a COMAU NJ130 2.0 which is a 6 DoF robot, and have coupled it with a 2-axis tilt-turn table (COMAU PTS-ORB-1000). This robotic system is used to perform Wire Arc Additive Manufacturing (WAAM) to create large-scale AlSi12 (Aluminum alloy) parts, which are used mainly for aerospace applications. Also, a monitoring system has been proposed which uses multivariate sensors to detect defects in the final part after printing.

A novel method for Laser-based Directed Energy Deposition was proposed by [45], for manufacturing dome-shaped structures in a single step. They used a 6-axis (6 DoF) Fanuc M-20iA robotic arm coupled with a 2-axis servo positioner H875 that has a 500 kg payload. The material used is SS 316 powder of size 15–45 μm.

Another paper by Ding et al. [43] focuses on the building of revolved parts by Laser-based Directed Energy Deposition using SS 431 L HC powder sized from 53 to 150 mm (a product of North America Hoganas High Alloys LLC). Here too an 8-axis robotic system is used, which consists of a 6-axis KUKA robot with a 2-axis tilt and rotary positioning system. A hybrid slicing method is proposed which maps the overhanging structures of the revolved part at a planar base. Finally, conventional process planning techniques are used to generate toolpaths for the mapped structure. Akbari et al. [46] used a similar 8-axis system for Laser-based Directed Energy Deposition but the material used for printing (Low alloy steel-AWS ER 100S-G) was in wire form.

Mcgee et al. [47] from the University of Michigan have used a 7 DoF KUKA robot coupled with a screw extruder, to perform material extrusion-based AM using thermoplastic elastomer (TPE) as the print material. The challenge was to design an elastic net geometry that could be extruded on a flat surface and could later be stretched into anti-elastic forms. Table 3.2 summarizes the specifications of the previously discussed robotic manipulators as well as the AM methods used.

The different papers reviewed for this particular category of robot systems (>6 DoF) reveal that Directed Energy Deposition (DED) is majorly performed using a 6 or greater degree of freedom robotic manipulator, coupled with a positioning system that adds extra degrees of freedom to the system (Figure 3.3).

3.2.3 COLLABORATIVE ROBOTS

Collaborative robots (Cobots) are robots that are meant to work in environments where humans are also present. Typically, these robotic arms have added safety features like rounded edges, can be made up of lightweight materials, and may have a limitation of speed or force. However, current trends also consider mobile robots with these safety features as cobots [48]. Application of cobots includes being used as information robots in public places, used as industrial robots for industrial tasks where human interaction takes place like assembly operations and lifting of heavy parts [1].

Wu et al. [49] and Danielsen Evjemo et al. [50] have used collaborative robots offered by Universal Robots for two different 3D printing methods using different materials. Wu et al. [49] used a UR3 robot coupled with an extrusion head for Fused Deposition Modeling (material extrusion). The material used for printing was

TABLE 3.2

Specifications of Robot Systems Having Greater Than 6 DoF Used in AM [41–47]

Robot	Payload	Horizontal reach	Max Speed (wrist angle)	Repeatability	Robot weight	AM Method	Material used	References
6-axis robot COMAU NJ130 2.0	130 kg	2050 mm		±0.07 mm	740 kg	WAAM	AlSi12 (Aluminum alloy)	[44]
Positioning system	**Payload**	**Max moment of flexure**	**Output Max Rotation Speed**	**Repeatability**	**Weight**			
2-axis tilt-turn table COMAU PTS-ORB-1000	1000 kg	2060 Nm	2.618 rad/sec	±0.06 mm	630 kg			
Robot	**Payload**	**Horizontal reach**	**Max Speed (wrist angle)**	**Repeatability**	**Robot weight**	**AM Method**	**Material used**	**References**
A 6-axis Fanuc M-20iA robotic arm with 2-axis servo positioner H875	20 kg	1811 mm	9.6 rad/s (J6)	±0.08 mm	250 kg	Laser-directed energy deposition	SS 316 L	[45]
KUKA 6-axis robot coupled 2-axis tilt and rotatory positioning system						Laser-based direct metal deposition (LBDMD)	Stainless steel powder 431 L HC sized from 53 to 150 mm (a product of North America Hoganas High Alloys LLC)	[43]
KUKA robot 6-axis along with a 2-axis positioning system						Laser/wire additive manufacturing	Low alloy steel(AWS ER 100S-G)	[46]
KUKA robot 7-axis, coupled with screw-assisted extruder						Extrusion probably	Thermoplastic elastomer (TPE)	[47]

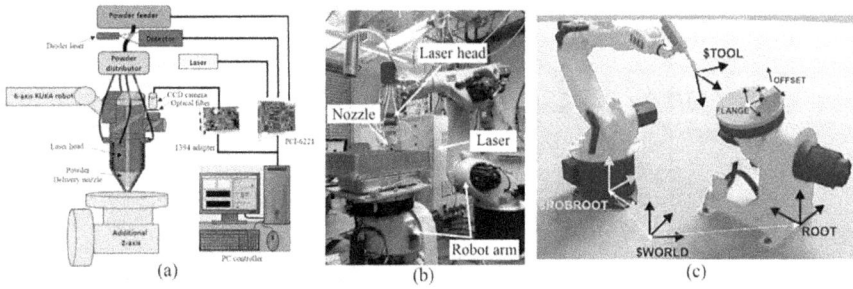

FIGURE 3.3 Setup of laser-based direct metal deposition system: (a) schematic, (b) photo, (c) coordinate systems [43].

Polylactic Acid (PLA) in filament form. A new algorithm was formulated by these researchers to divide models into separate support-free parts that can be printed one by one, and while avoiding collisions. On the other hand, Danielsen Evjemo et al. [50] also used a UR robot, specifically, the UR5 cobot to perform AM using the welding method of Cold Metal Transfer (CMT). The paper focuses on the possibility of printing large-scale structures using this process.

Bhatt et al. [29] have proposed a robotic cell for AM that uses two KUKA LBR iiwa 7 R800 collaborative robots. These robots are used to perform Sheet lamination-based AM of Acrylonitrile-Butadiene Styrene (ABS). The second robot can be replaced by a human operator also because it is mainly used for clamping and similar activities during the sheet lamination process. Table 3.3 summarizes the specifications of the previously discussed robotic manipulators as well as the AM methods used.

Felbrich et al. [51] have used Reinforcement Learning (RL) methods to introduce autonomy to robot-based AM. They have tested the proposed algorithm on a UR 10 robot setup for performing material extrusion of thermoplastic.

Industrial robots may be the better option for large-scale 3D printing, due to their high payload capacities. Also, the 3D printing process would not necessarily involve human intervention, and thus robots used need not be collaborative in nature (Figures 3.4 and 3.5).

3.2.4 Mobile Robots

Mobile robots are robots that can move around a place (achieve locomotion), unlike industrial robots that are stationary [54]. The study of mobile robots is a trending area of research in robotics these days. We can see mobile robots being used for material transportation in warehouses, medical facilities, and even in industries. Moreover, mobile robots are used in unconventional applications like land mine detection and other military operations [2]. These types of robots have also been used for AM by some researchers, mainly in the field of Building and Construction (B&C). For AM purposes, a robotic manipulator is attached to a mobile platform, so that the 3D printing can be carried out on a large scale, as the robot becomes capable of covering more area by moving around. Also, multiple mobile robots may be used. This increases the scalability of the AM process and is thus advantageous [55].

TABLE 3.3
Specifications of Collaborative Robots Used in AM [29,41,42,49–51]

Robot	Payload	Horizontal reach	Max Speed (wrist angle)	Repeatability	Robot weight	AM Method	Material used	References
UR3 robot arm 6 DoF, coupled with extrusion head	3 kg	500 mm	6.28 rad/sec	±0.1 mm	11 kg	FDM	Filament form polylactic acid (PLA)	[49]
UR5 robot arm	5 kg	850 mm	3.142 rad/sec	±0.1 mm	18.4 kg	Cold Metal Transfer (CMT) used for AM		[50]
2 KUKA LBR iiwa 7 R800, a 7 DoF robot arm	7 kg	800 mm	3.14 rad/sec	±0.1 mm	24 kg	Sheet lamination-based AM	ABS	[29]
UR 10 robot	10 kg	1300 mm	3.14 rad/sec	±0.1 mm	28.9 kg	FDM	Thermoplastic	[51]

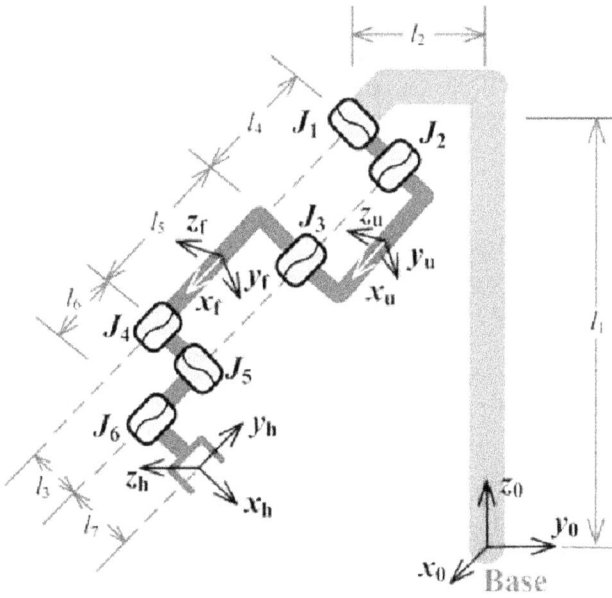

FIGURE 3.4 Schematic of UR5 collaborative robot [52].

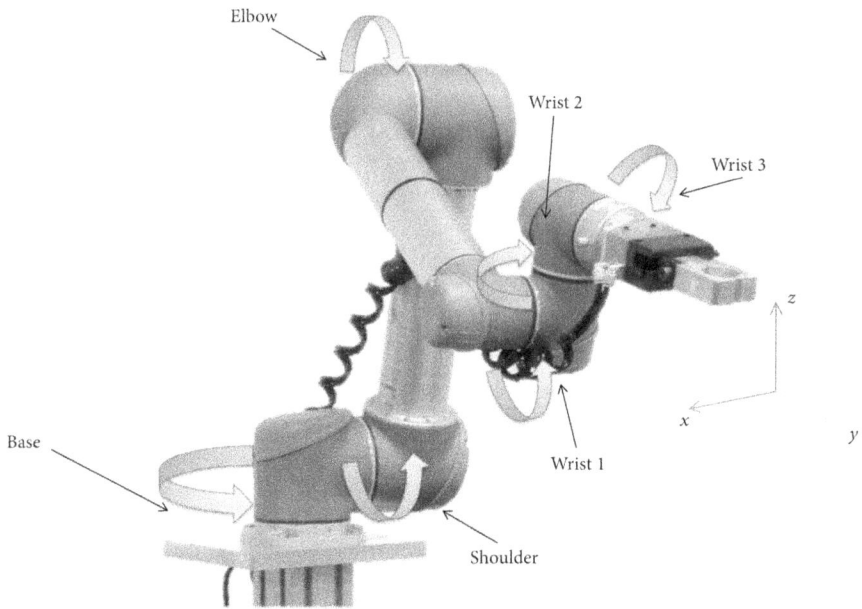

FIGURE 3.5 The UR5 collaborative robot [53].

Using mobile robots for AM provides some challenges as well. Mobile robots are majorly designed for transportation-based applications and thus may face positioning accuracies. Precision however is of utmost importance for performing AM. Thus, Lachmyer et al. [56] have proposed a laser-based approach to contour tracking during AM. A 7-axis Franka Emika manipulator is placed on Mobile Industrial Robot 100 (MiR 100) platform. A 2D line laser mounted to the end effector of the robot creates a depth profile for the component's surface. This setup is used to perform foam-based 3D printing, using PU foam as the building material. Zhang et al. [55] and Efe Tiryaki et al. [57] have also used Denso VS-087, which is a 6 DoF robotic arm (multiple mobile robots used for printing) with a holonomic mobile platform, Clearpath Ridgeback, for printing 3D concrete structures using Fused Deposition Modeling (FDM).

Hunt et al. [58] have introduced a unique method for mobile 3D printing, wherein flying robots are being used. In this experiment, a quadcopter (3DR ArduCopter Quad) is used to perform Fused Deposition Modeling of PU foam. Even though this system has advantages like 3D printing at high elevations, and in inaccessible as well as hazardous areas, it also brings some challenges like low payload carrying capacities.

Researchers from the University of Stuttgart have used fiber composites as the material in Fused Filament Fabrication (FFF)/Fused Deposition Modeling (FDM) for in-situ interior and urban scale fabrication [59]. They created custom mobile robots that were small in size and attached a lightweight extrusion system to each. Table 3.4 provides details related to the previously discussed robots.

It was observed that material extrusion (also called FDM/FFF) is generally the preferred AM technique when using mobile robots. The main reason could be the simplicity of this technique. A mobile robot may not be able to carry out other types of AM as precisely because of the complexity of the processes. Some difficulties are faced in performing 3D printing using multiple robots. First, the robots should move in such a way that collisions between them do not take place. For this, precise localization and mapping techniques need to be used; only then can the printing be precise [55] (Figures 3.6 and 3.7).

3.2.5 PARALLEL ROBOTS

Parallel robots consist of sets of serial chains, that are used to guide an effector plate to precise locations in 3D space [60]. Unlike serial manipulators (robotic manipulators discussed in previous sections), the end effector of these robots (the effector plate) is connected to their base with the help of several linkages that work in harmony. Some applications of parallel robots include being used for flight simulators, automobile simulators, assembly of PCBs, and high-precision milling operations [61]. These robots are generally less likely to be used in AM, as it limits the achievable workspace for 3D printing. However, we came across some papers that propose using these.

Barnett et al. [62] have used a cable-suspended robot for 3D foam printing of a statue, using PU foam as the build material. A Rostock max V2.0 delta robot 3D

TABLE 3.4

Specifications of Mobile Robots Used in AM [41,42,55–59]

Robot	Payload	Horizontal reach	Max Speed (wrist angle)	Repeatability	Robot weight	AM Method	Material used	References
7-axis Franka Emika manipulator	3 kg					Foam 3D printing	PU foam	[56]

Mobile robot	Payload	Running Time/ Running distance	Max Speed	Turning Radius	Weight			
Mobile Industrial Robot 100 (MiR 100) platform	100 kg	10 hrs/ 20 Km	1500 mm/s	520 mm	62.5 kg			

Robot	Payload	Horizontal reach	Max Speed (wrist angle)	Repeatability	Robot weight	AM Method	Material used	References
3DR ArduCopter Quad						FDM	two-part polyurethane (PU) foam	[58]
6 DOF robotic arm with holonomic mobile platform		900 mm		±0.02 mm		FDM (3D concrete printing)	OPC, ASTM Type 1 + Fly Ash + Silica fume + River Sand + Water	[55]

Robot	Payload	Horizontal reach	Max Speed (wrist angle)	Repeatability	Robot weight	AM Method	Material used	References
A 6-DoF industrial robot manipulator (Denso VS-087)	7 kg	905 mm	10.577 rad/sec	±0.03 mm	51 kg	FDM (3D concrete printing)	OPC, ASTM Type 1 + Fly Ash + Silica fume + River Sand + Water	[57]

(Continued)

TABLE 3.4 (*Continued*)
Specifications of Mobile Robots Used in AM [41,42,55–59]

Mobile robot	Payload	Running Time/ Running distance	Max Speed	Turning Radius	Weight
A holonomic mobile base (Clearpath Ridgeback)	100 kg		1100 mm/s		135 kg

Robot	Payload	Horizontal reach	Max Speed (wrist angle)	Repeatability	Robot weight	AM Method	Material used	References
Small mobile robots and a lightweight extrusion system						FFF/FDM	Filament form fiber composite	[59]

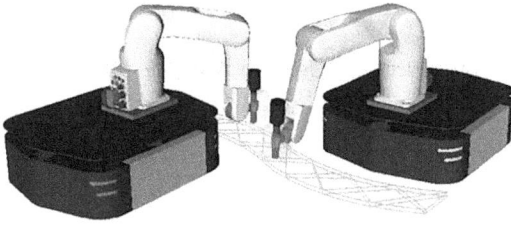

(a) Concurrent printing of one large-scale structure by two mobile robot printers.

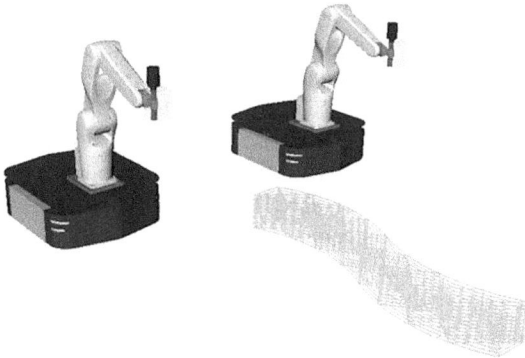

(b) Robot printers back to home station upon completion of printing.

FIGURE 3.6 Simulation of multi-robot printing process [55].

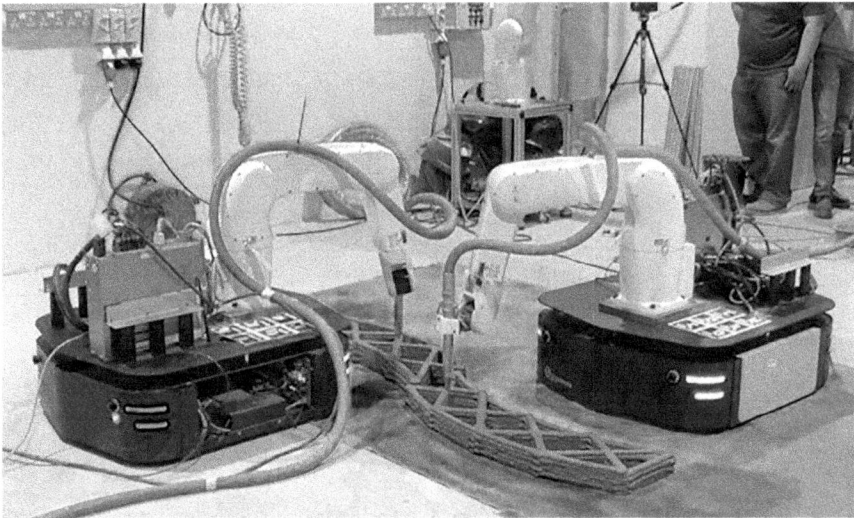

FIGURE 3.7 Concurrent printing of a large, single-piece, concrete structure by two mobile robot printers [55].

TABLE 3.5

Specifications of Parallel Robots Used in AM [41,42,60,62,63]

Robot	Max Speed	AM Method	Material used	References
Parallel robot (Cable-suspended robot)		Foam 3D printing	Polyurethane foam (object material)	[62]
Parallel robot (Delta robot; Rostock MAX V2.0)	300 mm/sec	Curved layer FFF/FDM	Polylactide (PLA) thermoplastic	[60]
Cable-driven Parallel Robot		FFF	TUDALIT TF10	[63]

printer, with an extrusion head, mounted at the robot's end effector was used by Allen et al. [60] for performing curved layer FFF of Polylactide thermoplastic. Specifications for both cases of robot-based AM are given in Table 3.5.

Hahlbrock et al. [63] created a cable-driven parallel robot for large-scale construction. The reachability of serial robots is quite less, and thus a cable-driven parallel robot can scale higher as well as vast regions, and is thus preferred in construction. The researchers performed material extrusion using this printer. Advantages of using a cable-driven robot for 3D printing are that its mechanical setup is flexible and its work capacity can be increased if needed. Also, the printer's working can be customized according to the accuracy needed to extrude the material, as well as the sequence of extrusion. In this project, the same CAD setup that is used for serial manipulators was applied to parallel robot, and this made incorporation of a parallel robot for 3D printing easier, even though this is still a niche research field (Figures 3.8 and 3.9).

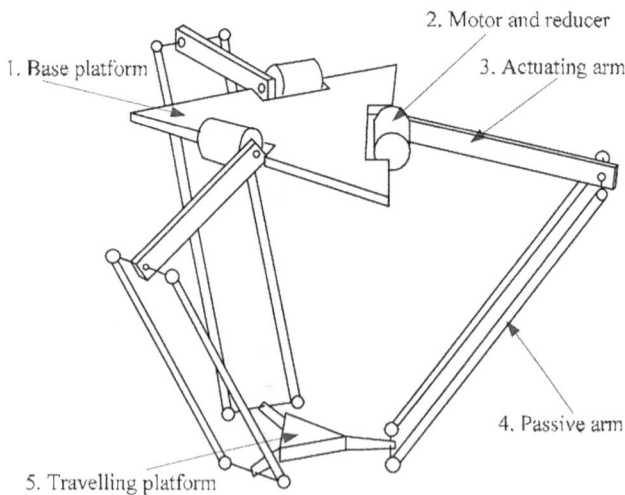

FIGURE 3.8 Schematic of Delta robot [64].

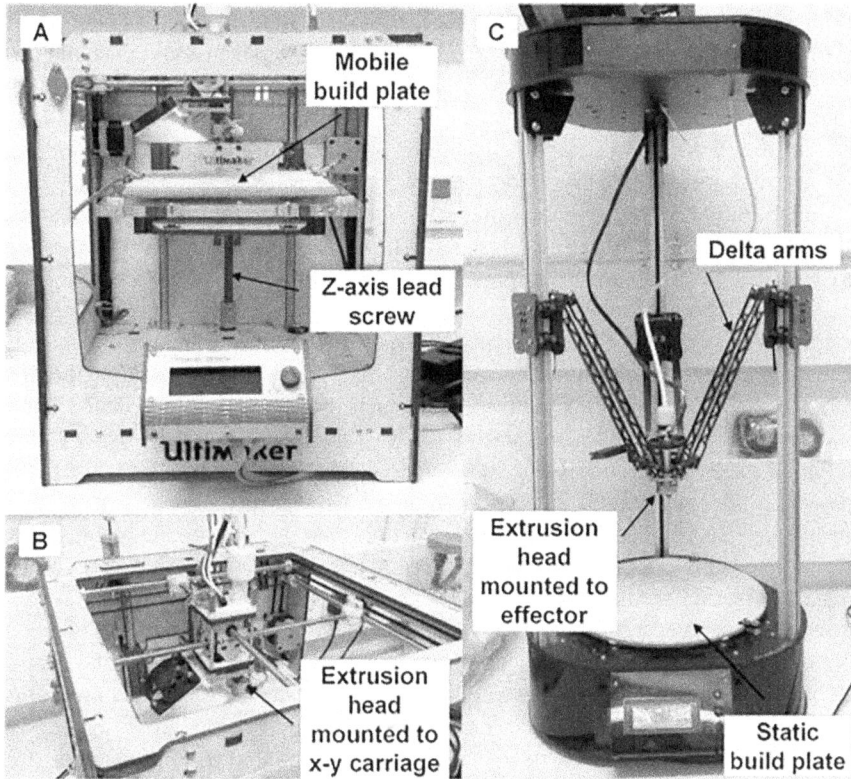

FIGURE 3.9 A comparison of Cartesian and delta-based FFF systems: (a) overview of a consumer Cartesian FFF system detailing the z-axis and mobile build plate, (b) detail of Cartesian x–y carriage, (c) overview of a delta style FFF detailing key components [60].

3.3 CONCLUSION AND FUTURE WORK

A total of 42 papers were reviewed and distinguished on the basis of the type of robot used for AM. We then compared the specifications of all these robots and the AM techniques for which they were used. The robots are majorly classified into two main types: serial and parallel. Within serial robots, we could find 6 DoF robotic manipulators, robotic AM systems with greater than 6 DoF, collaborative robots, and mobile robots. Serial robots are the preferred type for use in large-scale applications of AM because they have a bigger workspace (open loop chain) as compared to the confined area in which a parallel robot can work (closed loop chain). It was observed that Fused Deposition Modeling (material extrusion) was majorly performed using 6-degree-of-freedom robotic arms. The reason behind the extensive use of FDM could be its inexpensive setup. Moreover, in FDM, the extruder operates in a vector-based manner which is common in machining tools [38].

Directed Energy Deposition AM processes like WAAM, Laser-directed Energy Deposition, GMAW-based, or GTAW-based AM are better performed with a setup wherein a 6 DoF manipulator is coupled with a 2-axis positioning system. The positioning system helps to negate the offset of the build part with respect to the tool, as the tool head changes its orientation as and when there is a movement of the build part.

The use of mobile robots in AM is still being explored. It was found out that mobile robots are being used for Robot-based AM majorly in the field of building and construction. However, its capabilities in fields like healthcare-associated AM and automotive industry-associated AM are yet to be explored.

Collaborative robots are not suitable for being used in industrial applications of AM where build time is to be short. Cobots are designed to work in environments where humans are present and thus work at low speeds as compared to industrial robots. Nevertheless, cobots can be used for prototyping purposes or for small-scale 3D printing applications where human intervention may be needed to check the part-building qualities. Also, these may be used in the field of healthcare for in-situ 3D printing. Not much research has been carried out in this area of AM in healthcare, and this is a vast field to be explored in the future.

Parallel robots are very useful when large-scale applications like construction come into the picture. Of course, mobile robots might be used here as well, however, when compared to serial manipulators that are not mobile, cable-driven robots have a larger workspace and more mobility and can also be scaled according to the application requirement.

It was also observed that a cost-based comparison of robotic AM techniques is yet to be done. There is future scope for comparing and combining cost-effective AM techniques with precise yet economical robots and then concluding to find the best combination which balances cost as well as precision. Robot-based AM has many future aspects yet to be explored, and is a trend that will offer support-free and precise manufacturing of desired products, and will expand the scope of AM from large-scale construction to micro-scale healthcare applications.

ACKNOWLEDGMENT

The authors would like to express gratitude toward Pandit Deendayal Energy University (PDEU) for supporting our study.

REFERENCES

[1] Wikipedia. Cobot, https://en.wikipedia.org/wiki/Cobot.
[2] Yadav A, Prakash A, Kumar A, et al. Design of remote-controlled land mine detection troops safety robot. *Mater Today Proc* 2022; 56: 274–277.
[3] Gao M, Zhou H, Yang Y, et al. An intelligent master–slave collaborative robot system for cafeteria service. *Rob Auton Syst* 2022; 154: 104121.
[4] Duan B, Xiong L, Guan X, et al. Tele-operated robotic ultrasound system for medical diagnosis. *Biomed Signal Process Control* 2021; 70: 102900.
[5] Tian Y, Chen C, Sagoe-Crentsil K, et al. Intelligent robotic systems for structural health monitoring: Applications and future trends. *Autom Constr* 2022; 139: 104273.
[6] Fan Y, Hu Y, Jiang L, et al. Intelligent disinfection robots assist medical institutions in controlling environmental surface disinfection. *Intell Med* 2021; 1: 19–23.
[7] Gibson I, Rosen D, Stucker B. *Additive Manufacturing Technologies*. New York, NY: Springer New York, 2015. Epub ahead of print 2015. DOI: 10.1007/978-1-493 9-2113-3.

[8] Wikipedia. 3D Printing, https://en.wikipedia.org/wiki/3D_printing#Applications.

[9] Campbell I, Bourell D, Gibson I. Additive manufacturing: Rapid prototyping comes of age. *Rapid Prototyp J* 2012; 18: 255–258.

[10] Bhatt PM, Malhan RK, Shembekar A v, et al. Expanding capabilities of additive manufacturing through use of robotics technologies: A survey. *Addit Manuf* 2020; 31: 100933.

[11] Alsharhan AT, Centea T, Gupta SK. Enhancing Mechanical Properties of Thin-Walled Structures Using Non-Planar Extrusion Based Additive Manufacturing. In: *Volume 2: Additive Manufacturing; Materials*. American Society of Mechanical Engineers, 2017. Epub ahead of print 4 June 2017. DOI: 10.1115/MSEC2017-2978.

[12] Urhal P, Weightman A, Diver C, et al. Robot assisted additive manufacturing: A review. *Robot Comput Integr Manuf* 2019; 59: 335–345.

[13] Leal-Junior A, Frizera-Neto A. Optical fiber sensors applications for human health. In: *Optical Fiber Sensors for the Next Generation of Rehabilitation Robotics*. Elsevier, 2022, pp. 263–286.

[14] BLUEPRINT LAB. 'Degrees of Freedom' Vs 'Functions' of a Robotic Arm, https://blueprintlab.com/blog/degrees-of-freedom-vs-functions-of-a-robotic-arm/ (2021).

[15] Yoon YJ, Yon M, Jung SE, et al. Development of three-nozzle extrusion system for conformal multi-resolution 3D printing with a robotic manipulator. In: *Proceedings of the ASME Design Engineering Technical Conference*. American Society of Mechanical Engineers (ASME), 2019. Epub ahead of print 2019. DOI: 10.1115/DETC2019-98069.

[16] Bhatt PM, Kabir AM, Malhan RK, et al. A Robotic Cell for Multi-Resolution Additive Manufacturing. In: *2019 International Conference on Robotics and Automation (ICRA)*. IEEE, 2019, pp. 2800–2807.

[17] Ishak I, Fisher J, Larochelle P. *Robot Arm Platform for Additive Manufacturing: Multi-Plane Printing*. In: *29th Florida Conference on Recent Advances in Robotics, FCRAR 2016*, Miami, Florida, May 12–13, 2016.

[18] Ishak I, Larochelle P. *Robot Arm Platform for Additive Manufacturing: 3D Lattice Structures*. In: *30th Florida Conference on Recent Advances in Robotics,* Florida Atlantic University, Boca Raton, Florida, May 11–12, 2017.

[19] Ishak I bin, Fisher J, Larochelle P. *Robot Arm Platform for Additive Manufacturing Using Multi-plane Toolpaths*, http://proceedings.asmedigitalcollection.asme.org/pdfaccess.ashx?url=/data/conferences/asmep/90694/ (2016).

[20] Felsch T, Klaeger U, Steuer J, et al. Robotic system for additive manufacturing of large and complex parts. In: *2017 22nd IEEE International Conference on Emerging Technologies and Factory Automation (ETFA)*. IEEE, 2017, pp. 1–4.

[21] Felbrich B, Wulle F, Allgaier C, et al. A novel rapid additive manufacturing concept for architectural composite shell construction inspired by the shell formation in land snails. *Bioinspir Biomim*; 13. Epub ahead of print 1 March 2018. DOI: 10.1088/1748-3190/aaa50d.

[22] Li L, Haghighi A, Yang Y. A novel 6-axis hybrid additive-subtractive manufacturing process: Design and case studies. *J Manuf Process* 2018; 33: 150–160.

[23] Kubalak JR, Mansfield CD, Pesek TH, et al. *Design and Realization of a 6 Degree of Freedom Robotic Extrusion Platform*. *Solid Freeform Fabrication 2016: Proceedings of the 26th Annual International Solid Freeform Fabrication Symposium – An Additive Manufacturing Conference Reviewed Paper*.

[24] Zhang GQ, Spaak A, Martinez C, et al. Robotic additive manufacturing process simulation - towards design and analysis with building parameter in consideration. In: *2016 IEEE International Conference on Automation Science and Engineering (CASE)*. IEEE, 2016, pp. 609–613.

[25] Dine A, Vosniakos GC. On the development of a robot-operated 3D-printer. In: *Procedia Manufacturing*. Elsevier B.V., 2018, pp. 6–13.

[26] Yuan PF, Meng H, Yu L, et al. Robotic Multi-dimensional Printing Based on Structural Performance. In: *Robotic Fabrication in Architecture, Art and Design 2016*. Springer International Publishing, 2016, pp. 92–105.

[27] Keating S, Oxman N. Compound fabrication: A multi-functional robotic platform for digital design and fabrication. *Robot Comput Integr Manuf* 2013; 29: 439–448.

[28] Zheng H, Cong M, Dong H, et al. CAD-based automatic path generation and optimization for laser cladding robot in additive manufacturing. *Int J Adv Manuf Technol* 2017; 92: 3605–3614.

[29] Bhatt PM, Kabir AM, Peralta M, et al. A robotic cell for performing sheet lamination-based additive manufacturing. *Addit Manuf* 2019; 27: 278–289.

[30] Stevens AG, Oliver CR, Kirchmeyer M, et al. Conformal robotic stereolithography. *3D Print Addit Manuf* 2016; 3: 227–235.

[31] Bonaccorso F, Cantelli L, Muscato G. An arc welding robot control for a shaped metal deposition plant: Modular software interface and sensors. *IEEE Trans Ind Electron* 2011; 58: 3126–3132.

[32] Ding D, Shen C, Pan Z, et al. Towards an automated robotic arc-welding-based additive manufacturing system from CAD to finished part. *CAD Comput Aided Des* 2016; 73: 66–75.

[33] Evjemo LD, Langelandsvik G, Gravdahl JT. Wire arc additive manufacturing by robot manipulator: towards creating complex geometries. In: *IFAC-PapersOnLine*. Elsevier B.V., 2019, pp. 103–109.

[34] Lin M, San L, Ding Y. Construction of Robotic Virtual Laboratory System Based on Unity3D. *IOP Conf Ser Mater Sci Eng* 2020; 768: 072084.

[35] Bhatt PM, Peralta M, Bruck HA, et al. *Robot Assisted Additive Manufacturing of Thin Multifunctional Structures*, http://www.asme.org/about-asme/terms-of-use (2018).

[36] Gosselin C, Duballet R, Roux P, et al. Large-scale 3D printing of ultra-high performance concrete – A new processing route for architects and builders. *Mater Des* 2016; 100: 102–109.

[37] Laarman J, Jokic S, Novikov P, et al. *Chapter Title: Anti-gravity Additive Manufacturing Book Title: Fabricate 2014 Book Subtitle: Negotiating Design & Making*, https://about.jstor.org/terms.

[38] Brooks BJ, Arif KM, Dirven S, et al. Robot-assisted 3D printing of biopolymer thin shells. *Int J Adv Manuf Technol* 2017; 89: 957–968.

[39] Mostafavi S, Bier H. Materially informed design to robotic production: A robotic 3D printing system for informed material deposition. In: *Robotic Fabrication in Architecture, Art and Design 2016*. Springer International Publishing, 2016, pp. 338–349.

[40] Zhang G. *Robotic Additive Manufacturing Along Curved Surface ï¿½ a Step towards Free-Form Fabrication*. 2015. Epub ahead of print 2015. DOI: 10.0/Linux-x86_64.

[41] RobotWorx. Industrial Robots, https://www.robots.com/robots.

[42] RoboDK. RoboDK Library, https://robodk.com/library.

[43] Ding Y, Dwivedi R, Kovacevic R. Process planning for 8-axis robotized laser-based direct metal deposition system: A case on building revolved part. *Robot Comput Integr Manuf* 2017; 44: 67–76.

[44] Reisch R, Hauser T, Kamps T, et al. Robot based wire arc additive manufacturing system with context-sensitive multivariate monitoring framework. In: *Procedia Manufacturing*. Elsevier B.V., 2020, pp. 732–739.

[45] Kaji F, Jinoop AN, Zimny M, et al. Process planning for additive manufacturing of geometries with variable overhang angles using a robotic laser directed energy deposition system. *Addit Manuf Lett* 2022; 2: 100035.

[46] Akbari M, Ding Y, Kovacevic R. *Process Development for A Robotized Laser Wire Additive Manufacturing*, http://proceedings.asmedigitalcollection.asme.org/ pdfaccess.ashx?url=/data/conferences/asmep/93277/ (2017).

[47] Mcgee W, Velikov K, Thün G, et al. *Chapter Title: Infundibuliforms: Kinetic Systems, Additive Manufacturing for Cable Nets and Tensile Surface Control*, https://about.jstor.org/terms.

[48] Jason Walker. A Guide to Cobots & 6 Trends Impacting Industrial Collaborative Robots, https://waypointrobotics.com/blog/cobots-trends/.

[49] Wu C, Dai C, Fang G, et al. RoboFDM: A robotic system for support-free fabrication using FDM. In: *2017 IEEE International Conference on Robotics and Automation (ICRA)*. IEEE, 2017, pp. 1175–1180.

[50] Danielsen Evjemo L, Moe S, Gravdahl JT, et al. Additive manufacturing by robot manipulator: An overview of the state-of-the-art and proof-of-concept results. In: *2017 22nd IEEE International Conference on Emerging Technologies and Factory Automation (ETFA)*. IEEE, 2017, pp. 1–8.

[51] Felbrich B, Schork T, Menges A. Autonomous robotic additive manufacturing through distributed model-free deep reinforcement learning in computational design environments. *Constr Robot* 2022; 6: 15–37.

[52] Gong D, Zhao J, Yu J, et al. Motion mapping of the heterogeneous master–slave system for intuitive telemanipulation. *Int J Adv Robot Syst* 2018; 15: 172988141774813.

[53] Tang G, Webb P. The design and evaluation of an ergonomic contactless gesture control system for industrial robots. *J Robot* 2018; 2018: 1–10.

[54] Wikipedia. Mobile Robot, https://en.wikipedia.org/wiki/Mobile_robot.

[55] Zhang X, Li M, Lim JH, et al. Large-scale 3D printing by a team of mobile robots. *Autom Constr* 2018; 95: 98–106.

[56] Lachmayer L, Recker T, Raatz A. Contour tracking control for mobile robots applicable to large-scale assembly and additive manufacturing in construction. *Proc CIRP* 2022; 106: 108–113.

[57] Efe Tiryaki M, Zhang X, Pham Q-C. *Printing-while-moving: A New Paradigm for Large-scale Robotic 3D Printing*, https://www.researchgate.net/publication/ 327835498.

[58] Institute of Electrical and Electronics Engineers. 3D Printing with Flying Robots. p. 6822.

[59] Yablonina M, Prado M, Baharlou E, et al. *Chapter Title: Mobile Robotic Fabrication System for Filament Structures*, https://www.jstor.org/stable/j. ctt1n7qkg7.32.

[60] Allen RJA, Trask RS. An experimental demonstration of effective curved layer fused filament fabrication utilising a parallel deposition robot. *Addit Manuf* 2015; 8: 78–87.

[61] Wikipedia. Parallel Manipulator.

[62] Barnett E, Gosselin C. Large-scale 3D printing with a cable-suspended robot. *Addit Manuf* 2015; 7: 27–44.

[63] Hahlbrock D, Braun M, Heidel R, et al. Cable Robotic 3D-printing: Additive manufacturing on the construction site. *Constr Robot*. Epub ahead of print 22 October 2022. DOI: 10.1007/s41693-022-00082-3.

[64] Lu X, Liu M. Optimal design and tuning of PID-type interval type-2 fuzzy logic controllers for delta parallel robots. *Int J Adv Robot Syst* 2016; 13: 96.

4 Triply Periodic Minimal Surfaces

Properties, Applications, and Challenges

Aarya Patel, Falak Patel, and Vishvesh Badheka

4.1 INTRODUCTION

In the age of innovation and rapid mobility, the demand for non-disruptive digital technologies such as quantum computing, Internet of Things (IoT), and next-generation satellites has surged. The rise in the demands calls upon the requirement for effective, compact, and lightweight technologies that can be implemented in applications for better and faster performance. Majorly, there are two control parameters that are vital for any system's performance, namely, Structural and Thermal Response to the acting loads. However, when the devices are implemented in actual applications, there needs to be an optimum balance between physical and thermal properties for the system to function optimally. Thermal control devices such as Heat sinks and Heat Exchangers are crucial components to ensure the smooth functioning of these technologies across varied domains ranging from aerospace, medical, defense, and automobile. Traditionally, excess heat is dissipated using rectangular fins but due to the prolonged severe usage of these technologies, the conventional finned surfaces are not that effective, and it may affect the life cycle of components. To sync with heavy usage and demand, non-conventional ways to develop heat sinks must be adopted. One such technique involves replacing the heat sinks conventional linear geometries with architectured cellular materials. These involve lattice, non-periodic nodal surfaces, and Infinite Periodic Minimal Surfaces. Among them, Triply Periodic Minimal Surfaces (TPMS) offer the best customization properties and tailored designs which can be varied as per the applications. The uniformity of TPMS geometries in all three directions makes them easy to generalize and manufacture as compared to other lattice geometries. The paper presents an overview of the basics of minimal surfaces and their possible applications in industry.

4.2 PROCESSES

Minimal Surface Geometries find their utility in various domains. It is due to its unique properties and geometric design that makes them a perfect fit for designs and

DOI: 10.1201/9781003484325-4

emerges as the best possible solution for the problem statement. But before implementing these surfaces it becomes very important to understand different types of TPMS geometries, their unique properties, design analogies, and how each of them vary with applications. An overview of the basics of Minimal surfaces and their features is showcased here.

4.2.1 INTRODUCTION TO MINIMAL SURFACES

Minimal surfaces are a class of complex geometries that have zero mean curvature that is local area minimizing and has the smallest area inside a given boundary. In a nutshell, minimal surfaces increase the total surface area while maintaining the same control volume by locally minimizing the area, hence the name. The phenomenon is depicted in Figure 4.3 [1]. It can do this because of its intricate internal curvatures, which are controlled by numerous differential and trigonometric equations. The discovery of minimal surfaces ways back in the 19th century but has recently gained much attention due to extensive research carried out in this domain due to the advent of additive manufacturing technologies through which the fabrication can be made feasible. Beginning with catenoids or helicoids, which were the first to be discovered around the world, Celso Costa's discovery of the Costa Surface in 1982, which was followed by other discoveries in the field of minimal surfaces, was a watershed point in the research on minimal surfaces at the beginning of the 18th century [2].

By using the equilibrium conditions, the governing equation of the surface is obtained. The surface's free energy is reduced at thermodynamic equilibrium. The governing integral can then be solved using variational calculus by applying the boundary conditions to determine the governing parameter [3]. Energy minimization results in surface minimization and the minimization of the governing integral (most minor radius in this case). Each point on minimal surfaces is a saddle point with equal and opposite principal curvature, which is a property of minimal surfaces. There are various types of minimal surfaces, and each has a unique set of governing equations and variables. One such family is the infinite surfaces family, which includes TPMS [4].

TPMS are those that have minimized areas under a particular boundary constraint and are periodic in all XYZ directions (TPMSs). They are crystalline in nature and repeat their geometry in all three dimensions which is why they are referred to as triply periodic in nature. Schwartz P, Schwartz D, Neovios, Gyroid, and Fischer Surfaces are some of the minimal surfaces from the TPMS family [5]. TPMS are porous structures that have the potential to be used to generate multifunctional materials for a variety of technological applications. They are smooth, continuous, finitely extending surfaces that divide space into two congruent interwoven zones. Each section is periodic in 3D and has a volume fraction of 50%. Due to these characteristic traits, TPMSs are introduced as promising candidates for providing high stiffness, strength, and energy-absorbing capability.

They are commonly referred to as Bio-Inspired Surfaces as they take inspiration from nature and can be easily visualized in nature. A basic example of it is

Soap Films. It involves the dipping of two wire rings into a soap solution and following that a connecting surface is naturally formed between the two rings. That topology is termed a Minimal Surface. To realize different roles and adapt to natural environments, biological entities have evolved to contain complex and varied hierarchical structures on the macro, meso, and micro scales. Many biological systems, including skeletons and scaffolds, have been found to contain triply periodic minimum surfaces and related materials. Gyroid and Diamond structures, two examples of TPMS topologies, were found in the butterfly wing scales and the weevil Lamprocyphus Augustus [6]. Figure 4.1 displays images of C. rubi butterfly wing scales taken using optical microscopy and scanning electron microscopy

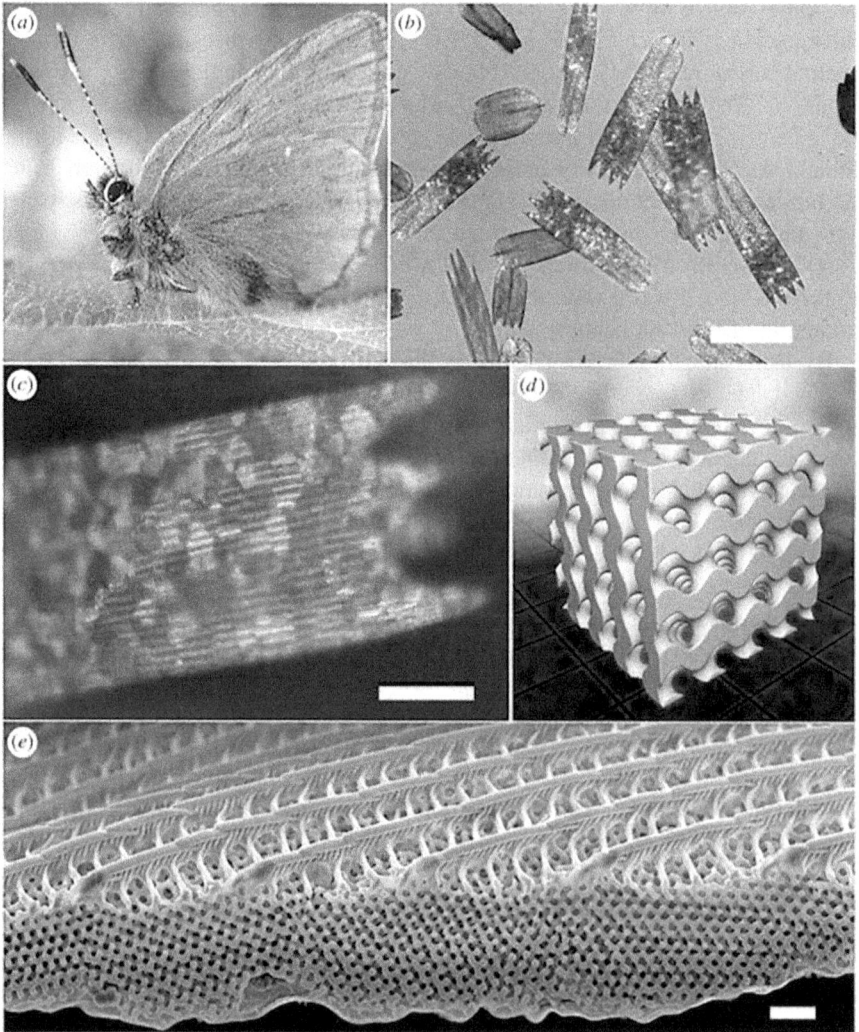

FIGURE 4.1   Nature-inspired minimal surface.

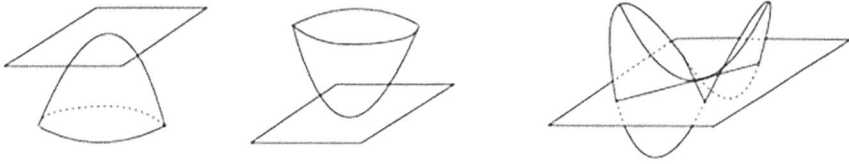

FIGURE 4.2 Zero mean curvature of minimal surfaces.

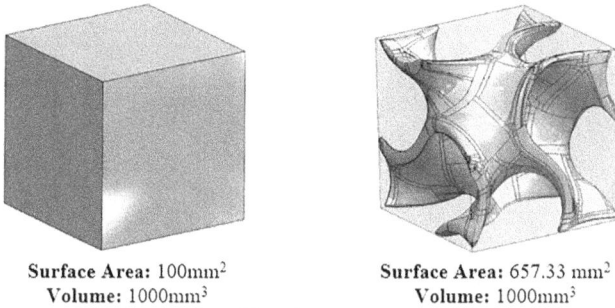

Surface Area: 100mm²
Volume: 1000mm³

Surface Area: 657.33 mm²
Volume: 1000mm³

FIGURE 4.3 Gyroid surface area comparison.

(SEM). The surface of the scales is covered in interconnected ridges and ribs that provide mechanical support and serve as interference reflectors. Inherent characteristics of the intertwined designs include a lack of sharp edges and improved fluid dynamics. A steady enlargement or contraction of a flow route has a far lower pressure drop than a rapid contraction or enlargement, according to fluid dynamics. The curved shape assists smooth flow through the channels and reduces pressure drop theoretically [7] (Figure 4.2).

4.2.1.1 Types of TPMS Geometries

TPMS are single and continuous curves and can be approximated by a trigonometric parametric equation. They are broadly classified into two categories, namely, solid-based and sheet-based. A solid-based network structure is generated by solidifying the volume enclosed by the minimal surface at a particular value of 'c'. In contrast, in a sheet network structure, the volume enclosed by the intersection of two minimal surfaces is evaluated at a value of 'c'. Apart from the type of selection, the value of 'c' also regulates the volume fraction (V_f) of the generated surfaces. The value of V_f is obtained by the ratio of the volume occupied by surface geometry to the total bulk volume. This value leads to the establishment of another critical parameter which is porosity. It can be calculated using $1-V_f$. Equations 1, 2, and 3 are the general mathematical representation, and Table 4.1 depicts the visual representations of Gyroid, Schwarz Diamond, and Schwarz Primitive TPMS structures, respectively. The surface morphology is governed by 'c' in the respective equations.

TABLE 4.1
Governing Equations of TPMS

Surface Geometry	Surface Depiction		Governing Equation
	Solid Based	Sheet Based	
Gyroid			$\sin(x)*\cos(y) + \sin(y)\cos(z) + \sin(z)*\cos(x) = c$
Diamond			$\sin(x)\sin(y)\sin(z) + \sin(x)\cos(y)\cos(z) + \cos(x)\sin(y)\cos(z) + \cos(x)\cos(y)\sin(z) = c$
Schwarz P			$\cos(x) + \cos(y) + \cos(z) = c$
Split P			$1.1(\sin(2x)\sin(z)\cos(y) + \sin(2y)\sin(x)\cos(z) + \sin(2z)\sin(y)\cos(x)) = c$ $- 0.2(\cos(2x)\cos(2y) + \cos(2y)\cos(2z) + \cos(2z)\cos(2x))$ $- 0.4(\cos(2x) + \cos(2y) + \cos(2z))$

4.2.2 Design Aspects of Minimal Surfaces

The surfaces can be modeled using parametric software and any functional algorithm. In the present study for visualization, the TPMS is modeled using nTopology Software, a widely used parametric design software for modeling complex structures. Additionally, it makes it feasible for manufacturing through additive manufacturing techniques. It is used primarily for parameter-driven designs and their optimizations. It consists of several governing parameters for any surface and can be varied to change the morphology of surfaces to understand its effect on results before finalizing the design. Some include the unit cell size, wall thickness, and bias length [8]. In the equations depicted in Table 4.1, there is a parameter 'c' which is on the RHS of the equation. The value of c can be either negative, zero, or positive and the range of c is decided according to the dimensions of design and application. For instance, if we are designing a TPMS-based heat sink, then the main parameter is overall exposed surface area. To finalize the value of 'c', the overall surface area is compared to obtain an optimum value of 'c' for maximum surface area to enhance the heat transfer rate. The solid-based and sheet-based networks are also compared based on surface area and porosity. It is found that the sheet-based networks had a higher surface area and hence a higher surface area to volume ratio, so it is considered for major applications. Gyroid, Diamond, and Schwarz P are considered in this example. A summary of the investigated cases is presented in Table 4.2. It also depicts the overall surface area porosity of solid and sheet-based networks with variation in 'c' values. Notably, it is found that a 'c' value increases, the surface area and porosity percentage decrease.

TABLE 4.2
Design Parameters of Minimal Surfaces

Structure	Parameter	Volume Fraction (%)	Porosity (%)	Surface Area (mm^2)
Gyroid Sheet	C=0	8.7	91.2	2656.26
	C=1	9.8	90.2	2610.70
Gyroid Solid	C=0	58.69	41.30	2422.60
	C=1	60.68	39.31	2609.10
Diamond Sheet	C=0	7.9	92.1	3262
	C=1	9.1	90.9	3189.42
Diamond Solid	C=0	48.65	51.35	2728.98
	C=1	51.42	48.58	2917.65
Schwarz P Sheet	C=0	14.1	85.9	2140.68
	C=1	16.5	83.5	1975.97
Schwarz P Solid	C=0	48.67	51.33	2896.45
	C=1	52.45	47.55	3118.33

4.2.3 MANUFACTURING ASPECTS OF TPMS

Minimal surfaces have applications in various domains ranging from structural to thermal problems. For instance, currently, heat sinks and heat exchangers are being designed by using TPMS structures as their core in place of a complete solid structure. The existing methods of manufacturing conventional heat sinks requires excessive tooling in form of milling, drilling, and casting which is not applicable to fabricate non-conventional lattice-based geometries due to its complex intertwined intrinsic surfaces. However, with the rise of additive manufacturing, it has now become possible to manufacture the topology of these surfaces and blend it according to the design requirements. The materials and methods are selected according to the application in which these surfaces would be implemented.

4.2.4 MECHANICAL FEATURES OF TPMS

Mechanical strength is always related to bulky designs with high-density and high-strength materials. But as the demand for lightweight structures increases, porous substances, and lattice structures come into action as they provide an excellent balance between strength and weight. Amongst the lattice structures minimal surfaces are the most popular option as they are porous in nature and at the same time do not compromise on strength aspects. There are some unique geometrical aspects of TPMS surfaces that make them as a frontrunner in applications.

4.2.4.1 Interconnected Uniform Geometry

To sustain compressive and tensile loads, the stress acting should be well distributed so that it does not concentrate at a single point or area. For that reason, trusses, and uniform lattices such as octet and honeycomb structures are highly regarded as they dissipate the stress uniformly and avoid any sort of buckling failure. TPMS geometries being periodic in nature repeat their geometric pattern in all three axes in a uniform manner. However, when these surfaces are used in designs to replace the solid cores, multiple arrays of unit cells are stacked together according to the profile of the fitting curve of design. These surfaces are interconnected to each other via side edge-to-edge connection and form a tight enclosed space within the domain. So, when it is subjected to loads, the surfaces in contact act as a channel and propagate the stress down the domain uniformly and avoiding any case of rupture of unit. Figure 4.4 depicts how the unit cells connect to each other and how the stress is being distributed.

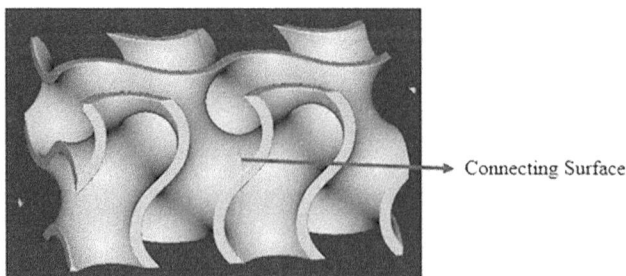

FIGURE 4.4   Interconnectivity of surfaces.

4.2.4.2 Tailored Parameters

Every application demands different mechanical properties such as Strength, Modulus of Elasticity, Density, etc. And that calls upon a demand for variable properties of surfaces alongside the dimensions. There are mainly three governing parameters of topology, namely, unit cell size, wall thickness, and relative density. By a proper combination of these parameters, one can achieve an optimum design that fits best for that application. The unit cell size governs the distribution and packaging of surfaces in three directions, namely, x, y, and z. The smaller the unit cell size in an array, the denser its packaging is. The study carried out by Zhilong C. on Investigations on porous media customized by TPMS suggests a parameter, namely, effective bearing area. It refers to the solid zone area in a specific cross-section [9]. A ratio of these areas is normally used for comparison. The higher the ratio, lesser are the chances of failure. It is mainly due to more solid zone area in that cross-section. This approach is very important to avoid any mechanical failures in form of buckling or crippling. And it also helps in improving the strength while achieving weight reduction (Table 4.3 and 4.4).

TABLE 4.3
Variation of Unit Cell Size in Gyroid Element

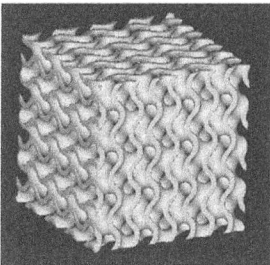
Gyroid_5mm

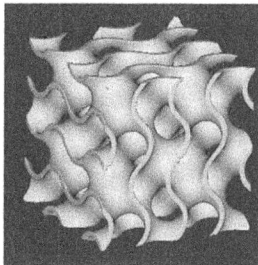
Gyroid_10mm

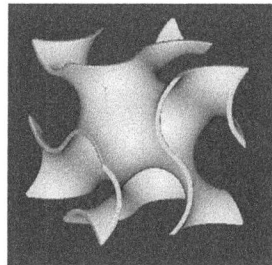
Gyroid_20mm

TABLE 4.4
Variation of Thickness in Gyroid Element

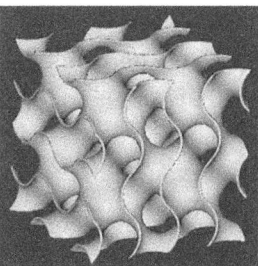
T=0.5 mm

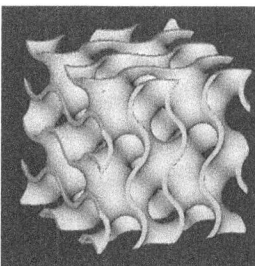
T=1 mm

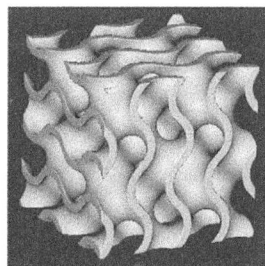
T=1.5 mm

4.2.5 Thermal Features of TPMS

There are normally three modes of heat transfer, namely, conduction, convection, and radiation. The common factor in all these modes is the overall exposed surface area. The heat transfer rate is directly proportional to the overall surface area. Conventionally, to solve the overheating issues, Linear geometry-based heat sinks are used in the electronic packages to increase the heat dissipation rate and maintain the operating temperature under the safety limit. They include rectangular, pinned, or circular fins. Sometimes, perforations are provided on the surfaces so that the surface area can be maximized by keeping the control volume as constant. In case of conventional ones, the only surface area that is exposed is that of the cylindrical surface, but in the case of perforated, the perforations also contribute to the surface area. However, with the rising needs of more efficient and effective heat sinks, the perforations are no longer seen as the sustainable and long-term solution to the problem. There must be a solution that bridges the gap between the usage and effectiveness of the heat sinks. One such non-conventional approach that can be implemented to bridge the gap is the inclusion of minimal surfaces geometry which increases the overall surface area by an average of 40% and meanwhile keeping the control volume as the constant parameter. The study carried out by Zhilong C. on heat transfer correlations of TPMS structures suggests that when the TPMS-based heat sinks were subjected to cooling medium, the periodic acceleration and deceleration and change in flow direction broke the boundary layer to enhance the heat transfer rate [9]. The large surface area to volume ratio, smooth surfaces, and complex geometry qualify TPMS-based lattices to be investigated as potential high-performance heat sinks. Also, being mathematically realized, the level of complexity of TPMS-based heat sinks can be controlled to optimize the pressure drop while maintaining good thermal performance. A study done by Jaisree Iyer on heat transfer and pressure drop characteristics of heat exchangers depicts how a tradeoff between pressure drop and heat transfer rate can be done [10]. With the aid of minimal surfaces, it becomes possible to achieve the same heat transfer rate as compared to the conventional fins at same pressure drop but with lesser control volume dimensions. So, the reduction in space and weight is achieved without compromising on heat transfer characteristics and pressure drop.

4.3 TPMS POTENTIAL APPLICATIONS

Structures with TPMS possess unique properties like smooth surfaces and highly interconnected pores and have found applications in many different industries. An overview of applications of TPMS in different industries is showcased here.

4.3.1 Mechanical Applications

The mechanical performance of TPMS structures has been studied thoroughly and owing to that many varying mechanical applications of TPMS structures have been explored in different industries.

The longer linear elastic stage of TPMS has enabled us to use these structures as energy and impact absorbers [11–13]. As the weight of TPMS structures is less than a solid block of the same envelope volume, they have found its application in the development of lightweight structures to save material and energy. These developed structures can eventually find their application in automobiles and aerospace. TPMS with its mechanical vibration band gap has found its application as a vibration isolator.

Recently the sandwich panels were constructed using TPMS cores. A comparative study between honeycomb, lattice, and TPMS sandwich panels proved that TPMS showcased better results. It was seen that the highest strength, modulus, and stiffness-to-weight ratio were obtained by TPMS cores [14,15].

Functional engineering components can also be fabricated using TPMS structures. A study showed that turbine blades composed of TPMS structure had optimized stress and deformation performance [16]. Another study suggested that compared to traditional leaf flexure hinges better compliance and compliance ratio was obtained in TPMS-based hinge structure. It was also determined that P surface was most suitable for Flexure Hinges [17].

4.3.2 THERMAL APPLICATIONS

The heat transfer performance of TPMS is excellent owing to its high-volume-specific area. Therefore, the main application of TPMS in thermal domain is in heat exchangers. Li et al attempted to use bio-inspired TPMS structures to make heat exchangers. It was observed that heat transfer rate of TPMS was higher than printed circuit heat exchangers. Also, the heat transfer coefficient of TPMS was 16%–120% higher [18,19].

Compared to applications of TPMS in other domain the exploration of application of TPMS in thermal domain is very less. The research and development of TPMS in thermal domain is in its infancy. It is an interesting problem which will require more attention to further its reach than just in heat exchangers.

4.3.3 BIOMEDICAL APPLICATIONS

The uncanny ability of TPMS to mimic natural structures has unlocked vast applications of TPMS structures in Biomedical field. The porous structures can be used as medical implants and in tissue engineering scaffolds. TPMS surface makes it easier for the cells to attach and grow and the high-volume-specific surface area and interconnected pores can provide space for the exchange of nutrients and removal of waste.

A comparative study suggested that in comparison to typical salt-leached scaffolds the permeability of TPMS scaffolds is ten times higher [20]. Another study discussed early osteo-integration of Ti6Al4V TPMS scaffolds. The results from this experiment suggested that a stable interface between implant and surrounding bone was observed after five weeks. In micro-CT bone formation around all the pores of TPMS was observed [21].

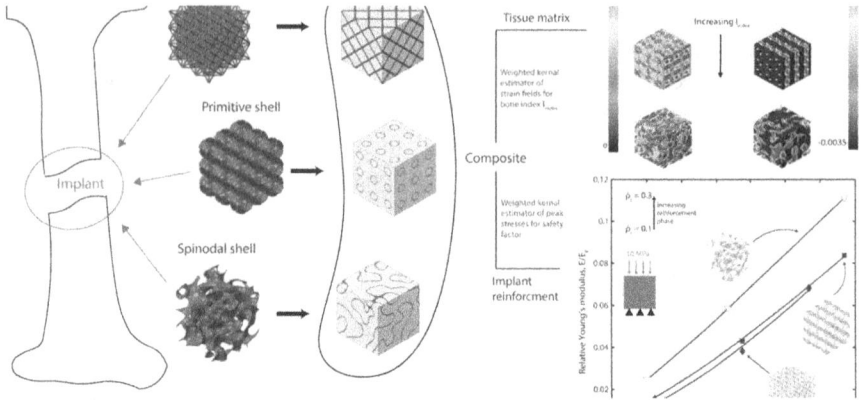

FIGURE 4.5 Comparisons between lattice and TPMS scaffolds.

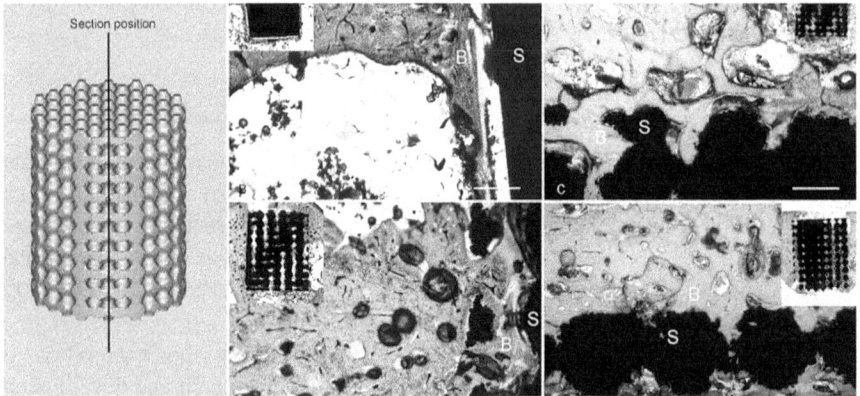

FIGURE 4.6 Early osteo-integration of Ti6Al4V.

Recently TPMS were employed to replicate natural bones. Zhu et al employed TPMS structures as meniscal implants. In comparison with the commercial solid meniscal implants, the TPMS can prevent the transfer of higher magnitude of compressive and shear stress on articular cartilage. TPMS can also be used as implant for craniofacial bones and can help repair craniofacial bone abnormalities [22]. Comparative study between lattice and TPMS scaffolds said that more bone growth was observed in TPMS scaffolds, and they were less susceptible to fragile breaks. It was noted that TPMS provide stress shielding effect making them optimum for scaffold designing. Alabort et al. confirmed the possibility of applying TPMS as metallic bones by 3D printing [23] (Figures 4.5 and 4.6).

4.3.4 CHEMICAL APPLICATIONS

The interconnected structures with high-volume-specific surface areas of TPMS can increase the efficiency of chemical reactions as it provides a higher contact area.

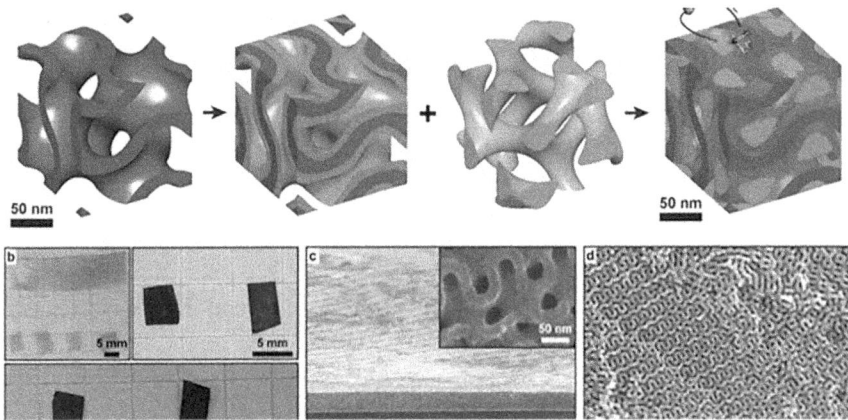

FIGURE 4.7 TPMS battery.

TPMS structures can be used to fabricate catalyst support structures and reactors to improve the efficiency of the process.

Lei et al used TPMS for catalytic structure and compared the results obtained with commercial solution. It was noted that the hydrogen production performances were improved in TPMS catalytic structures [24]. TPMS structures are also widely utilized as spacers in membrane distillation [25–29]. According to a study anode and cathode fabricated from different channels of TPMS structures can provide more stable open circuit voltage and reversible discharge voltage. It was also noted that to TPMS structures provided more internal space. This gives TPMS unique applications in energy-storing devices. When microreactors with TPMS structures were employed for CO_2 methanation it was found that it provided 14% improvement in CO_2 conversion when compared with honeycomb-based structures [30] (Figure 4.7).

4.4 CONCLUSION

TPMS with their unique geometric properties and orientation find their utility and potential in a diverse set of applications. They are promising solutions to various existing problems in varied industries ranging from biomedical to automobile and aerospace. They are known to exist in nature and were discovered way back in the 19th century, but recently it has garnered attention due to the advent of Additive Manufacturing methods as it made possible to fabricate these complex intricate geometries. The study has been going on currently on exploring different aspects of minimal surfaces by designing them for specific applications and then performing the suitable tests for testing its integrity. However, it is still in its rudimentary phase and a lot of potential of these surfaces will be explored as time goes on. However, there are certain challenges that needs to be overcome. The fabrication of these surfaces is possible only via additive manufacturing and it is currently expensive and time consuming. Different methods of AM should be explored for finding the optimum parameters and methods for fabricating the geometry so that it can be manufactured easily. The next stage is the post-processing and validation of surface geometries and to test its integrity whether it confides with the design or not before implementing into applications. A standard process needs to be carried out to test the surface integrity. Barring these gaps, TPMS are an excellent group of geometries and has a lot of potential solutions to the existing problems and can be implemented in novel and specific applications.

ACKNOWLEDGMENT

The authors would like to express gratitude toward Pandit Deendayal Energy University (PDEU) for supporting our study.

REFERENCES

[1] Ma, Q., Zhang, L., Ding, J., Q, Shuo, Fu, J., Zh, M., & Song, X. (2021). Elastically-isotropic open-cell minimal surface shell lattices with superior stiffness via variable thickness design, Additive Manufacturing, 47, 1–18. 10.1016/j.addma.2021.102293

[2] Meeks Iii, William H., & Perez, J. (2011). The classical theory of minimal surfaces. Bulletin of the American Mathematical Society, 48, 325–407.

[3] Sebastein J, P. Callens, Christoph H, Alina K, Amir A. Decoupling minimal surface metamaterial properties through multi-material hyperbolic tilings, Advanced Functional Materials (2021), 1–15, 10.1002/adfm.202101373

[4] Diab W, Mete B, Rashid K, Jorgen S, Nahil A, Iwona J. Mechanical properties of 3D printed polymeric cellular materials with triply periodic minimal surface architectures, Materials and Design (2017), 255–267. 10.1016/j.matdes.2017.03.018

[5] Li, W., Yu, G., & Yu, Z. (2020). Bioinspired heat exchangers based on triply periodic minimal surfaces for supercritical CO_2 cycles, Applied Thermal Engineering, doi: 10.1016/j.applthermaleng.2020.115686

[6] Pérez, J. A new golden age of minimal surfaces. Notices of the AMS. 2017. pp 347–358

[7] Daneshmehr, A. Using Minimal Surface theory to design bone tissue scaffold and validate it with SLS 3D printer. 27th Annual International Conference of Iranian Society of Mechanical Engineers-ISME2019. 2019.

[8] Hannah S. Paul M. Steven M. (2022). Development of an architecture-property model for triply periodic minimal surface structures and validation using material extrusion additive manufacturing with polyetheretherketone (PEEK). Journal of the Mechanical Behavior of Biomedical Materials, 133, 105345–105358.

[9] de Oliveira, A. R., de Andrade Mendes Filho, A., Masoumi, M., & Del Conte, E. G. (2021). Compression and energy absorption of maraging steel primitive scaffolds produced by powder bed fusion. The International Journal of Advanced Manufacturing Technology, 116(3–4), 1271–1283. 10.1007/s00170-021-07514-4

[10] Zhao, M., Zhang, D. Z., Liu, F., Li, Z., Ma, Z., & Ren, Z. (2020). Mechanical and energy absorption characteristics of additively manufactured functionally graded sheet lattice structures with minimal surfaces. International Journal of Mechanical Sciences, 167, 105262. 10.1016/j.ijmecsci.2019.105262

[11] Yin, H., Zheng, X., Wen, G., Zhang, C., & Wu, Z. (2021). Design optimization of a novel bio-inspired 3D porous structure for crashworthiness. Composite Structures, 255, 112897. 10.1016/j.compstruct.2020.112897

[12] Peng, C., Fox, K., Qian, M., Nguyen-Xuan, H., & Tran, P. (2021). 3D printed sandwich beams with bioinspired cores: Mechanical performance and modelling. Thin-Walled Structures, 161, 107471. 10.1016/j.tws.2021.107471

[13] Shaer, A. W., & Harland, D. J. (2021). An investigation of the strength and stiffness of weight-saving sandwich beams with CFRP face sheets and seven 3D printed cores. Composite Structures, 257, 113391. 10.1016/j.compstruct.2020.113391

[14] Alkebsi, E. A. A., Ameddah, H., Outtas, T., & Almutawakel, A. (2021). Design of graded lattice structures in turbine blades using topology optimization. International Journal of Computer Integrated Manufacturing, 34(4), 370–384. 10.1080/0951192x.2021.1872106

[15] Jiansheng Pan, A., Jianwei Wu, B., Yin Zhang, C., Hui Wang, D., & Jiubin Tan, E. (2021). Design and analyze of flexure hinges based on triply periodic minimal surface lattice. Precision Engineering, 68, 338–350. 10.1016/j.precisioneng.2020.12.019

[16] Melchels, F. P., Barradas, A. M., van Blitterswijk, C. A., de Boer, J., Feijen, J., & Grijpma, D. W. (2010). Effects of the architecture of tissue engineering scaffolds on cell seeding and culturing. Acta Biomaterialia, 6(11), 4208–4217. 10.1016/j.actbio.2010.06.012

[17] Li, L., Shi, J., Zhang, K., Yang, L., Yu, F., Zhu, L., Liang, H., Wang, X., & Jiang, Q. (2019). Early osteointegration evaluation of porous Ti6Al4V scaffolds designed based on triply periodic minimal surface models. Journal of Orthopaedic Translation, 19, 94–105. 10.1016/j.jot.2019.03.003

[18] Alabort, E., Barba, D., & Reed, R. C. (2019). Design of metallic bone by additive manufacturing. Scripta Materialia, 164, 110–114. 10.1016/j.scriptamat.2019.01.022

[19] Paré, A., Charbonnier, B., Tournier, P., Vignes, C., Veziers, J., Lesoeur, J., Laure, B., Bertin, H., De Pinieux, G., Cherrier, G., Guicheux, J., Gauthier, O., Corre, P., Marchat, D., & Weiss, P. (2019). Tailored three-dimensionally printed triply periodic calcium phosphate implants: A preclinical study for craniofacial bone repair. ACS Biomaterials Science & Engineering, 6(1), 553–563. 10.1021/acsbiomaterials.9b01241

[20] Castillo, E. H. C., Thomas, N., Al-Ketan, O., Rowshan, R., Abu Al-Rub, R. K., Nghiem, L. D., Vigneswaran, S., Arafat, H. A., & Naidu, G. (2019). 3D printed spacers for organic fouling mitigation in membrane distillation. Journal of Membrane Science, 581, 331–343. 10.1016/j.memsci.2019.03.040

[21] Thomas, N., Sreedhar, N., Al-Ketan, O., Rowshan, R., Abu Al-Rub, R. K., & Arafat, H. (2018). 3D printed triply periodic minimal surfaces as spacers for enhanced heat and mass transfer in membrane distillation. Desalination, 443, 256–271. 10.1016/j.desal.2018.06.009

[22] Sreedhar, N., Thomas, N., Al-Ketan, O., Rowshan, R., Hernandez, H., Abu Al-Rub, R. K., & Arafat, H. A. (2018). 3D printed feed spacers based on triply periodic minimal surfaces for flux enhancement and biofouling mitigation in RO and UF. Desalination, 425, 12–21. 10.1016/j.desal.2017.10.010

[23] Sreedhar, N., Thomas, N., Al-Ketan, O., Rowshan, R., Hernandez, H. H., Abu Al-Rub, R. K., & Arafat, H. A. (2018b). Mass transfer analysis of ultrafiltration using spacers based on triply periodic minimal surfaces: Effects of spacer design, directionality and voidage. Journal of Membrane Science, 561, 89–98. 10.1016/j.memsci.2018.05.028

[24] Thomas, N., Sreedhar, N., Al-Ketan, O., Rowshan, R., Abu Al-Rub, R. K., & Arafat, H. (2019). 3D printed spacers based on TPMS architectures for scaling control in membrane distillation. Journal of Membrane Science, 581, 38–49. 10.1016/j.memsci.2019.03.039

[25] Thomas, N., Swaminathan, J., Zaragoza, G., Abu Al-Rub, R. K., Lienhard V, J. H., & Arafat, H. A. (2021). Comparative assessment of the effects of 3D printed feed spacers on process performance in MD systems. Desalination, 503, 114940. 10.1016/j.desal.2021.114940

[26] Lei, H. Y., Li, J. R., Wang, Q. H., Xu, Z. J., Zhou, W., Yu, C. L., & Zheng, T. Q. (2019). Feasibility of preparing additive manufactured porous stainless steel felts with mathematical micro pore structure as novel catalyst support for hydrogen production via methanol steam reforming. International Journal of Hydrogen Energy, 44(45), 24782–24791. 10.1016/j.ijhydene.2019.07.187

[27] Baena-Moreno, F. M., González-Castaño, M., Navarro de Miguel, J. C., Miah, K. U. M., Ossenbrink, R., Odriozola, J. A., & Arellano-García, H. (2021). Stepping toward efficient microreactors for CO2 methanation: 3D-printed gyroid geometry. ACS Sustainable Chemistry & Engineering, 9(24), 8198–8206. 10.1021/acssuschemeng.1c01980

[28] Li, W., Yu, G., & Yu, Z. (2020). Bioinspired heat exchangers based on triply periodic minimal surfaces for supercritical CO2 cycles. Applied Thermal Engineering, 179, 115686. 10.1016/j.applthermaleng.2020.115686

[29] Attarzadeh, R., Rovira, M., & Duwig, C. (2021). Design analysis of the "Schwartz D" based heat exchanger: A numerical study. International Journal of Heat and Mass Transfer, 177, 121415. 10.1016/j.ijheatmasstransfer.2021.121415

[30] Sreedhar, N., Thomas, N., Al-Ketan, O., Rowshan, R., Abu Al-Rub, R. K., Hong, S., & Arafat, H. A. (2020). Impacts of feed spacer design on UF membrane cleaning efficiency. Journal of Membrane Science, 616, 118571. 10.1016/j.memsci.2020.118571

5 Ultrasonic Additive Manufacturing Process

Process Parameters, Defects, Properties, and Applications

Bhumi K. Patel and Vishvesh Badheka

5.1 INTRODUCTION TO ULTRASONIC ADDITIVE MANUFACTURING

Additive manufacturing is a revolutionary innovation with a wide range of industrial applications that conventional production methods cannot match. Dawn White pioneered ultrasonic additive manufacturing (UAM) also known as ultrasonic consolidation (UC) in 1999 [1]. UAM is a solid-state additive manufacturing technology that utilizes ultrasonic metal welding (USW) to consecutively join metal foils together, layer by layer, and CNC machining to remove material to attain the required geometry throughout the additive build-up process [2,3]. Figure 5.1 shows a schematic representation of the process and its components.

The method employs an ultrasonic transducer to activate a horn or sonotrode to generate a scrubbing motion while a downward force is applied, similar to ultrasonic metal welding [5]. Up to a frequency of 20 kHz, this scrubbing motion or welder vibration amplitude occurs [6]. Shearing occurs as a result of the ultrasonic vibration, which removes surface oxides via friction and allows direct metal-metal contact [4]. In addition, the dynamic interfacial tensions between the contacting surfaces are formed because of the shearing process [1]. During the operation, the ultrasonic power supply cautiously modulates the vibration amplitude [5]. Similarly, the motion system can control the downward force and linear travel speed [5]. A heated anvil is sometimes used to modulate the process temperature [5]. The circular design of the horn (tool or sonotrode) enables continuous material deposition along the part's surface [5]. In Figure 5.2, steps are shown to achieve high-strength bonding with the help of UAM.

Since UAM operates at a much lower temperature than melting point temperature, many application areas are enabled. Generally, UAM operates at 0.3–0.5 Tm [6]. Many defects caused by high temperature are eradicated because of low-temperature application [7]. Due to system power restrictions (2 kW),

DOI: 10.1201/9781003484325-5

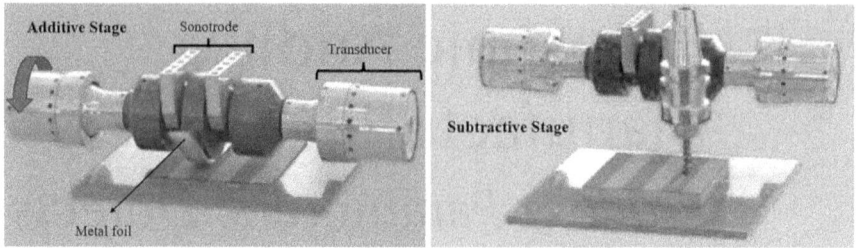

FIGURE 5.1 Components of UAM technique [4].

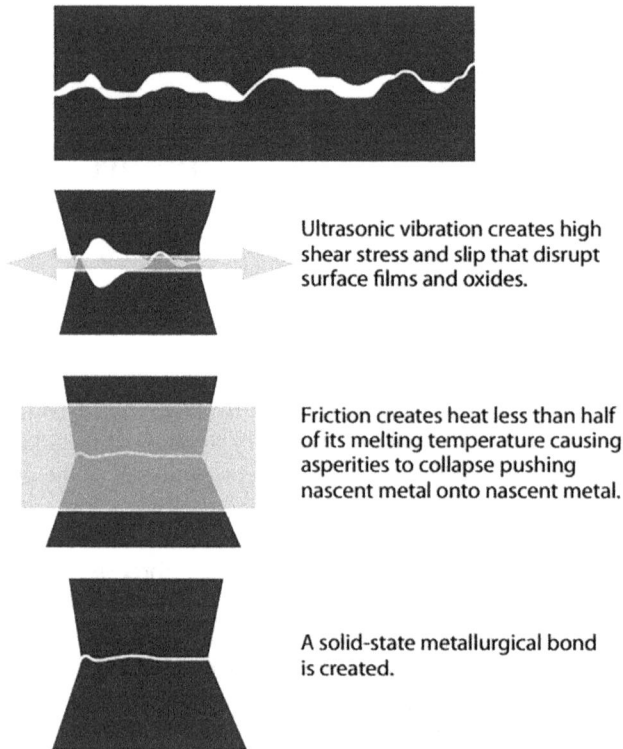

Ultrasonic vibration creates high shear stress and slip that disrupt surface films and oxides.

Friction creates heat less than half of its melting temperature causing asperities to collapse pushing nascent metal onto nascent metal.

A solid-state metallurgical bond is created.

FIGURE 5.2 Steps utilized to obtained high-strength solid-state bonding with UAM [4].

UAM systems were first confined to connecting soft materials such as aluminum alloys [8–10]. This problem has been solved by designing high-powered (9 kW) UAM systems [11]. With the help of UAM process dissimilar materials are easily bonded together while retaining the properties of each building material. So far, many combinations for similar and dissimilar materials are feasible e.g., Al–Al [12–16], Ti–Al [9,17,18], steel-Ni [19], steel-Ta [20], Al–Cu [21], Al–NiTi [22], Al–SiC [23,24], Al with embedded dielectric materials [25], low carbon steel [26,27], Al-shape memory alloy [28]. One of the joining combinations of various

	Al	Be	Cu	Ge	Au	Fe	Mg	Mo	Ni	Pd	Pt	Si	Ag	Ta	Sn	Ti	W	Zr
Al Alloys	•	•	•	•	•	•	•	•	•	•	•	•	•	•	•	•	•	•
Be Alloys	•	•				•										•		
Cu Alloys	•		•	•	•	•	•	•	•			•	•			•	•	•
Ge				•							•							
Au					•	•		•	•	•	•	•				•	•	•
Fe Alloys						•		•	•	•			•	•		•	•	•
Mg Alloys							•						•			•		
Mo Alloys								•	•		•		•			•	•	•
Ni Alloys									•	•	•		•			•	•	
Pd										•			•	•				
Pt Alloys											•	•		•		•	•	
Si												•	•					
Ag Alloys													•	•				•
Ta Alloys														•	•	•		
Sn															•			
Ti Alloys																•	•	
W Alloys																	•	
Zr Alloys																		•

• Material pair proven for ultrasonic welding

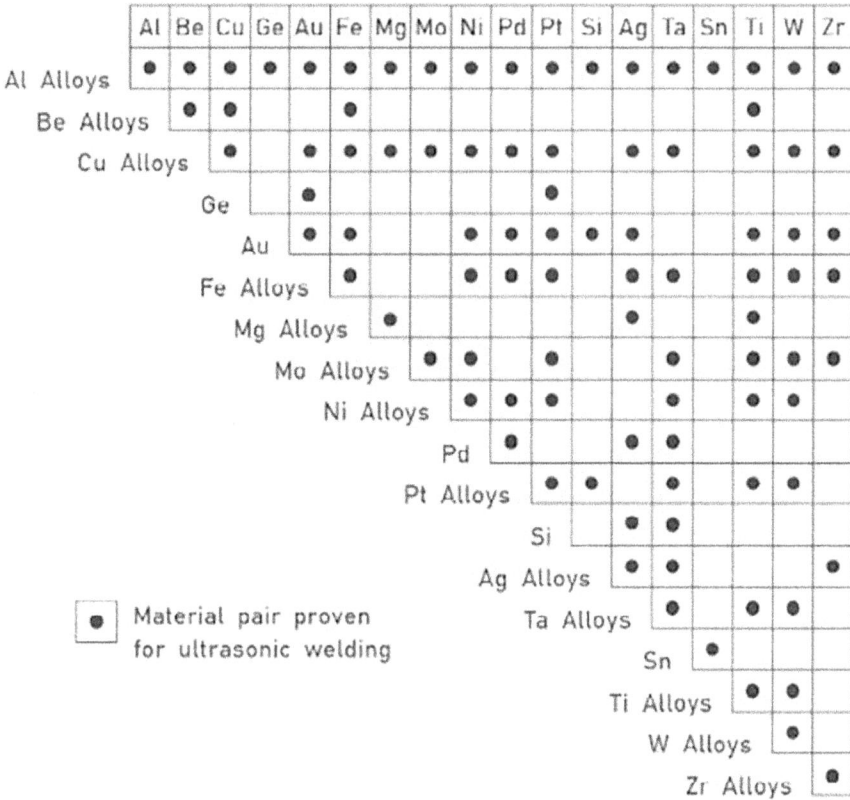

FIGURE 5.3 Possible joining of various materials with the help of UAM [4].

material with the help of UAM is shown in Figure 5.3. Fabrication of components with embedded sensors [29,30], actuators [31,32], and parts for thermal management [11] also feasible with the help of UAM. As we can see from Figure. 5.1, UAM is referred to as a hybrid process, as both additive and subtractive stages are involved.

5.1.1 Process Associated with Ultrasonic Additive Manufacturing

UAM works on the basic principle of ultrasonic waves. The ultrasonic sound waves create vibration between the consecutive layers of the metal sheets. Friction is generated between the plates as a result of the vibration. The plates are welded together by the combined action of applied pressure force and friction. As shown in Figure 5.4, In the UAM, the first step is to attach the base foil to the build platform with the help of ultrasonic vibrations that are produced by the sonotrode. The second step is to weld thin foil sheets until a predetermined depth is reached. Once the intended depth has been attained, the third step is to cut the block into the needed shape with a suitable milling tool. The fourth step is to repeat this process. In the fifth step, a smaller milling tool further shapes the part and completes any

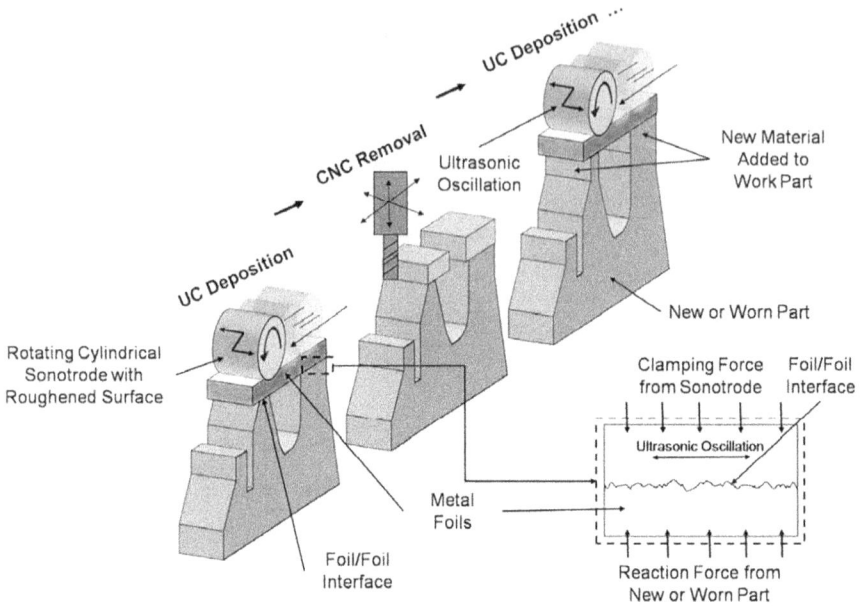

FIGURE 5.4 Schematic representation of Ultrasonic additive manufacturing process [35].

internal milling required for the finished part. The entire UAM cycle is then repeated until the necessary component is obtained and the part is removed from the build platform in the final step [5,6,33,34].

5.2 PROCESS PARAMETERS OF UAM

The process parameters used throughout the UAM process play an important role in the formation of high-strength and high-quality bonds. Normal force (N), amplitude (μm), speed (mm/s), and temperature (°C) are some of the most significant process parameters. From the research percentage contribution these process parameters are shown in Figure 5.5. In order to fabricate components with utmost build strength, these parameters must be optimized.

The sonotrode is one of the important components of the UAM process. It is a disc-shaped horn that transmits ultrasonic vibrations to the foils that are to be welded. The prime objective of the sonotrode is to apply downward force to the sheets while keeping them fixed [36]. The normal force is applied via the rough surface of the sonotrode. Sonotrodes are often composed of titanium, aluminum, or steel, and come in a variety of surface finishes to accommodate various purposes. Generally, a steel sonotrode with an average surface roughness (Ra) of 5.2 μm [25] is utilized for low-amplitude applications requiring hardness. Because of low vibration loss and great strength, Titanium is the most used material whereas aluminum is utilized to decrease wear and is frequently coated with chrome or nickel [36]. The travel speed and motion in all three directions are regulated by CNC controller.

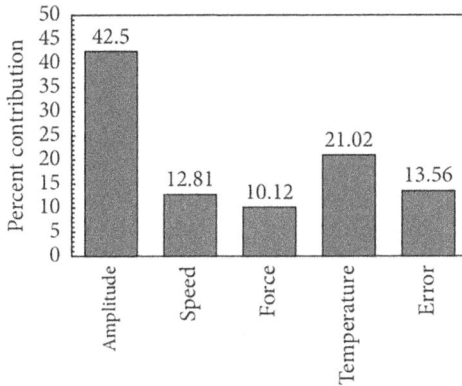

FIGURE 5.5   Percent contribution of process parameters during UAM process.

5.2.1 NORMAL FORCE (N)

Sonotrode applies force in a downward direction which is referred to as the normal force, and the objective of the normal force is to manage adequate coupling between the foils, allowing vibrations to be successfully transferred. Usually, it is in the range from 100 N to 9000 N [6] but from the few research works, it can be concluded that in-low power systems the normal force was restricted to 2000 N [8,11,37] and 20,000 N in the case of the high-power system [11,38]. Due to persistent pressure, the direct contact between the building materials was maintained during fabrication, and in addition to that plastic flow has been enhanced [37]. Dynamic stresses are formed as a result of the interaction of force and ultrasonic vibrations, which are vital for oxide removal and plastic flow [37]. The higher normal force will create a larger bonding, resulting in a better quality weld, but in contrast to that, excessive stresses and deformation can occur when pressures are too high [39–41]. Similarly, the low value of normal force results in poor weld quality and insufficient deformation. The input energy is a function of the normal force, hence to acquire higher bond quality and strength, optimum parameters should be utilized [6].

5.2.2 AMPLITUDE (MM)

Vibration amplitude is the distance covered by the sonotrode, measured from the outer diameter of the sonotrode while moving along the x-y plane, which is the driving factor to the interface thickness, density of the bonded area, and the size of the deformed area [6]. For the various studies, it can be stated that the amplitude lies between 10 and 50 μm [42]. The most significant and crucial parameter in bond quality is amplitude [43]. The amount of plastic deformation between the building components is determined by the amplitude [44]. An increase in the amplitude will result in improved scrubbing action, which disperses oxides and impurities more efficiently away from the interface, enhancing the interface's strength by increasing the density of metallic bonding [15,28,43,45]. A higher value of the amplitude is not advisable because the flaws may be incorporated into the structure which results in the breaking of the bonds [43].

5.2.3 SPEED (M/S)

The travel speed refers to the sonotrode linear motion across the workpiece during the UAM process [26]. It can be kept in the range of 1–100 mm/s [26]. The speed controls how long the material under the sonotrode is compressed and oscillated [36]. Additionally, the amount of energy given to the workpiece is inversely related to the speed. In an instance, Lowering the speed increased the energy delivered to the build material, resulting in higher degrees of strain hardening [46]. This happens because when the speed is kept low, it offers more time for scrubbing the interface, resulting in more ultrasonic energy being delivered to the interface for welding [43]. As an outcome, improvement of bonding between two foils, dispersion of oxides, and impurities at the contact is attained [43]. When the speed was raised, insufficient bonding occurs [28].

5.2.4 TEMPERATURE (°C)

According to numerous experts, substrate temperature is another process parameter that might affect bond quality [15,21,38,41,47,48]. But the substrate temperature has less influence on the UAM process compared to other process parameters. The temperature of the substrate/part can be raised above the ambient temperature during processing. The use of a higher temperature during processing results in denser and stronger bonding [49]. Build materials influenced the increase in temperature, so careful selection is advisable, especially in the case of dissimilar joining.

5.2.5 LINEAR WELD DENSITY

The quality of the bond created is related to the amount of UAM energy used during processing, which may be measured using the linear weld density (LWD) method [13,30]. As illustrated in an equation, LWD is the length of a given interface that appears to be bonded divided by the overall length of the interface inspected [13]. Total interface length refers to the apparent length measured, whereas bonded interface length refers to the actual length.

$$\%LWD = \frac{bonded \quad interface \quad length}{total \quad interface \quad length} \times 100[6] \qquad (5.1)$$

LWD must be kept as high as possible in order to maintain good bond quality. LWD on UAM parts normally ranges from 45% to 95% [13]. Lower values correspond to lower UAM energies, resulting in weaker bonds [30]. Parameters like speed, amplitude, substrate temperature, and normal force influenced the LWD [8].

5.2.6 PROCESS PARAMETERS USED IN SEVERAL RESEARCH WORK BY VARIOUS AUTHORS

Table 5.1 shows various process parameters used for build materials.

TABLE 5.1
Various Process Parameters For Different Build Materials

Sr. No.	Build Material	Foil Thickness (mm)	Force(N)	Speed (mm/s)	Amplitude (µm)	Temperature (°C)	References
1	Al 1100-Commercially pure Ti	0.127	3500	25.4	41.55	93.3333	[50]
2	SAE4130 steel	0.127	6000	21	30.87	204	[26]
3	Al2024 foil	0.15	2500	46	24	120	[48]
4	Al6061 H-18	0.15	5000	85	35	75	[14]
5	Al6061 H-18	0.15	5000	84.7	35	75	[3]
6	NiTi-Al6061	NiTi diameter – 0.28 thickness of Al6061 – 0.152	6000	84.6	32.76	22	[51]
7	NiTi SMA-Al3003-H18	0.1	138 kPa	$\leq$34.5	8.4–14.3	70	[28]
8	Al3003 H18-Al3003 H14	0.1524	1000	53	26	149	[15]
9	SiC-Al3003	SiC diameter – 0.1 Al3003 thickness - 0.15	1700	34	20	149	[15]
10	Al3003 H18	0.15	5600	35.6	26	NA	[52]
11	C11000 copper-Al3003 H18	0.15	6700	30	36	25	[38]
12	Al3003 H18-CP Ti	CP Ti – 0.075 Al3003 H18 – 0.150	1750–2000	10.58–23.70	16–28	148.889	[9]
13	Al3003 O-Al3003 H18	0.1	1400	42.3	19	149	[53]
14	Al3003 H18-Al1050 H14	0.1	1600	20	25	NA	[25]
15	Al3003-Al3003 H18	0.15	1450–1900	28–40	10–19	24–149	[8]
16	CF-Al6061 H18	Al6061 H18 – 0.1524	5000	84.67	32	22	[54]
17	Ni-coated CF-Al6061 H18	Al6061 H18 – 0.1524	5000	84.67	32.5	22	[54]

(Continued)

TABLE 5.1 (Continued)

Various Process Parameters For Different Build Materials

Sr. No.	Build Material	Foil Thickness (mm)	Force(N)	Speed (mm/s)	Amplitude (μm)	Temperature (°C)	References
18	CF-4130 steel/Ni201	4130 steel – 0.13 Ni201 – 0.025	7000	33.83	40.63	93	[54]
19	CFRP-Ti	CFRP – 0.600 Ti – 0.100	100	NA	300	NA	[55]
20	1010 steel-95% pure Ta	99.5% pure Ti – 0.050	7000	15	36	NA	[20]
21	Al6061	0.152	4000–6000	84–106	28–34	80–95	[56]
22	Al6061 H-18	0.15	5600	35.6	31	NA	[57]
23	Ti–Al	0.127	500–2000	21–85	15–30	150	[58]
24	CFRP/Ti	CFRP – 0.600 Ti – 0.100	150	NA	10, 35, 50	24.85	[59]
25	CFRP/Ti	CFRP – 0.600 Ti – 0.100	NA	NA	300	24.85	[60]
26	Al-Au-Ni	0.15	4000	33	28	24.85	[61]
27	YSZ-Al Composites	YSZ – 0.040 Al – 0.152	4000	33.87	30	Al –204.4	[62]
28	FBGs - Al6061	NA	4000	84.67	30–32	25	[63]
29	Al-PVDF sensor-polyimide film	Al – 0.152 PVDF – 0.040 Polyimide – 2 ×0.050	4000	42	34	23	[64]
30	Uncoated Carbon steel	0.127	6500	17	32.4	204	[65]

5.3 DEFECTS

UAM can create high-quality functional components. However, they are prone to defects. Several studies have focused on identifying the best UAM process parameters and subsequently reducing flaws with post-weld treatments [3,15,26,66–70]. As shown in Figure 5.6 defects in UAM are known as type 1 and type 2 [37,70]. Type 1 defects can be further categorized into type 1a and type 1b [71]. Other than these, defects like voids, kiss bonds, etc. are also present in the UAM process. When the sonotrode has deteriorated, the surface roughness was poor, and it was unable to hold the top foil properly and efficiently pass ultrasonic energy to the welding interface, resulting in the poor-quality weld [72].

Both type 1 and type 2 faults in UAM components must be eradicated, either by process improvement or post-manufacturing healing, before UAM components may be employed in structural applications [3,26]. Heat treatment of UAM components after processing has been found to improve bond quality and reduce type 1 flaws, but it cannot repair type 2 defects [71]. It is critical to create a methodology to heal/repair the typical flaws in UAM due to the variety in processing circumstances that affect bond quality [71]. Because the primary benefit of UAM is solid-state processing, only solid-state procedures should be considered for the repair of flaws in UAM parts [71].

Friction deposition [73,74], friction lap-seam welding [75], and friction stir additive manufacturing (FSAM) [76] are some of the additional friction-based solid-state additive manufacturing processes that have been recommended and demonstrated for diverse materials. Friction stir processing (FSP) is a solid-state method used for enhancing material microstructural properties and repairing defects at the same time [77]. Fine-grain microstructure, surface, and bulk composites, as well as repairing crack and surface defects, have all been proven using FSP [71].

When a soft and a tougher material are bound together with UAM, they can be deposited in one of two ways, in which one is direct and another one is indirect. In the indirect type, there are two ways: first, the deposition of the soft layer on top of the harder layer, and second, the deposition of the harder layer on top of the soft layer [72]. From the research, it was found that in terms of bonding density and joint quality, strategy 2 was found to be superior to strategy 1 [72]. Good welding density can be accomplished by utilizing the second approach, whereas many

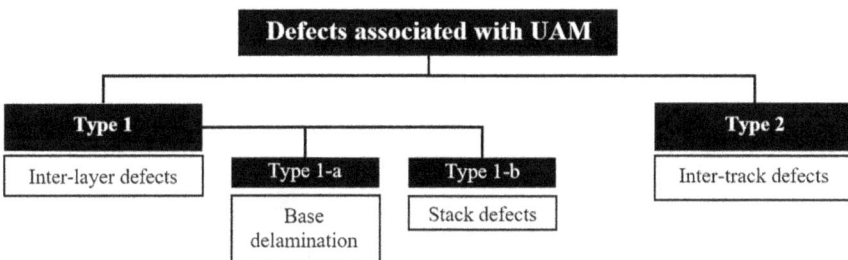

FIGURE 5.6 Classification of defects.

defects occurred between the harder layer and the previously as-deposited soft layer in the case of strategy 1 [19,78,79].

5.3.1 TYPE 1

A type 1 defect is called an inter-layer defect. Type 1 defects can be divided into two categories: type 1a for base/build delamination and type 1b for defects within the stack [71]. Because the base/build interface is prone to delamination, this distinction is critical [80]. Because the UAM component is tightly attached to the base, various research works show that type 1a defects are widespread, even if they are not physically evident to the naked eye. Though bonding is great at the base/build interface, some part of it is delaminated [71].

5.3.1.1 Type 1a

Base delamination is the term for this type of flaw. Type 1a defect is shown in Figure 5.7. With layer build-up, the base interface is the most vulnerable to delamination. The base is also a component of the structure of interest for various applications. In this instance, the quality of the base interface must be measured and monitored. Minor delamination is permissible for applications where the base is not a part of the structure and if the structure stays firmly attached to the base [71]. Understanding the mechanism of base interface delamination is critical in both circumstances.

Type 1a defect evolution can be categorized into three types [71]: 1) Because of insufficient energy intake, poor bonding takes place. This kind of defect generally occurs largely on the UAM weld's inner side while fabrication of the first few layers. 2) Because of the UAM stack vibrations, this kind of defect begins at the edge. 3) Because too much energy is applied, this kind of defect arises and results in a fractured base/build interface. If the optimal value of the amplitude has been selected then there might be a slight possibility to reduce type 1a defect, however, this is an unviable solution because it will cause type 1b defect [71]. Schematic representation of this defect is shown in Figure 5.8.

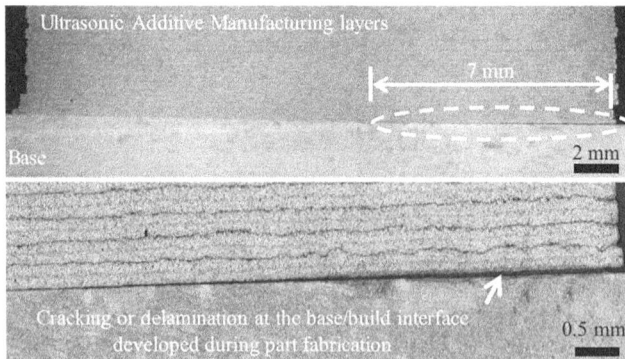

FIGURE 5.7   Base declamation (Type 1a) defect.

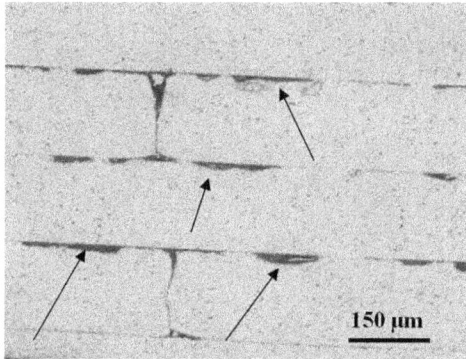

FIGURE 5.8   Schematic representation of Type 1a defect [81].

5.3.1.2   Type 1 b

Type 1b defects are associated with stack quality. Because substrate stiffness impacts ultrasonic energy input, the quality of a UAM stack is governed by its structure. Inadequate power input into the UAM component causes type 1b defects [71]. Figure 5.9 represents type 1b defect.

5.3.2   Type 2

Type 2 defects are known as inter-track defects. In complex systems containing features such as ribs with transverse foil orientation to the loading direction, the weakening effects of these defects are more severe. In the presence of these defects, those elements are more prone to failure, sooner than other components with foils in the longitudinal direction to the applied load [81] (Figures 5.10 and 5.11).

5.3.3   Voids

Surface asperities of the original foils and those on top of the newly deposited foil caused by compression and shear deformation induced by the sonotrode are

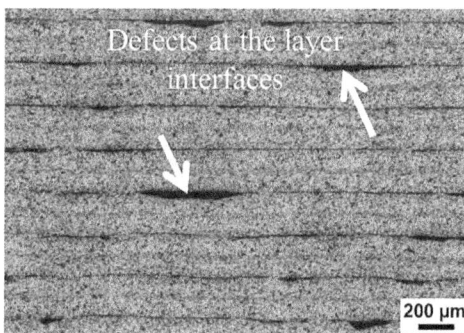

FIGURE 5.9   Illustration of type 1b defect [71].

FIGURE 5.10 Schematic representation of type 2 defects [71].

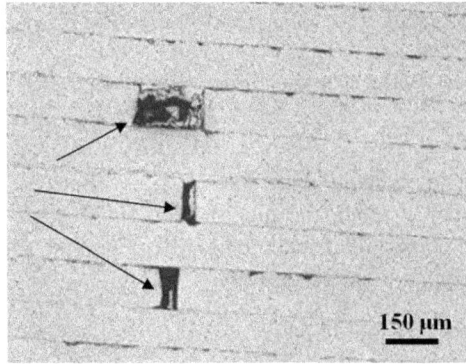

FIGURE 5.11 Inter-track defect [81].

considered to be the cause of void formation at the weld interface [8,45,53,57,82]. The shattered oxide layers on the original foil surface can potentially cause welding problems at the interface [72]. The inconsistent contact pressure from the sonotrode can also lead to voids [72]. Last, the building pattern or methodology can also affect voids [72,82].

The voids at the UAM interfaces were classified as follows: linear, parabola like, and point like [8]. Large micro-valleys on the rough surface caused by the sonotrode during the preceding deposition are what cause the parabola-like voids to form [8,72]. It can also form due to insufficient material flow into the micro-valleys during the subsequent deposition [72].

5.3.4 REPAIRING OF DEFECTS

Friction stir processing has been used to heal defects. In the FSP process rotating tool is being used to accomplish microstructural refinement, densification, and homogeneity. FSP tool design, process parameters, and the active cooling-heating cycle are responsible for the mechanical properties and microstructure of the processed region [71]. In FSP, depth is adjustable and has a negligible effect on the shape of the processed region, and that is why it is a better suitable technique for repairing defects [71].

Employing standard joining procedures to repair flaws in UAM components diminishes the benefits of using the UAM technique. Both types of flaws have been fixed by FSP. Repairing type 1a flaws during component production is possible, but it necessitates the removal of the base plate and some setup time whereas the type 2 flaws can be repaired in between UAM layers by changing the weld head [71]. Changing the UAM process parameters is a better way to tackle type 1b defects.

For a defect-free surface, the surface roughness of the sonotrode should be kept high as possible but not too high. If the value of surface roughness is kept too low, energy is lost due to slippage between the sonotrode and the top foil [72]. When the surface roughness is excessively great, bigger voids can form at the welding interfaces.

5.3.4.1 Repairing Type 1 Defect

Repairing of the type 1 defects is only required when the base is an important component otherwise base can be removed once the manufacturing is done. With the help of FSP, type 1 defects can be repaired, and enhanced mechanical properties and fine microstructure of the final part have also been observed [71].

For the base/build crack repair, the supporting blocks in Figure 5.12 are not required. Often, just securing the base plate is enough to ensure a proper repair. Extra component support is necessary in order to heal very severe delamination [71]. Kissing bonds at the base/build interface are the most commonly seen delamination in UAM.

As seen in Figure 5.13, all the defects are completely healed after using FSP. Not only does FSP heal all the defects, but it also enhances properties.

5.3.4.2 Repairing Type 2 Defects

Repairing type 2 defects has been challenging, and only a little research has been done until now [70]. When a UAM component consists of a single track, the component's building height is limited. The building height can be considerably

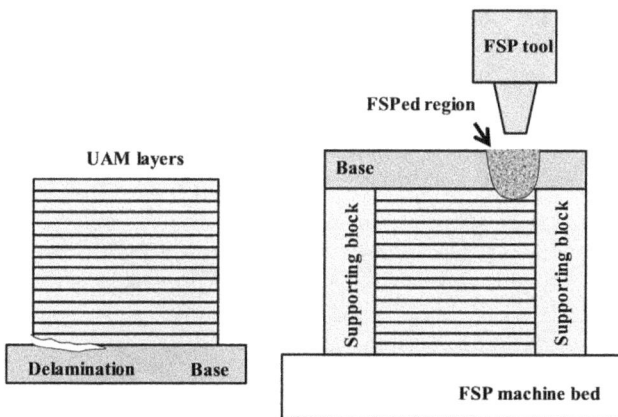

FIGURE 5.12 Schematic illustration of FSP used for repairing type 1 defect [71].

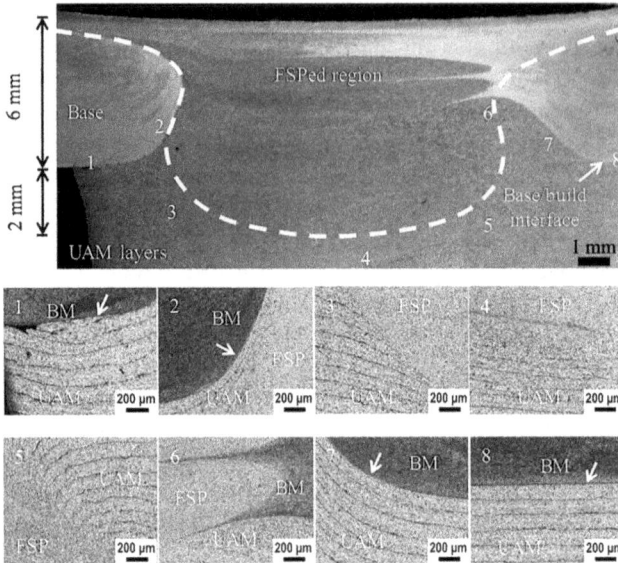

FIGURE 5.13 Optical micrographs of a sample after FSP are used to repair the type 1 defects.

expanded by using additional tracks. It is vital to fix type 2 flaws for these reasons. To ensure strong inter-track bonding, the most common way to deal with type 2 flaws is to change the overlap between consecutive foils as shown in Figure 5.14 [71].

The lack of bond formation between consecutive tracks arranged side-by-side is the cause of type 2 defect formation. The overlapping region has the tendency to reduce type 2 flaws, but it does not provide any guarantee that they will be eliminated completely [71]. For the complete elimination of type 2 flaws, precautions should be taken during fabrication, but again, perfect results will not be attainable while fabricating hundreds of inter-tracks. The acceptable expectation of bond quality with type 2 faults may be dependent on various localized geometrical parameters that are very difficult to manage [71].

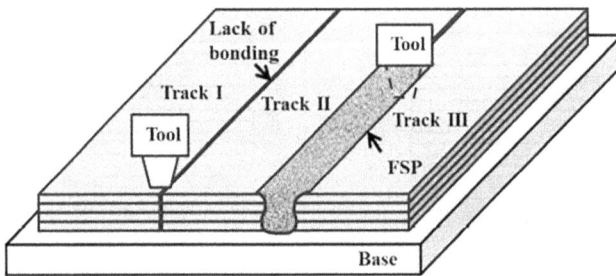

FIGURE 5.14 FSP used for the repair of type 2 defect.

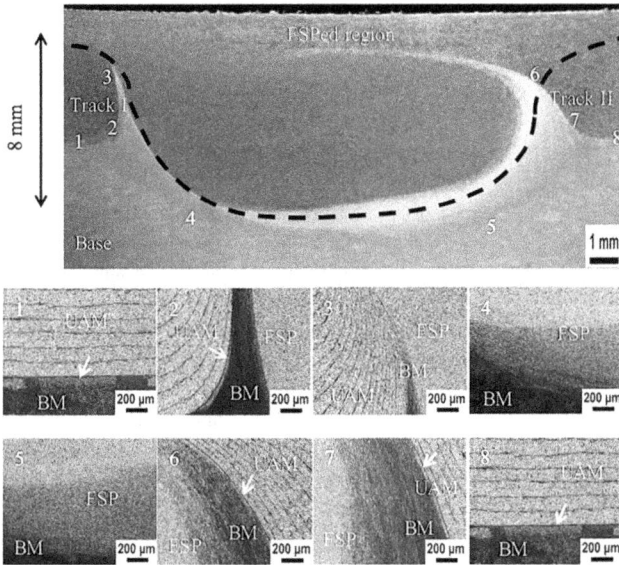

FIGURE 5.15 Optical micrographs of a sample after FSP are used to repair inter-track defects.

From Figure 5.15, it can be seen that cracks are completely healed and as a result of the FSP uniformity, better quality, enhanced mechanical property, fine microstructure, etc. were attained. Despite the many benefits of FSP, there is one limitation: that at the edges this methodology is not applicable [71].

5.3.4.3   Repairing of Voids

The voids can be eliminated by introducing enough plastic deformation/flow to fill the gaps created at the bonding interface, shattering the oxides into as small of bits as possible, and then redistributing them to the materials around the interface [72]. For reducing defects, selected process parameters should be optimal. In addition to that, heating will also reduce void defects [72].

5.4   MECHANICAL PROPERTIES

Factors affecting mechanical properties of the UAM part (Table 5.2):

1. Characteristic of the interface region between layers: UAM is a layer-by-layer process like any other AM technique, hence the quality and characteristics of the interface between the layers will influence the mechanical properties of the as-built parts [6]. Generally, between metal foils or tapes, interfacial bonding occurred [13]. Metal-metal contact under pressure causes plastic deformation, oxide dispersion, and asperity collapse [26]. A strong continuous interfacial bonding is essential for improved mechanical properties and defect-free weld [26].

TABLE 5.2
Mechanical Properties

Sr. No.	Property	Description
1	Shear Strength	• To achieve precise result within the layered region, special tooling is used [26]. • Happens at joining of the first layer and the base plate [87] • Poor quality in the z-direction i.e., building direction [14,66] • Higher temperature will result in better quality and consistency of the part [87] • Temperature is inversely proportional to shear strength [87] • Post-treatment methods have been employed for satisfying results [6] • Attain improved strength by employing HIP and SPS techniques [17,87] • After SPS treatment – ductile failure and Before SPS treatment – brittle failure (Because of voids) [15,17]

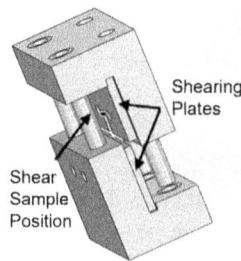

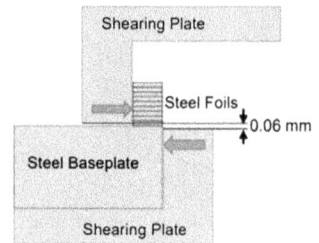

[87]   [87]

Sr. No.	Property	Description
2	Hardness	• Dependent on the process parameters, post-processing, and microstructure [26] • Hardness is inversely proportional to the fine microstructure. Finer the structure, better the hardness [6]. • According to research thickness of dielectric does not affect the hardness [25]. • From some of the work, the HIP process improves the hardness whereas according to a few research works it is showing the opposite result [6,87]. It's possible that this is because of the longer heating period, which resulted in large grain growth and lower hardness [6]. Both the results are contradicting one another and that is why more solid research is required. • SPS process always has a positive impact on the hardness [26]. It is because of the material's short heating time and retention of structure [88,89]. • The hardness values of the layers closest to the base plate are lower. The SPS process, in which the top layer comes into contact with a graphite punch and some carburization occurs, is responsible for this phenomenon [26].

TABLE 5.2 *(Continued)*
Mechanical Properties

Sr. No.	Property	Description
3	Peel Strength	• Used to determine the bond strength of tapes, glues, and individual layers [13,40,46,47]. • In the case of dielectric material, hardness affects the peel strength. The hardness of the dielectric material is proportional to the peel strength [25]. • Among all the process parameters amplitude has a significant impact on the peeling load [15,90] and it was found that the higher the amplitude, the higher the peeling load [38,47,53] due to fewer voids generation [91]. • Two reasons can contribute to a reduced peeling strength: 1) a lower joining strength of each micro-bond 2) a lesser number of micro-bonds. The LWD results, on the other hand, are only interested in the number of micro-bonds, not their strength. This is why a low peeling strength is often accompanied by a low LWD, but this is not always the case [92].

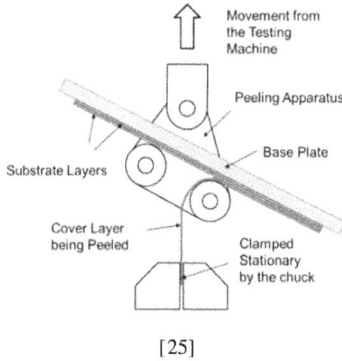

[25]

4	Tensile Strength	• Strength is weaker in x-y-z-direction compared to the as-built part. Among the x, y, and z axes, parts loaded along the Z-axis showed a considerable drop in strength [3]. • In the Z-direction, brittle failure has been observed due to defects or voids and without any failure, parts are failed because of inadequate bonding [3,6]. • In the case of aluminum/iron-based amorphous composite, brittle and ductile both type of failure have been observed [93,94]. • By using heat treatment, tensile properties have been improved but more work is required in this area to get a clear idea.

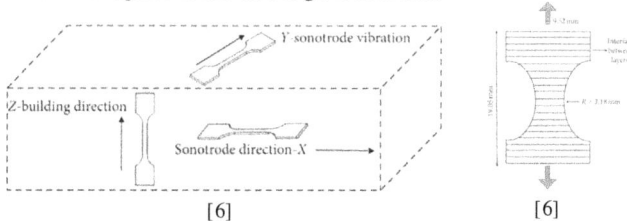

[6]   [6]

5	Fatigue Life	• More research is needed in this area to get proper results and relations.

2. Materials and building direction: From the different research work, it can be concluded that poor build strength is achieved when the load is normal to z-direction, whereas in x and y directions the same properties have been obtained [14,66]. Different materials have different characteristics and, in that way, they influence the overall mechanical properties of the UAM part.

3. Post-processing treatment: Another parameter that will influence the mechanical properties is post-processing treatment. By using post-processing, structural homogeneity and enhanced mechanical properties have been attained in the final fabricated UAM component [7,14,17,26,66,83]. Until now, many processes like spark plasma sintering [7,17,26,83], hot isostatic pressing [7,26], solutionizing, and aging [14] have been employed and have shown significant results.

4. Surface roughness: Another factor that should be taken into consideration is surface roughness because all the defects or cracks start to form at the surface itself [84]. The rough surface will result in restricting fatigue behavior whereas better surface quality will provide promising fatigue behavior [84–86].

5. Presence of voids/defects: Defects can be commonly seen at the interface, but lack of bonding [13] and fusion [85] are internal defects that will reduce the mechanical properties. Due to these defects, stress concentration takes place around the defect and diminishes the load-bearing capacity [6].

5.5 APPLICATIONS

- UAM method is used in manufacturing injection molding dies [94].
- UAM is suitable for the incorporation of second-phase material and is also used for joining dissimilar materials [94].
- UAM is suitable for metal to polymer joining [55,59,60].
- Components with complex geometries that are difficult to create using traditional fabrication techniques can benefit from UAM [94].
- Embed sensors and circuitry
  - UAM is a one-step solution for creating improved electrical applications, such as embedded sensors and RFID, as well as tamperproof devices [95,96]. Sensors may be inserted anywhere in a metal structure with UAM, making it useful in fields such as healthcare, industrial control, and even the Internet of Things [97].
- Parts with embedded channels are manufactured with the help of UAM method [94].
- Fiber Embedment
  - Many types of optical fibers can be deposited without being damaged since UAM operates at low processing temperatures. Silicon carbide, structural fibers, and optical fibers embedded in aluminum matrices are the most prevalent. Fibers can also be inserted and embedded between materials that are not the same and metal structures can be used to optically transfer data and energy. Figure 5.16 shows SiC fibers embedded within Al.

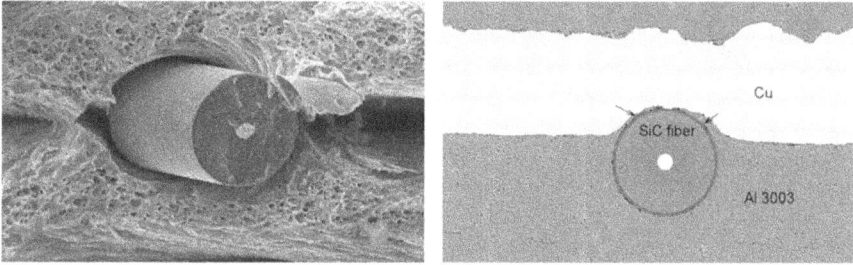

FIGURE 5.16 SiC fiber embedded within A1 [94].

- Internal channel
  - Morden Fabrisonic machines can manufacture parts from aluminum and copper using very high-power UAM, which helps to discharge heat while also reducing weight [97]. Within metallic materials, UAM can develop complicated internal structures like honeycomb structures, internal pipes or channels, and enclosed cavities [94].
- A combination of Carbon Fibre Reinforced Polymers (CFRP) and titanium (Ti) can be used in the automobile and aerospace industries [55].
- UAM was successfully used in nuclear component manufacturing by embedding neutron-absorbing materials into aluminum directly with UAM [98].
- Aerospace Industry
  - NASA put UAM heat exchangers manufactured by the Fabrisonic to the test as a replacement for heat exchangers made with standard technologies. It boasts a 30% weight reduction and a 30% improvement in performance. A curving channel is carved into the layered metal and then encased under further layers to construct the heat exchanger. New fiber optic sensors can detect metal strain or flaws and anticipate failures ahead of time. These Fabrisonic sensors aid in the detection of flaws and performance difficulties in commercial aircraft in NASA aeronautics testing [99–101]. Example of aluminum HX is shown in Figure 5.17.
- Thermal Management
  - Many engineering systems require the ability to extract or deliver heat at specific points across a structure. Heat sinks on the outside of the structure near the heat source are commonly used to accomplish this. Designing the structure with internal channels at the source and pumping fluid via those channels is a more effective strategy. There are numerous issues with the standard procedure. Additive manufacturing technology can be used to create improved thermal management designs with conformal cooling and heating channels. Because of its capacity to weld incompatible materials and embed components such as cooling tubes or heat-wicking materials, UAM is highly suited for the fabrication of complicated parts among various technologies [102]. One of the applications of this area is shown in Figure 5.18.

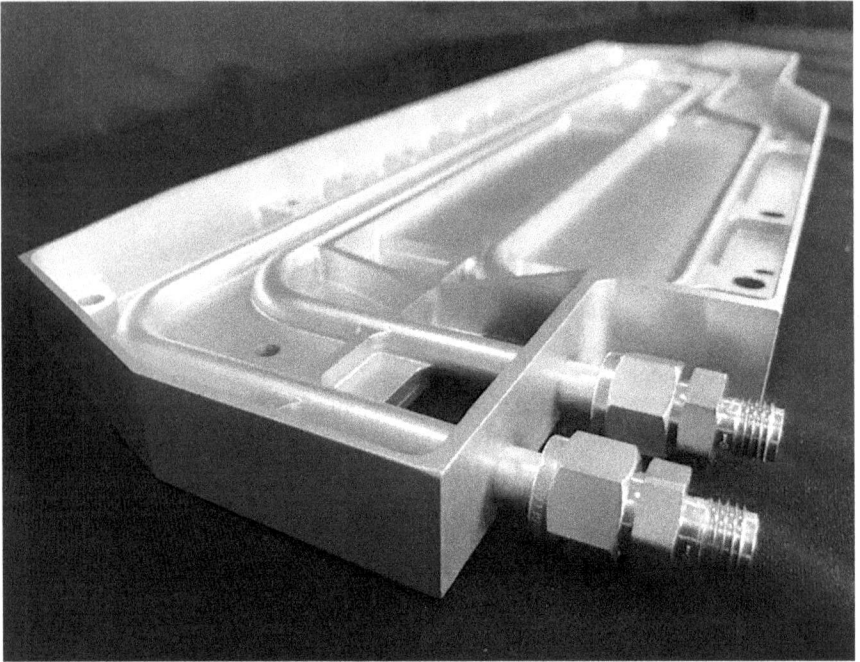

FIGURE 5.17 Aluminum HX manufactured by Fabrisonic.

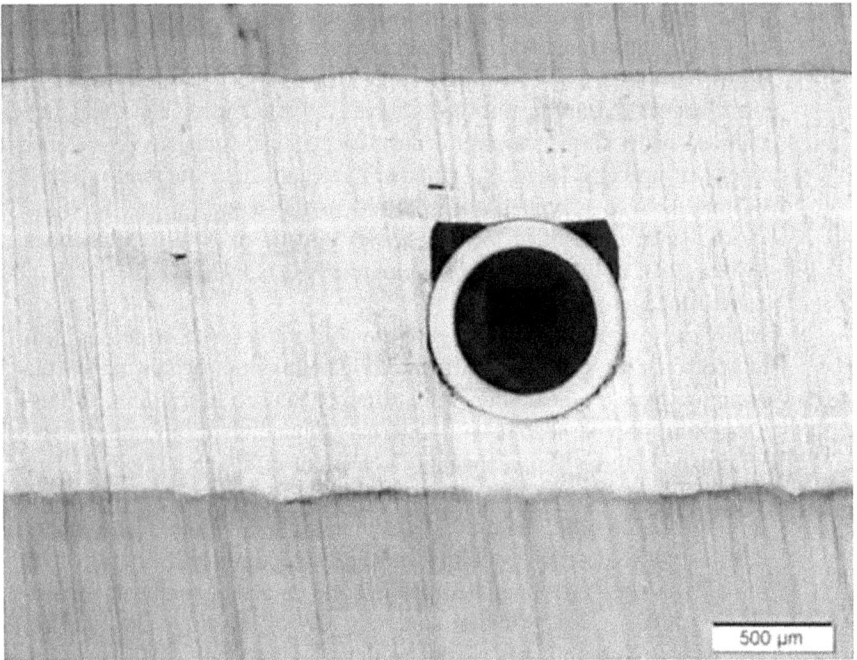

FIGURE 5.18 Section view of the part including copper, aluminum, and a cooling channel via UAM [102].

5.6 CHALLENGES AND FUTURE SCOPE

The quality and size of builds, as well as the spectrum of incompatible material combinations that can be additively welded, are limited due to a lack of thorough scientific understanding of the UAM process and how it influences building attributes [43]. Eliminating interfacial defects, more thorough research on type 2 defects, direct comparison of the various mechanical properties, and lack of high-temperature applications are still a concern for UAM [6]. Standard ASTM specimens are typically utilized for testing and usually necessitate massive structures. As a result, the sub-sized specimen was created and the problem with it is related to the effect of the scaled size. One of the difficulties is achieving significant heights without producing delamination in the bottom layers and a lack of compliance in the structure. Unsupported geometries are difficult to produce with UAM, and as a result, poor bonding has been obtained [3]. It is quite difficult to detect the kissing bond defect in UAM [103]. It is still difficult to compare the work of different study groups because of the lack of established and approved mechanical property evaluation methodologies [6]. When working with strong steel and nickels, the foil material has a tendency to adhere to the steel UAM equipment and hard-faced UAM tooling with Stellite-like alloys is being investigated as a solution to the tool stick problem [5]. UAM is not yet compatible with mass-production lines because only small batch sizes and one-off items are fabricated with the existing UAM equipment [43].

Though there are lots of limitations, UAM has a wide range of applications. Fabricating heat exchangers with complex internal geometries utilizing high thermally conductive metals, for example, has proven difficult, but Fabrisonic, on the other hand, has successfully produced copper and aluminum heat exchangers with UAM [99–101]. It has also been accomplished to build layers of tantalum, molybdenum, and titanium for radiation shielding in structural panels [6]. There are many concerns also that should be taken care of. UAM technique should focus on improving the understanding of UAM mechanisms, the bonding density and joint quality of hard materials, and active elements [72]. More research work is required for a better understanding of mechanical properties. There is yet to be developed a credible UAM process model that can reliably anticipate joint outcomes. The future of UAM should focus on repairing type 2 defects and examine processing modifications to reduce or eliminate fiber damage in order to maximize the performance in aerospace applications. UAM has applications in a variety of fields, including the electronics industry, aerospace industry, and automotive industry.

5.7 CONCLUSION

UAM is successfully employed to join similar materials, dissimilar materials, and metal to the polymer. Because of the no melting attribute, delicate application areas like fiber embedment and sensor embedment enables. High-power UAM enables high-temperature application and a considerable amount of time is required in order to achieve considerable results. FSP is used for repairing type 1 and type 2 defects.

Sonotrode roughness and building height are important parameters that can create voids. In order to get a better quality weld, process parameters such as force, amplitude, temperature, and speed should be selected carefully. Amplitude is the most vital parameter and to some extent properties, defects, bonding strength, etc., are dependent on it. By using post-treatment, mechanical properties can be enhanced and defects can be reduced. Until now, HIP and SPS have been commonly used post-treatment methods. Standardized testing and comparing methods should be incorporated because properties of UAM parts such as fatigue and creep are still lacking. UAM has shown great potential in a wide range of applications and in the future, it will be proven a great technique than other AM methods due to its unique characteristics. In conclusion, research is still going on to develop new tooling, understand and compare different properties, create defects-free joints, increase application area, and create an accurate process model.

ACKNOWLEDGMENT

The authors would like to express gratitude toward Pandit Deendayal Energy University (PDEU) for supporting our study.

REFERENCES

[1] D. R. White, Ultrasonic consolidation of aluminum tooling. Advanced Materials Processing, vol. 161, pp. 64–65, 2003.

[2] Dehoff, R., & Babu, S. (2010). Characterization of interfacial microstructures in 3003 aluminum alloy blocks fabricated by ultrasonic additive manufacturing. Acta Materialia, 58(13), 4305–4315. 10.1016/j.actamat.2010.03.006

[3] Sridharan, N., Gussev, M., Seibert, R., Parish, C., Norfolk, M., Terrani, K., & Babu, S. S. (2016). Rationalization of anisotropic mechanical properties of Al-6061 fabricated using ultrasonic additive manufacturing. Acta Materialia, 117, 228–237. 10.1016/j.actamat.2016.06.048

[4] Fabrisonic, 3D Metal Printing without Melting, Fabrisonic, Columbus, OH, USA, 2019, https://fabrisonic.com/ultrasonic-additive-manufacturing-overview/

[5] Hehr, A., & Norfolk, M. (2019). A comprehensive review of ultrasonic additive manufacturing. Rapid Prototyping Journal, 26(3), 445–458. 10.1108/rpj-03-2019-0056

[6] Gujba, A. K., & Medraj, M. (2020). Power ultrasonic additive manufacturing: Process parameters, microstructure, and mechanical properties. Advances in Materials Science and Engineering, 2020, 1–17. 10.1155/2020/1064870

[7] Miriyev, A., Kalabukhov, S., & Frage, N. (March 2016). "Ultrasonic additive manufacturing of dissimilar material systems: method, postprocessing and proper-ties," in Proceedings of the Fraunhofer Direct Digital Manufacturing Conference (DDMC2016), Berlin, Germany.

[8] Janaki Ram, G., Yang, Y., & Stucker, B. (2006). Effect of process parameters on bond formation during ultrasonic consolidation of aluminum alloy 3003. Journal of Manufacturing Systems, 25(3), 221–238. 10.1016/s0278-6125(07)80011-2

[9] Obielodan, J., Stucker, B., Martinez, E., Martinez, J., Hernandez, D., Ramirez, D. A., & Murr, L. (2011). Optimization of the shear strengths of ultrasonically consolidated Ti/Al 3003 dual-material structures. Journal of Materials Processing Technology, 211(6), 988–995. 10.1016/j.jmatprotec.2010.12.017

[10] Johnson, K., Edmonds, H. C., Higginson, R. L., & Harris, R. A. (2011). New discoveries in ultrasonic consolidation nano-structures using emerging analysis techniques. Proceedings of the Institution of Mechanical Engineers, Part L: Journal of Materials: Design and Applications, 225(4), 277–287. 10.1177/095442 0711413656

[11] Graff, K. F., Short, M., & Norfolk, M. (2010). "Very high power ultrasonic additive manufacturing (VHP UAM) for advanced materials," International Conference on Additive Manufacturing. https://doi.org/10.26153/tsw/15165

[12] Kelly, G. S., Just, M. S., Advani, S. G., & Gillespie, J. W. (2014). Energy and bond strength development during ultrasonic consolidation. Journal of Materials Processing Technology, 214(8), 1665–1672, .

[13] Schick, D. E., Hahnlen, R., & Dehoff, R. (2010). Microstructural characterization of bonding interfaces in aluminum 3003 blocks fabricated by ultrasonic additive manufacturing. Welding Journal, 89(5), 105–115.

[14] Gussev, M. N., Sridharan, N., Norfolk, M., Terrani, K. A., & Babu, S. S. Effect of post weld heat treatment on the 6061 aluminum alloy produced by ultrasonic additive manufacturing, Netherlands. 10.1016/j.msea.2016.12.083

[15] Hopkins, C. D., Wolcott, P. J., Dapino, M. J., Truog, A. G., Babu, S. S., & Fernandez, S. A. (2011). Optimizing ultrasonic additive manufactured al 3003 properties with statistical modeling. Journal of Engineering Materials and Technology, 134(1). 10.1115/1.4005269

[16] Sridharan, N., Gussev, M., Seibert, R., Parish, C., Norfolk, M., Terrani, K., & Babu, S. S. (2016b). Rationalization of anisotropic mechanical properties of Al-6061 fabricated using ultrasonic additive manufacturing. Acta Materialia, 117, 228–237. 10.1016/j.actamat.2016.06.048

[17] Wolcott, P. J., Sridharan, N., Babu, S. S., Miriyev, A., Frage, N., & Dapino, M. J. (2016). Characterisation of Al–Ti dissimilar material joints fabricated using ultrasonic additive manufacturing. Science and Technology of Welding and Joining, 21(2), 114–123. 10.1179/1362171815y.0000000072

[18] Sridharan, N., Wolcott, P., Dapino, M., & Babu, S. (2016). Microstructure and texture evolution in aluminum and commercially pure titanium dissimilar welds fabricated using ultrasonic additive manufacturing. Scripta Materialia, 117, 1–5. 10.1016/j.scriptamat.2016.02.013

[19] Kuo, C. H., Sridharan, N., Han, T., Dapino, M. J., & Babu, S. S. (2019). Ultrasonic additive manufacturing of 4130 steel using Ni interlayers. Science and Technology of Welding and Joining, 24(5), 382–390. 10.1080/13621718.2019.1607486

[20] Sridharan, N., Norfolk, M., & Babu, S. S. (2016). Characterization of steel-Ta dissimilar metal builds made using very high power ultrasonic additive manufacturing (VHP-UAM). Metallurgical and Materials Transactions A, 47(5), 2517–2528. 10.1007/s11661-016-3354-5

[21] Truog, A. G. (2012). Bond Improvement of Al/Cu Joints Created by Very High Power Ultrasonic Additive Manufacturing. Master's thesis, The Ohio State University.

[22] Hehr, A., & Dapino, M. J. (2015). Interfacial shear strength estimates of NiTi–Al matrix composites fabricated via ultrasonic additive manufacturing. Composites Part B: Engineering, 77, 199–208. 10.1016/j.compositesb.2015.03.005

[23] Li, D., & Soar, R. C. (2009). Characterization of process for embedding SiC fibers in Al 6061 O matrix through ultrasonic consolidation. Journal of Engineering Materials and Technology, 131(2). 10.1115/1.3030946

[24] MARSHALL, D. B., & OLIVER, W. C. (1987). Measurement of interfacial mechanical properties in fiber-reinforced ceramic composites. Journal of the American Ceramic Society, 70(8), 542–548. 10.1111/j.1151-2916.1987.tb05702.x

[25] Li, J., Monaghan, T., Masurtschak, S., Bournias-Varotsis, A., Friel, R., & Harris, R. (2015). Exploring the mechanical strength of additively manufactured metal structures with embedded electrical materials. Materials Science and Engineering: A, 639, 474–481. 10.1016/j.msea.2015.05.019

[26] Levy, A., Miriyev, A., Sridharan, N., Han, T., Tuval, E., Babu, S. S., Dapino, M. J., & Frage, N. (2018). Ultrasonic additive manufacturing of steel: Method, post-processing treatments and properties. Journal of Materials Processing Technology, 256, 183–189. 10.1016/j.jmatprotec.2018.02.001

[27] Sridharan, N., Wolcott, P., Dapino, M., & Babu, S. S. (2016b). Microstructure and mechanical property characterisation of aluminium–steel joints fabricated using ultrasonic additive manufacturing. Science and Technology of Welding and Joining, 22(5), 373–380. 10.1080/13621718.2016.1249644

[28] Kong, C., Soar, R., & Dickens, P. (2004). Ultrasonic consolidation for embedding SMA fibres within aluminium matrices. Composite Structures, 66(1–4), 421–427. 10.1016/j.compstruct.2004.04.064

[29] Petrie, C. M., Sridharan, N., Subramanian, M., Hehr, A., Norfolk, M., & Sheridan, J. (2019). Embedded metallized optical fibers for high temperature applications. Smart Materials and Structures, 28(5), 055012. 10.1088/1361-665x/ab0b4e

[30] Monaghan, T., Capel, A., Christie, S., Harris, R., & Friel, R. (2015). Solid-state additive manufacturing for metallized optical fiber integration. Composites Part A: Applied Science and Manufacturing, 76, 181–193. 10.1016/j.compositesa.2015.05.032

[31] Hahnlen, R., & Dapino, M. J. (2010). Active metal-matrix composites with embedded smart materials by ultrasonic additive manufacturing. SPIE Proceedings. 10.1117/12.848853

[32] Hahnlen, R., Dapino, M. J., Short, M., & Graff, K. (2009). Aluminum-matrix composites with embedded Ni-Ti wires by ultrasonic consolidation. SPIE Proceedings. 10.1117/12.817036

[33] Natarajan, J., Cheepu, M., & Yang, C. (2021). Advances in Additive Manufacturing Processes. Bentham Science Publishers.

[34] Friel, R., & Harris, R. (2013). Ultrasonic additive manufacturing – A hybrid production process for novel functional products. Procedia CIRP, 6, 35–40. 10.1016/j.procir.2013.03.004

[35] Schwope, L. A., Friel, R., Johnson, K. E., & Harris, R. (2009). Field repair and replacement part fabrication of military components using ultrasonic consolidation cold metal deposition. Loughborough University. Conference contribution. https://hdl.handle.net/2134/14489

[36] Friel, R. (2015). Power ultrasonics for additive manufacturing and consolidating of materials. Power Ultrasonics, 313–335. 10.1016/b978-1-78242-028-6.00013-2

[37] Yang, Y., Janaki Ram, G., & Stucker, B. (2009). Bond formation and fiber embedment during ultrasonic consolidation. Journal of Materials Processing Technology, 209(10), 4915–4924. 10.1016/j.jmatprotec.2009.01.014

[38] Sriraman, M., Babu, S., & Short, M. (2010). Bonding characteristics during very high power ultrasonic additive manufacturing of copper. Scripta Materialia, 62(8), 560–563. 10.1016/j.scriptamat.2009.12.040

[39] Friel, R. J., & Harris, R. A. (2009). A nanometre-scale fibre-to-matrix interface characterization of an ultrasonically consolidated metal matrix composite. Proceedings of the Institution of Mechanical Engineers, Part L: Journal of Materials: Design and Applications, 224(1), 31–40. 10.1243/14644207jmda268

[40] Kong, C. Y., Soar, R. C., & Dickens, P. M. (2005). A model for weld strength in ultrasonically consolidated components. Proceedings of the Institution of Mechanical Engineers, Part C: Journal of Mechanical Engineering Science, 219(1), 83–91. 10.1243/095440605x8315

[41] Yang, Y., Janaki Ram, G. D., & Stucker, B. E. (2007). An experimental determination of optimum processing parameters for Al/SiC metal matrix composites made using ultrasonic consolidation. Journal of Engineering Materials and Technology, 129(4), 538–549. 10.1115/1.2744431

[42] Natarajan, J., Cheepu, M., & Yang, C. (2021b). Advances in Additive Manufacturing Processes. Bentham Science Publishers.

[43] Badiru, A. B., Valencia, V. V., & Liu, D. (2017). Additive Manufacturing Handbook: Product Development for the Defense Industry (Systems Innovation Book Series) (1st ed.). CRC Press.

[44] Powers, J. J., & Jones, J. B. (1956). Ultrasonic welding. Welding Journal, 35(8), 761–766.

[45] Li, D., & Soar, R. (2009a). Influence of sonotrode texture on the performance of an ultrasonic consolidation machine and the interfacial bond strength. Journal of Materials Processing Technology, 209(4), 1627–1634. 10.1016/j.jmatprotec.2008.04.018

[46] Kong, C., Soar, R., & Dickens, P. (2003). Characterisation of aluminium alloy 6061 for the ultrasonic consolidation process. Materials Science and Engineering: A, 363(1–2), 99–106. 10.1016/s0921-5093(03)00590-2

[47] Kong, C., Soar, R., & Dickens, P. (2004b). Optimum process parameters for ultrasonic consolidation of 3003 aluminium. Journal of Materials Processing Technology, 146(2), 181–187. 10.1016/j.jmatprotec.2003.10.016

[48] He, X. H., Shi, H. J., Zhang, Y. D., Yang, Z. G., Wilkinson, C. E., Neal, A. L., & Norfolk, M. (2015). Mechanical properties and microstructure of Al/Al laminated structure produced via ultrasonic consolidation process. Materials Science and Technology, 31(15), 1910–1918. 10.1179/1743284715y.0000000045

[49] George, J., & Stucker, B. (2006). Fabrication of lightweight structural panels through ultrasonic consolidation. Virtual and Physical Prototyping, 1(4), 227–241. 10.1080/17452750601106799

[50] Wolcott, P. J., Sridharan, N., Babu, S. S., Miriyev, A., Frage, N., & Dapino, M. J. (2016b). Characterisation of Al–Ti dissimilar material joints fabricated using ultrasonic additive manufacturing. Science and Technology of Welding and Joining, 21(2), 114–123. 10.1179/1362171815y.0000000072

[51] Hehr, A., Pritchard, J., & Dapino, M. J. (2014). Interfacial shear strength estimates of NiTi-Al matrix composites fabricated via ultrasonic additive manufacturing. SPIE Proceedings. 10.1117/12.2046317

[52] Fujii, H. T., Sriraman, M. R., & Babu, S. S. (2011). Quantitative evaluation of bulk and interface microstructures in Al-3003 alloy builds made by very high power ultrasonic additive manufacturing. Metallurgical and Materials Transactions A, 42(13), 4045–4055. 10.1007/s11661-011-0805-x

[53] Friel, R., Johnson, K., Dickens, P., & Harris, R. (2010). The effect of interface topography for Ultrasonic Consolidation of aluminium. Materials Science and Engineering: A, 527(16–17), 4474–4483. 10.1016/j.msea.2010.03.094

[54] Guo, H., Gingerich, M. B., Headings, L. M., Hahnlen, R., & Dapino, M. J. (2019). Joining of carbon fiber and aluminum using ultrasonic additive manufacturing (UAM). Composite Structures, 208, 180–188. 10.1016/j.compstruct.2018.10.004

[55] James, S., & Dang, C. (2020). Investigation of shear failure load in ultrasonic additive manufacturing of 3D CFRP/Ti structures. Journal of Manufacturing Processes, 56, 1317–1321. 10.1016/j.jmapro.2020.04.026

[56] Wolcott, P. J., Hehr, A., & Dapino, M. J. (2014). Optimized welding parameters for Al 6061 ultrasonic additive manufactured structures. Journal of Materials Research, 29(17), 2055–2065. 10.1557/jmr.2014.139

[57] Shimizu, S., Fujii, H., Sato, Y., Kokawa, H., Sriraman, M., & Babu, S. (2014). Mechanism of weld formation during very-high-power ultrasonic additive manufacturing of Al alloy 6061. Acta Materialia, 74, 234–243. 10.1016/j.actamat.2014.04.043

[58] Hopkins, C. D., Dapino, M. J., & Fernandez, S. A. (2010). Statistical characterization of ultrasonic additive manufacturing Ti/Al composites. Journal of Engineering Materials and Technology, 132(4). 10.1115/1.4002073

[59] James, S., & de la Luz, L. (2019). Finite element analysis and simulation study of CFRP/Ti stacks using ultrasonic additive manufacturing. The International Journal of Advanced Manufacturing Technology, 104(9–12), 4421–4431. 10.1007/s00170-019-04228-6

[60] James, S., Sonate, A., Dang, C., & de la Luz, L. (2018). Experimental and simulation study of ultrasonic additive manufacturing of CFRP/Ti stacks. Volume 4: Processes. 10.1115/msec2018-6647

[61] Pagan, M., Petrie, C., Leonard, D., Sridharan, N., Zinkle, S., & Babu, S. S. (2021). Interdiffusion of elements during ultrasonic additive manufacturing. Metallurgical and Materials Transactions A, 52(3), 1142–1157. 10.1007/s11661-020-06131-2

[62] Deng, Z., Gingerich, M. B., Han, T., & Dapino, M. J. (2018). Yttria-stabilized zirconia-aluminum matrix composites via ultrasonic additive manufacturing. Composites Part B: Engineering, 151, 215–221. 10.1016/j.compositesb.2018.06.001

[63] Schomer, J. J. (2017). OhioLINK ETD: Schomer, John J. Embedding Fiber Bragg Grating Sensors through Ultrasonic Additive Manufacturing. https://etd.ohiolink.edu/apexprod/rws_olink/r/1501/10?clear=10&p10_accession_num=osu1483670362650083

[64] Ramanathan, A. K., Gingerich, M. B., Headings, L. M., & Dapino, M. J. (2021). Metal structures embedded with piezoelectric PVDF sensors using ultrasonic additive manufacturing. Manufacturing Letters. 10.1016/j.mfglet.2021.08.001

[65] Han, T., Kuo, C. H., Sridharan, N., Headings, L. M., Babu, S. S., & Dapino, M. J. (2020). Effect of weld power and interfacial temperature on mechanical strength and microstructure of carbon steel 4130 fabricated by ultrasonic additive manufacturing. Manufacturing Letters, 25, 64–69. 10.1016/j.mfglet.2020.07.006

[66] Sridharan, N., Gussev, M. N., Parish, C. M., Isheim, D., Seidman, D. N., Terrani, K. A., & Babu, S. S. (2018). Evaluation of microstructure stability at the interfaces of Al-6061 welds fabricated using ultrasonic additive manufacturing. Materials Characterization, 139, 249–258. 10.1016/j.matchar.2018.02.043

[67] Hehr, A., Wolcott, P. J., & Dapino, M. J. (2016). Effect of weld power and build compliance on ultrasonic consolidation. Rapid Prototyping Journal, 22(2), 377–386. 10.1108/rpj-11-2014-0147

[68] Ward, A. A., & Cordero, Z. C. (2020). Junction growth and interdiffusion during ultrasonic additive manufacturing of multi-material laminates. Scripta Materialia, 177, 101–105. 10.1016/j.scriptamat.2019.10.004

[69] Robinson, C. J., Ram, G. J., & Stucker, B. E. (2011). Role of substrate stiffness in ultrasonic consolidation. International Journal of Rapid Manufacturing, 2(3), 162. 10.1504/ijrapidm.2011.043457

[70] Obielodan, J. O., Ram, G. D. J., Stucker, B. E., & Taggart, D. G. (2009). Minimizing defects between adjacent foils in ultrasonically consolidated parts. Journal of Engineering Materials and Technology, 132(1). 10.1115/1.3184033

[71] Nadimpalli, V. K., Karthik, G. M., Janakiram, G. D., & Nagy, P. B. (2020). Monitoring and repair of defects in ultrasonic additive manufacturing. The International Journal of Advanced Manufacturing Technology, 108(5–6), 1793–1810. 10.1007/s00170-020-05457-w

[72] Li, D. (2021). A review of microstructure evolution during ultrasonic additive manufacturing. The International Journal of Advanced Manufacturing Technology. 10.1007/s00170-020-06439-8

[73] Dilip, J. J. S., Babu, S., Rajan, S. V., Rafi, K. H., Ram, G. J., & Stucker, B. E. (2013). Use of friction surfacing for additive manufacturing. Materials and Manufacturing Processes, 28(2), 189–194. 10.1080/10426914.2012.677912

[74] Longhurst, W. R., Cox, C. D., Gibson, B. T., Cook, G. E., Strauss, A. M., Wilbur, I. C., & Osborne, B. E. (2016). Development of friction stir welding technologies for in-space manufacturing. The International Journal of Advanced Manufacturing Technology, 90(1–4), 81–91. 10.1007/s00170-016-9362-1

[75] Kalvala, P. R., Akram, J., & Misra, M. (2016). Friction assisted solid state lap seam welding and additive manufacturing method. Defence Technology, 12(1), 16–24. 10.1016/j.dt.2015.11.001

[76] Sharma, A., Bandari, V., Ito, K., Kohama, K., M., R., & B.V., H. S. (2017). A new process for design and manufacture of tailor-made functionally graded composites through friction stir additive manufacturing. Journal of Manufacturing Processes, 26, 122–130. 10.1016/j.jmapro.2017.02.007

[77] Mishra, R., & Ma, Z. (2005). Friction stir welding and processing. Materials Science and Engineering: R: Reports, 50(1–2), 1–78. 10.1016/j.mser.2005.07.001

[78] Janaki Ram, G., Robinson, C., Yang, Y., & Stucker, B. (2007). Use of ultrasonic consolidation for fabrication of multi-material structures. Rapid Prototyping Journal, 13(4), 226–235. 10.1108/13552540710776179

[79] Kaya, R., Cora, M. N., Acar, D., & Koç, M. (2018). On the formability of ultrasonic additive manufactured Al-Ti laminated composites. Metallurgical and Materials Transactions A, 49(10), 5051–5064. 10.1007/s11661-018-4784-z

[80] Nadimpalli V. K., Na J. K., Bruner D. T., King B. A., Yang L., Stucker B. E. (2016) In-situ non-destructive evaluation of ultrasonic additive manufactured components. 27th Annual International Solid Freeform Fabrication Symposium, p 1557–1567

[81] Obielodan, J. O. (2010). Fabrication of Multi-material Structures Using Ultrasonic Consolidation and Laser-engineered Net Shaping. Utah State University.

[82] Wolcott, P., Hehr, A., Pawlowski, C., & Dapino, M. (2016). Process improvements and characterization of ultrasonic additive manufactured structures. Journal of Materials Processing Technology, 233, 44–52. 10.1016/j.jmatprotec.2016.02.009

[83] Miriyev, A., Levy, A., Kalabukhov, S., & Frage, N. (2016). Interface evolution and shear strength of Al/Ti bi-metals processed by a spark plasma sintering (SPS) apparatus. Journal of Alloys and Compounds, 678, 329–336. 10.1016/j.jallcom.2016.03.137

[84] Gujba, A., & Medraj, M. (2014). Laser peening process and its impact on materials properties in comparison with shot peening and ultrasonic impact peening. Materials, 7(12), 7925–7974. 10.3390/ma7127925

[85] DebRoy, T., Wei, H., Zuback, J., Mukherjee, T., Elmer, J., Milewski, J., Beese, A., Wilson-Heid, A., De, A., & Zhang, W. (2018). Additive manufacturing of metallic

components – Process, structure and properties. Progress in Materials Science, 92, 112–224. 10.1016/j.pmatsci.2017.10.001

[86] Uhlmann, E., Fleck, C., Gerlitzky, G., & Faltin, F. (2017). Dynamical fatigue behavior of additive manufactured products for a fundamental life cycle approach. Procedia CIRP, 61, 588–593. 10.1016/j.procir.2016.11.138

[87] Han, T., Kuo, C. H., Sridharan, N., Headings, L. M., Babu, S. S., & Dapino, M. J. (2020a). Effect of preheat temperature and post-process treatment on the micro-structure and mechanical properties of stainless steel 410 made via ultrasonic additive manufacturing. Materials Science and Engineering: A, 769, 138457. 10.1016/j.msea.2019.138457

[88] Al-Aqeeli, N., Abdullahi, K., Hakeem, A. S., Suryanarayana, C., Laoui, T., & Nouari, S. (2013). Synthesis, characterisation and mechanical properties of SiC reinforced Al based nanocomposites processed by MA and SPS. Powder Metallurgy, 56(2), 149–157. 10.1179/1743290112y.0000000029

[89] Al-Aqeeli, Nasser, Gujba, Abdullahi, Suryanarayana, C., Laoui, Tahar, & Saheb, Nouari. (2013). Structure of mechanically milled CNT-reinforced Al-alloy nano-composites. Materials and Manufacturing Processes, 28. 10.1080/10426914. 2012.746703.

[90] Sriraman, M. R., Gonser, M., Foster, D., Fujii, H. T., Babu, S. S., & Bloss, M. (2011). Thermal transients during processing of 3003 Al-H18 multilayer build by very high-power ultrasonic additive manufacturing. Metallurgical and Materials Transactions B, 43(1), 133–144. 10.1007/s11663-011-9590-6

[91] Sojiphan, K., Sriraman, M. R., & Babu, S. S. (2010). Stability of microstructure in al3003 builds made by very high power ultrasonic additive manufacturing. 2010 International Solid Freeform Fabrication Symposium, 22(2), 362–371. 10.26153/tsw/15203

[92] Li, P., Wang, Z., Diao, M., Guo, C., Wang, J., Zhao, C., & Jiang, F. (2021). Effect of processing parameters on bond properties and microstructure evolution in ultrasonic additive manufacturing (UAM). Materials Research Express, 8(3), 036507. 10.1088/2053-1591/abe9d3

[93] Wang, Y., Yang, Q., Liu, X., Liu, Y., Liu, B., Misra, R., Xu, H., & Bai, P. (2019). Microstructure and mechanical properties of amorphous strip/aluminum laminated composites fabricated by ultrasonic additive consolidation. Materials Science and Engineering: A, 749, 74–78. 10.1016/j.msea.2019.01.039

[94] Applications of ultrasonic additive manufacturing. (2015). Inside Metal Additive Manufacturing. https://www.insidemetaladditivemanufacturing.com/blog/applications-of-ultrasonic-additive-manufacturing#:%7E:text=Multi%2Dmaterials%20manufacturing&text=By%20depositing%20various%20metal%20foils,property%20changes%20for%20various%20applications.

[95] Bournias-Varotsis, A., Friel, R. J., Harris, R. A., & Engstrøm, D. S. (2018). Ultrasonic Additive Manufacturing as a form-then-bond process for embedding electronic circuitry into a metal matrix. Journal of Manufacturing Processes, 32, 664–675. 10.1016/j.jmapro.2018.03.027

[96] Ultrasonic Additive Manufacturing (UAM): Part One: Total Materia Article. (2021). Total Materia. https://www.totalmateria.com/page.aspx?ID=CheckArticle&site=ktn&LN=ES&NM=497

[97] Team, S. (2021, April 27). Ultrasonic Additive Manufacturing Explained. Spatial. https://blog.spatial.com/ultrasonic-additive-manufacturing

[98] Hehr, A., Wenning, J., Terrani, K., Babu, S. S., & Norfolk, M. (2016). Five-axis ultrasonic additive manufacturing for nuclear component manufacture. JOM, 69(3), 485–490. 10.1007/s11837-016-2205-6

[99] Ultrasonic Welding Makes Parts for NASA Missions, Commercial Industry. (2021, May 28). NASA Jet Propulsion Laboratory (JPL). https://www.jpl.nasa.gov/news/ultrasonic-welding-makes-parts-for-nasa-missions-commercial-industry

[100] Ultrasonic Welding Makes Parts for NASA Missions, Commercial Industry | NASA Spinoff. (2021). NASA Spinoff. https://spinoff.nasa.gov/ultrasonic-welding-additive-manufacturing

[101] Creating Complex Internal Geometry. (2018). Fabrisonic. https://fabrisonic.com/creating-complex-internal-geometry/

[102] Dapino, M. J. (2014). Smart structure integration through ultrasonic additive manufacturing. Volume 2: Mechanics and Behavior of Active Materials; Integrated System Design and Implementation; Bioinspired Smart Materials and Systems; Energy Harvesting. 10.1115/smasis2014-7710

[103] Oosterkamp, A., Oosterkamp, L. D., & Nordeide, A. (2004). Kissing bond phenomena in solid-state welds of aluminum alloys. Welding Journal, 83, 225–231.

6 Solid-state Metal Additive Manufacturing Process
A Systematic Review

L. K. Yadav, J. P. Misra, and R. Tyagi

6.1 INTRODUCTION

Additive manufacturning (AM) is a revolutionary technology that produces three-dimensional (3D) components layer by layer based on a digital model. This unique property enables the fabrication of complex structures that are difficult to develop using conventional systems [1,2]. AM encompasses a diverse array of dynamic, adaptable, and extensively customizable fabrication techniques that possess the capability to generate an extensive assortment of materials, including but not limited to metals, ceramics, and polymers [3]. Among these materials, metallic materials are gaining the attention of many researchers and industries. AM has the potential to provide environmental benefits such as reduced waste, higher accuracy, reduced pollutants, and the ability to make parts on demand [4]. Metallic AM (MAM) is a rapidly increasing discipline that involves a wide range of operations for an even larger variety of metallic materials. The field of MAM, or AM, encompasses a diverse array of methodologies and substances, as visually depicted in Figure 6.1(b). The seven AM processes under consideration are material jetting, sheet lamination, binder jetting, powder bed fusion (PBF), material extrusion, directed energy deposition (DED), and VAT photopolymerization [5–7]. Binder jetting involves injecting binders as particles into plaster-based, elastomeric, metal, and ceramic materials [8]. This AM approach is fast, versatile, and easy to use [9]. In material jetting, photopolymer is injected layer by layer onto a substrate and cured by UV light [10]. This method is expensive and involves printing. It allows for improved precision, accuracy, and smooth surfaces in digital materials and flexible opaque/transparent components [11]. VAT photopolymerization cures polymers by visible light. Design analysis, tooling for cast prototypes, and medical model prototyping [12]. This method makes accurate, smooth volumetric materials. Material extrusion involves thermoplastic extrusion or layer-by-layer syringe injection. ABS, polycarbonate, and wax are suitable [13,14]. Accuracy, sluggish processing, abrupt shrinkage, and poor surface quality limit its use. The AM process that includes bonding and cutting sheets is sheet lamination, which is more

DOI: 10.1201/9781003484325-6

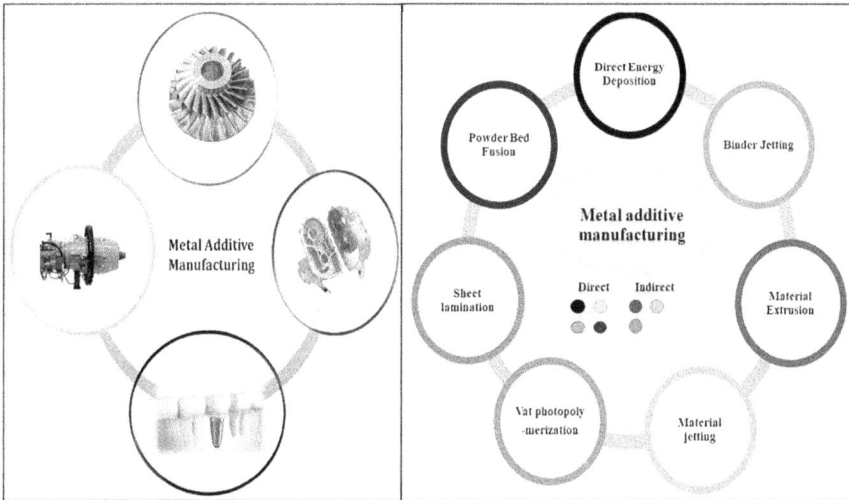

FIGURE 6.1   (a) Applications of metal AM, and (b) types of metal AM.

prevalent with plastics and metals [15]. MAM new avenues for enhancing manufacturing capability. Reduced lead times, new material exposure, minimum material waste, and the ability to produce unique intricate shapes and hard materials are all significant reasons for organizations and industries to adopt metal AM technology [16]. MAM technology is now being successfully employed in aerospace, defense, medical equipment, the automotive sector, and various other industries [17]. Figure 6.1(a) depicts the applications of the metal AM process.

6.1.1   METAL ADDITIVE MANUFACTURING

One of the newest emerging and innovative AM technologies is PBF. As it uses concentrated sources of energy such as gas, solid-state or fiber lasers, or electron beam to fuse metallic powders promulgated by a roller, this approach can address a wide range of materials, including metals and ceramics. It allows for the low-volume production of functional components in aerospace, medical, and tooling applications, as well as the development of disposable implants. A concentrated source of energy, such as a laser, electron beam, plasma, or arc, is used to fuse particles or wire in AM processes capable of metal deposition, known as DED. To build items layer by layer, injecting or feeding on a nozzle is utilized, followed by interaction with CAD/CAM modeling. Compared to rival technologies like PBF, this method delivers denser products with faster manufacturing speeds. According to contemporary research, however, the manufactured items lacked appropriate load-bearing capability due to structural integrity issues. Since the structure of these methods' 3D-printed materials is produced by layer-by-layer stacking of molten pools. PBF is a novel AM technology. DED fuses particles or wire using a laser, electron beam, plasma, or arc, shown in Figure 6.1(b). Injecting or feeding a nozzle is used to create layer-by-layer, followed by CAD/CAM modeling [18].

Microstructure regulation remains difficult despite recent advancements. Columnar grain formations grow in the build direction. Deformation analysis can help with fusion-based AM quality control [19,20]. The mechanical behavior of produced components can be influenced by non-equilibrium solidification problems such as porosity and cracking. Fusion-based metal AM inherently undermines porosity control, residual stress, and hot-cracking due to the liquid-phase bonding method. Significant temperature gradients and accelerated cooling rates all contribute to these challenges. Furthermore, despite recent improvements, microstructure regulation has remained a challenge. Because textured, columnar grain structures often grow along the build direction, to substantially tackle the quality control issues inherent in fusion-based AM, we need to apply deformation analysis.

6.1.2 SOLID-STATE AM TECHNOLOGY

In recent times, there has been significant advancement in deformation-based metal AM techniques, mostly propelled by the progress made in solid-state welding. The focus of this study is SSAM techniques utilized for metals and alloys. These techniques may be classified into two subcategories based on the bonding mechanism employed for material joining [21]. SSAM may be classified into two main categories: sintering-based techniques and mechanical deformation-based approaches. Sintering-based solid-state refers to the process of subjecting a 3D-printed powder compact to homogeneous external heating, resulting in a significant increase in temperature close to the melting point. The use of sintering-based solid-state AM (SSAM) encompasses both binders jetting AM (BJAM) and selective powder deposition AM (SPAM). The Mechanical deformation-based AM technique involves the mechanical disruption of the oxide layer, which is accompanied by material bonding through severe plastic deformation. This deformation can be induced by various mechanisms such as ultrasonic scrubbing, friction, or the supersonic impact of powder particles onto a substrate or a previously deposited layer of the same material [22]. Ultrasonic AM (UAM), cold spray solid-state AM (CSAM), and additive friction stir deposition (AFSD) represent a selection of approaches in this domain. The techniques employed in these approaches utilize deformation bonding to achieve layer adhesion by the application of ultrasonic vibration, particle impact, and friction stirring, as documented in reference [23]. It is worth noting that these techniques do not include any melting processes. The potential of AFSD surpasses that of UAM and CSAM. UAM integrates the process of ultrasonic foil welding with computer numerical control (CNC) machining [24]. The AFSD technology offers high-precision printing capabilities in conjunction with freeform manufacturing processes [25]. Ultrasonic, Cold spray, and AFSD techniques all employ deformation bonding; however, it should be noted that the former two methods exclusively include localized plastic deformation [26]. In the context of AFSD, it is observed that plastic deformation has a global nature. Elevated temperatures cause deformation in each individual feed voxel. This creates a microstructure with minimal internal defects [27]. Figure 6.2 shows the SSAM categorization flowchart.

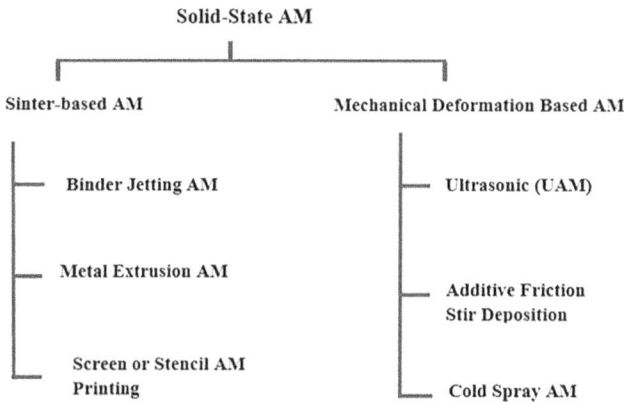

FIGURE 6.2 Classification of SSAM.

6.1.2.1 Sintering-based AM Process

Sintering-based AM technologies encompass many steps, including diverse processes for geometry assembly and consolidation. The initial procedure involves the consolidation of metal powder particles into a three-dimensional preform by the use of polymeric binders [28]. Subsequently, a procedure is implemented to eliminate organic binders, often known as debinding. The last component consolidation is sintering. Thermal sintering binds powder particles via atomic diffusion [29]. Increased by oxide reduction at high temperatures, especially in low-oxygen environments. Reduced powder surface area drives sintering. The sintering procedure improves physical and mechanical qualities [30]. Getting part sizes (sintering causes shrinkage). Additionally, the process of chemical homogenization involves the first combination of constituent powders, followed by sintering to get a uniform alloy.

6.1.2.1.1 Binder Jetting AM

Binder jetting is an AM that relies on sintering. It involves the creation of a three-dimensional structure by sequentially depositing thin layers of powder, as seen in Figure 6.3(a). In a controlled bed, it is common for the particle size to range from 50 to 200 μm. Subsequently, an organic binder, often dissolved in a solvent, is applied onto the intended regions by the use of a mobile printhead equipped with numerous nozzles. This phenomenon results in the adhesion of the binder-sprayed patches, while the adjacent powder particles remain unconsolidated [31]. BJAM has the capability to effectively handle a wide range of powdered materials, which may then be consolidated by various treatments such as sintering, container-less hot isostatic pressing (HIP), and heat treatment. The consolidation of BJAM pieces occurs during the sintering process, whereby their final mechanical properties are determined [32]. Sintering is a widely employed technique in which fine particles are utilized due to their significant surface area, which results in a substantial compressive driving force.

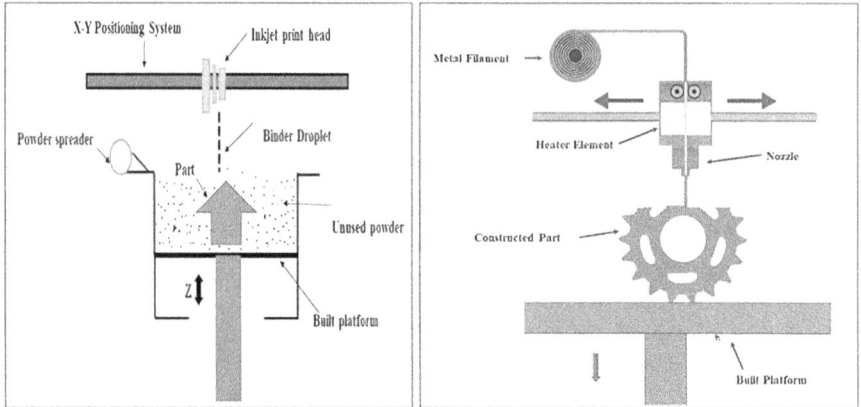

(a) Sinter-based Binder Jetting AM process   (b) Metal Extrusion AM method

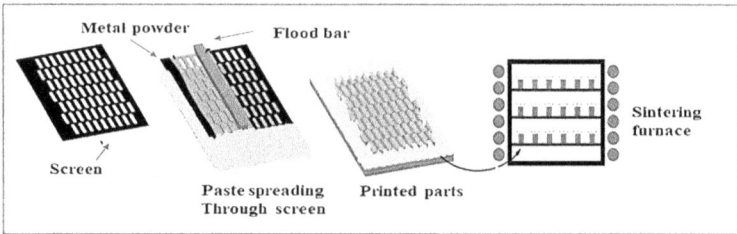

(c) Screen/stencil 3D printing

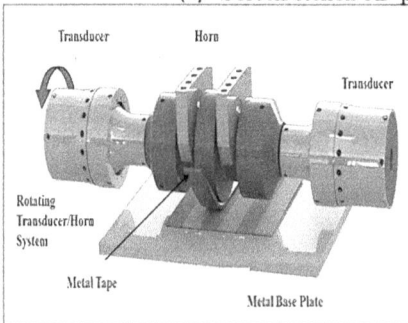

(a) Ultrasonic AM process   (e) Additive friction stir deposition

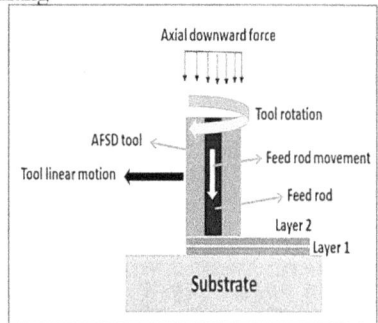

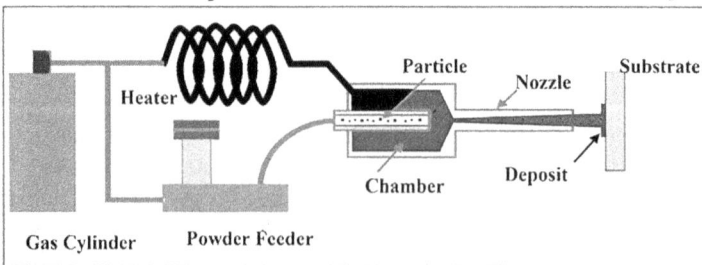

(f) Cold Spray AM process

FIGURE 6.3   Schematic outline of various solid-state AM process [28–64].

6.1.2.1.2 Material Extrusion-based AM

The introduction of material extrusion by Stratasys in 1989 marked the inception of fused filament fabrication (FFF), which has since been the prevailing method in this field. The process operates by sequentially depositing a thermoplastic polymer in layers through a heated nozzle with a narrow opening onto a substrate [33]. The development of extrusion-based metal AM was undertaken with the aim of producing metal components using FFF green parts [34]. Metallic alloys or powders, when combined with organic binders, serve as the feedstock material in Metal Extrusion Additive Manufacturing (MEAM) and Metal injection molding (MIM) processes. This approach uses flexible filaments or rods as raw material. This is inserted in a cartridge [35,36]. The binder system consists of several components, whereby the primary component may be extracted at an early stage, while a secondary binder retains the item until the sintering process, as seen in Figure 6.3(b). The principal binder can be effectively removed using several methods such as dissolution, heat wicking, supercritical extraction procedure, or sublimation. The secondary binder is thermally eliminated before the occurrence of closed porosity and powder necking. The feedstock is extruded layer by layer onto a print bed using a nozzle [37,38]. This process is remarkable because it uses several extrusion processes, including print heads supplied with different materials, to print on flexible substrates. Dense compounds need additional time for completion of the solvent and thermal debinding phases [39]. Binder dissolution and evaporation cause outer shell porosity. This sequential procedure takes longer with increasing thickness [40].

6.1.2.1.3 Screen and Stencil 3D Printing AM

The process of screen or stencil printing involves the use of pastes composed of metal or alloy powders, as well as biological or water-based solvent sludge. The depiction of SPAM printing may be observed in Figure 6.3(c) [41]. The screen is immediately above the component's substrate. Screens can be created to match component designs, but stencils are laser-cut from sheet metal. A printing squeegee bar pushes paste after the flood bar sprays it [42,43]. Then pushes the screen to force material through the mesh. During processing, paste shears [44] before printing the next layer, the paste-coated substrate is quickly dried to remove solvents. SPAM can print millions of minuscule pieces at high volumes [45].

6.1.2.2 Mechanical Deformation-based AM

The major difference between Mechanical Deformation-based AM techniques and sinter-based AM methods is in the mechanism of material bonding. It utilizes thermal energy to facilitate the bonding process, whilst Mechanical Deformation-based AM techniques employ kinetic energy to aid in the bonding process [46,47]. Kinetic energy can be imparted by the use of high-frequency ultrasonic waves, rotational friction accompanied by a normal force, or supersonic impact. The generation of heat occurs as an incidental outcome of the operation. The extent of the increase in local temperature is influenced by the physical and thermal characteristics of the specimen, as well as the material of the substrate [48]. In regions where significant plastic deformation takes place, the microstructures of components fabricated using MD-based processes exhibit the presence of recrystallized small grains [49,50].

6.1.2.2.1 Ultrasonic AM

UAM exfoliates metal foils with ultrasonic vibrations at ambient temperature [51]. This ultrasonic process combines metal foils into 3D shapes. Figure 6.3(d) shows how the motion system regulates downforce and linear travel speed. Ultrasonically produced parts are most substantial in the build direction due to porosity. Noncontinuous metallurgical bonding and interface oxide decomposition. Overlapping and staggered foils decrease or eliminate void formation [52,53]. Prevents seam alignment to improve mechanical quality. The increase in temperature beneath the sonotrode in UAM is directly proportional to the amplitude and the shear strength of the alloy being used. It has been shown that copper (Cu) exhibits a higher temperature rise compared to aluminum (Al) in UAM [54]. The phenomenon of grain refinement recrystallization is brought about by changes in the microstructure. The microstructural homogeneity is characterized by the transformation from equiaxed grains with a polished surface at the interface to elongated grains resulting from the process of foil rolling [55].

6.1.2.2.2 Additive Friction Stir Deposition

AFSD is the only metal AM technology that can offer equivalent or comparable mechanical qualities in the as-printed stage because of its thermomechanical processing method [56]. Combining friction stir with material feeding provides site-specific solid-state deposition. Solid feed rods feed material into a cylindrical tool head. During accumulation, it spins fast. Frictional heating occurs at the feed-rod and surface interface [57]. Compression from feeding equipment produces soft feed material below the tool head. Figure 6.3(e) shows how larger material can flow between the tool head and substrate. This is related to tool head shear force [58,59]. During deposition, the material undergoes severe shear strain and plastic deformation. The friction stir method's thermomechanical treatment achieves good metallurgical adhesion and bonding. This new cutting-edge technique is beneficial for solid-state bonding, coatings, maintenance, and AM. Solid-state processing may prevent porosity, hot-cracking, and other fusion flaws. Dense and homogenous components eliminate the need for additional processing [60].

6.1.2.2.3 Cold Spray Solid-state AM (CSAM)

This material has technical appeal in several sectors, including aerospace, automotive, marine, biomedical, industrial, and energy industries. Figure 6.3(f) depicts a high-pressure gas distribution network, a gas radiator, hopper, a control console and a cold spray gun. Cold-sprayed materials exhibit distinct microstructural characteristics in comparison to non-SSAM procedures like SLM, DED, WAAM, as well as conventional thermomechanical processes including rolling, forging, extrusion, and AFSD. During the cold spray process, the powder particles undergo significant strain rates, strain gradients, and localized increased temperatures, leading to the formation of intricate and diverse microstructures [63]. Gas type, temperature, pressure, operating distance, scanning speed, and spray angle are all important in cold spraying. Optimizing processing conditions reduces porosity and improves mechanical properties [64].

6.1.3 MICROSTRUCTURE

The microstructure of SSAM is influenced by both the bonding process and the temperature profile. The characteristics and coherence of heat production during component manufacture exhibit variations between the two primary SSAM techniques [65,66]. Table 6.1 presents the thermal and microstructural alterations observed throughout each operational procedure. Sintering procedures involve constant external heating, resulting in a uniform microstructure [67]. If the temperature profile is not set correctly, grain coarsening during sintering may be a process-related microstructure change. Typical microstructures have equiaxed, uniform grains without deformation-related roughness. The microstructure [68] of

TABLE 6.1

Thermal and Microstructural Variations in Parts Created Using the SSAM Technologies [72]

Type	Method	Max. Temp.	Bonding Mechanism	Microstructure Alteration	Microstructure Homogeneity
Sinter-based AM	BJAM	>0.75Tm	Atomic Diffusion	Minimal/grain \coarsening	Equiaxed in every direction
Sinter-based AM	MEAM	>0.75Tm	Atomic Diffusion	Minimal/grain \coarsening	Equiaxed in every direction
Sinter-based AM	SPAM	>0.75Tm	Atomic Diffusion	Minimal/grain \coarsening	Equiaxed in every direction
MD-AM	UAM	0.2–0.5 Tm	SPD	Recrystallization texture refining of grain	The transition from refined equiaxed grains to elongated grains with the rolling texture of the foil is characterized by the gradient in their respective properties.
MD-AM	AFSD	0.6–0.9Tm	SPD	Dynamic recrystallization and adiabatic shear Refinement of grains	Equiaxed, fine, and random in every direction
MD-AM	CSAM	<Tm	SPD	Recrystallization under dynamic conditions Refinement of grains	Fine granules at particle contacts

(a)   (b)

(c)

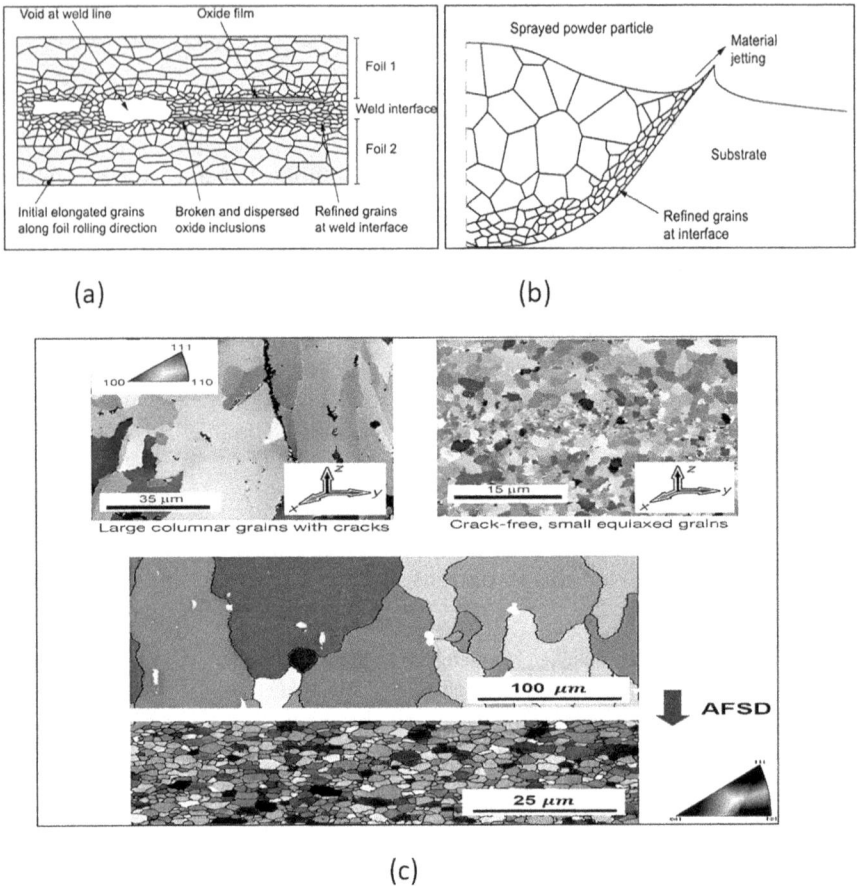

FIGURE 6.4 (a) Microstructure of UAM, (b) microstructure of cold spray AM assisted parts [65–68], and (c) microstructure of AFSD process [69].

Ultrasonic and cold spray additive manufactured parts are shown in Figures 6.4(a) and (b), respectively.

Various microstructures can develop in alloys containing elongated grains, such as some titanium-based alloys. These microstructures can arise from processes such as partial liquid-phase sintering, creation of grain boundary phases, precipitation of second phases, or irregular grain growth during sintering at high temperatures [70]. Figure 6.4(c) shows how the as-sprayed stacking fault energy affects the microstructure after cold spray and post-processing. Cold-spraying titanium has a high deposition efficiency, a significant porosity, and a heterogeneous microstructure due to titanium's weak heat conductivity and ductility [71,72]. Epitaxial solidification produces highly aligned, columnar grains in fusion-based metal AM. AFSD's thermomechanical processing yields refined, equiaxed microstructures [73]. Dynamic microstructure evolution generates discontinuous and continuous dynamic recrystallization [74]. Based on processing factors including peak temperature and strain rate, as well as intrinsic material attributes. High-strain-rate deformation inhibits strain hardening in UAM by causing adiabatic heating. Localized strain can increase void formation and brittle interface cracking [69].

The current study attempts to accomplish the following:

a. Outline research on solid-state AM.
b. Provide an overview of SSAM research and recent trends.

The evaluation is being done to address major research issues in the field of SSAM, such as (1) Who are the active researchers working in this field globally; (2) What are the major organizations and countries involved in this study; (3) Most cited documents and leading countries in the field of SSAM; (4) What are prominent journals and publications in this field?

6.1.4   BIBLIOMETRIC ANALYSIS

6.1.4.1   Publication Trend

When conducting research, it is customary to evaluate the time frame spanning from 2007 to 2022 in order to identify relevant papers. According to the publishing trend, Figure 6.5(a) illustrates a consistent upward trajectory in research within the field of SSAM. However, the rate of growth is not uniform.

The data exhibits a downward trend in 2014, 2015, 2016, and 2017, followed by a subsequent sharp increase over a span of two years. According to Figure 6.5(b), there has been a decline in the quantity of published articles over the years 2014 and 2015. However, there has been a rise in the production of scientific works during the same period. Between the years 2020 and 2021, a total of 17 and 25 works, respectively, achieved the highest ranking in terms of publications. The investigation of SSAM has emerged as a prominent subject of interest among scholars and academics. This emphasizes the importance of SSAM research to improve manufacturing practices, tool life, microstructure, and final surface quality.

6.1.4.2   Most Contributing Countries

This section examines nations with similar SSAM research. Table 6.2 provides nationwide paper and citation counts (based on the institutions of the authors). The USA (57) has the most published documents, followed by Canada (seven) and China (four). India has three papers on this topic. With a TLS of 77, the USA is the most successful country, followed by Canada and Iran (TLS of 19).

Figure 6.5(c) shows the geographical region-by-region location clusters in the global map made with Microsoft Excel. The number of published papers in a certain geographic region indicates that the study subject has been adapted and acknowledged. The area encompasses both established and emerging nations, therefore emphasizing its significance irrespective of varying economic conditions.

(a) Global publication trend

(b) Published articles (year-wise)

(c) Global distribution of published articles

(d) Top contributing countries

(e) Top contributing authors

(f) mutual citation network

(g) Research publications

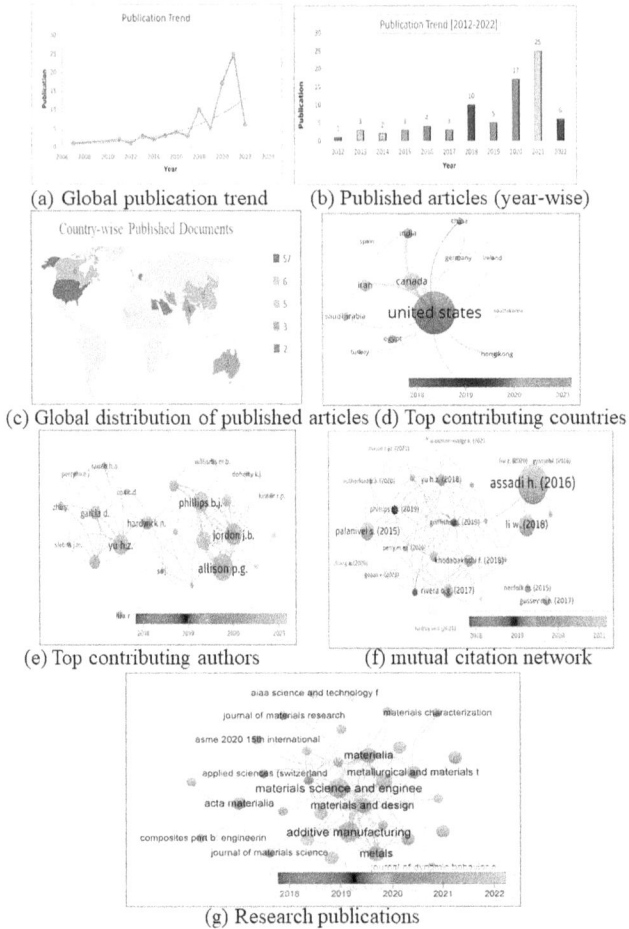

FIGURE 6.5   Bibliometric analysis of solid-state AM.

TABLE 6.2
Top Contributing Countries in the Area of SSAM

Country	Documents	Citations	TLS
USA	57	704	77
Canada	7	95	29
China	6	174	11
United Kingdom	5	53	0
Australia	3	4	3
India	3	32	16
Iran	3	61	19
Egypt	2	1	13
Germany	2	404	8
Saudi Arabia	1	1	13

Figure 6.5(d) depicts the connections between the most prominent countries in the field of SSAM. The size of nodes shows the citation count, as described in the previous sub-sections. There are 36 connections, showing that the network is well-connected

6.1.4.3 Author-based Analysis

The most prolific and influential Scopus authors are ranked by number of publications. Allison p.g has the most articles in Scopus. With about 16 publications, Allison pg. is followed in Scopus by Jordan and Yu, Hz who have 13 and 12 publications, respectively.

In the field of SSAM, a collective of 253 scholars have contributed to the publication of 82 scholarly articles. As a result, a threshold of three publications per author was established, leading to the exceeding of this restriction by 32 writers. The ranking is determined by the cumulative number of citations for authors who have an equivalent number of publications, as seen in Figure 6.5(e). The provided map illustrates a total of 24 circles, symbolizing individual researchers, wherein the proximity of the circles indicates partnerships among authors. The aforementioned circles have been partitioned into three distinct clusters, each of which symbolizes a distinct research community. The cartographic representation of lines on the map serves to indicate a correlation between two researchers, whereby the quantity of co-authored papers is symbolically shown by the number of lines.

6.1.4.4 Most Cited Documents

Figure 6.5(f) presents the top ten documents in the field of SSAM. From 2015 to 2019, publications with the subject *'Additive friction stir deposition: Cold spraying – A materials perspective'* had the most citations (403), followed by *SSAM and repairing by cold spraying: A review* with 143 citations.

This study investigates the available literature to gather comprehensive information on SSAM as a prospective approach for achieving functionally optimal Mg alloys. Figure 6.5(f) illustrates the network topologies of mutual citations among publications on the subject of SSAM, which were published in the Scopus database between the years 2007 and 2022.

6.1.4.5 Top Cited Journals

Among the 82 journals shown in Figure 6.5(g), the top contributing journals that align with the specified criteria are identified. The sources that possess the highest aggregate link strength are deemed to be the most significant.

Material science and engineering has the large volume of publications and total link strength whereas *AM* and *Materials and design* have a significant number of citations and total link strength as shown in Table 6.3. Figure 6.5(g) illustrates the interconnections established through citations among the many sites that publish publications related to SSAM in the Scopus database.

TABLE 6.3

Top Contributing Journals in Domain of SSAM

Source	Documents	Citations	TLS
Material Science and Engineering	5	179	58
Additive Manufacturing	5	39	39
Materialia	3	53	33
Material and Design	3	130	30
Metals	3	15	27
Scripta Materialia	1	57	26
Journal of Material Science and Technology	1	1	24
JOM	2	30	20
Metallurgical and Materials Transactions A	2	56	18
CIRP Journal of Manufacturing Science and Technology	2	36	6

6.1.5 CONTENT ANALYSIS

Industry 4.0 is comprised of disruptive technologies. The use of many technologies, IoT, Digital Manufacturing (DM), augmented reality, Machine or Artificial Intelligence systems (AI), virtual reality (VR) technology, and other similar advancements, has significantly transformed the operational processes of conventional manufacturing sector [75,76]. The increasing pace of industrial growth has led to a rising need for a novel class of engineered materials that possess distinct features. This phenomenon gives rise to novel production methodologies that exhibit enhanced efficiency and faster operations compared to the techniques devised in the initial and subsequent industrial revolutions [77]. The traditional component fabrication and validation paradigm have been dramatically altered by AM. This innovation has been at the forefront for the last three decades, and it now dominates the traditional manufacturing process to develop complex materials with minimal cost and effort [78]. It allows for both unitization and topological optimizations at the same time. AM is considered a pioneering methodology in the contemporary industrial landscape [79]. The process encompasses the production of diverse 3D things by incremental addition of material, irrespective of their dimensions or form (including complex structures), regardless of the material composition, such as steel, concrete, human tissue, plastic, and other variants. Metal Additive production (MAM) is a prominent constituent of AM processes, especially in cases where metals are employed in Layer-wise production techniques. [80]. The MAM methodology has been embraced as a preferred way to build metallic components for the automobile and aircraft sectors [81]. The development of lightweight materials composed of magnesium, titanium, aluminum, and other metals has increased the performance of the marine and aircraft industries [82]. The MAM method creates components with distinctive features such as complex shapes and minimal material usage. However, there have been several challenges in employing AM technology to manufacture metal items [83]. The major issues with today's modern liquid state-based AM technique include Shrinkage, internal cavities, internal porosity, inclusions, microstructural defects such as non-homogeneity, anisotropic material behavior, and insufficient mechanical properties [84]. Baker published the first patent literature on metal AM in 1926, which constructed components using electric arc melt/fusion processes to make 3-dimensional (3D) things. As a result, friction stir-based AM (AFSD and AFS) was established as an alternative solid-state (non-melting) approach that can deliver enhanced microstructural and mechanical qualities through grain refinement [85]. 'White' originally used and patented AFSD in 2002, while Thomas et al. presented a similar approach in 2005 [86]. In 2006, Airbus exhibited the ability to build structures faster while reducing material waste, making AFSD commercially viable. AFSD is an advanced AM technique characterized by its rapid expansion. This technique entails the sequential deposition of various materials in layers to fabricate a three-dimensional (3D) object based on digital data [87]. The system functions based on the idea of friction stir welding. In 2012, Kandasamy et al. presented the concept of Additive Friction Stir (AFS), a solid-state localized thermomechanical technique involving the placement of wrought metal onto metallic surfaces [88]. Al and Mg structures were manufactured using AFS in this work. To investigate the mechanism of powder consolidation and mechanical characteristics, microstructural and mechanical characterization were performed [89]. In order to create a metallurgical link between the powder particles and layers, shear-induced interfacial heating and severe plastic deformation are used [90]. He employed a solid-state localized thermomechanical process known as additive friction stir to combine and deposit atomized WE43 powder. The tensile properties of deposited WE43 alloy are presented and discussed [91]. Moreover, utilizing atomized powders with additive friction stir technology to produce near-net-shaped components is both energy-efficient and cost-effective. Kandasamy et al. published a paper in 2015 that described how ground and atomized WE43 and AZ91 powders were consolidated and deposited using additive friction stir, and their microstructural and mechanical characteristics were assessed [92]. WE43 and AZ91 alloys deposited using this SSAM technology have mechanical qualities that were superior to conventional powder metallurgy products and equivalent to wrought alloys [93]. Cold spray, also known as gas dynamic cold spray, was first introduced in the mid-1980s. In 2016, Faizan-Ur-Rab M. et al. presented a cold spray SSAM technique in which metal particles are accelerated toward a metallic or non-metallic substrate using high-velocity gas [94]. The microstructure of cold-sprayed materials differs from those produced by non-SSAM techniques such as selective laser melting (SLM), DED, wire arc AM (WAAM), as well as classic thermomechanical processes like rolling, forging, and extrusion. Additionally, it also differs from materials made using AFSD [95]. Blenheim J et al. (2018) provide a proposal for a new AM process for aluminum alloys. Based on Bonding (HYB) technology, Hybrid Metal Extrusion as well as the benefits and disadvantages of each approach [96]. In recent years, the emergence of a diverse array of SSAM technologies has been facilitated by the advancement of the AFSD process. The AFSD technique involves the stacking of a base material and a powder material, followed by the addition of an AFSD additive during the friction stir deposition process. Both techniques have a common concept, but with distinct processing approaches. In a recent work conducted by Martin L.P. et al. in 2022, a solid-state volumetric repair was successfully accomplished via the use of advanced friction stir deposition technique on an aluminum alloy 6061-T6. This technique involved filling grooves that were etched onto the surface of a plate made of 6061-T651 alloy [97]. The groove-filling technique is a valuable approach for the replacement of damaged or deteriorated material that has been selectively removed by precise grinding. The microstructural characteristics of AFSD are very captivating. Consequently, the microstructural characteristics of the AFSD and AFS processes undergo alterations at different magnitudes. The friction deposition method of A) utilizing friction stir welding (FSW) equipment was pioneered by Dilip et al.. Yuqing et al. investigated the properties of AA7075-O produced by FSAM [98]. Nine layers of composites were fabricated by the utilization of friction stir lap welding technique on a horizontal milling machine employing GH4169 steel tools in order to enhance the materialistic properties of aluminum 1060 sheets. In their study, Yuqing and Mondal employed Am and FSP as indicated by their citation [99]. The authors introduced two distinct categories of AM techniques for producing layered composite materials: fusion-based AM and friction-based AM. In their study, Khodabakhshi et al. (year) conducted an analysis to evaluate the advantages and drawbacks associated with FSAM, as documented in reference [100]. Selective laser melting (SLM) was employed to deposit aluminum 7075 powder into the groove, followed by additional processing using friction stir processing (FSP) with the aid of an H13 steel tool. The Advanced Friction Stir Welding (AFSD) technique has superior performance compared to existing SSAM methods, including friction stir deposition and AFS technology. Moreover, this methodology is employed across several fields of engineering and medical. However, it is important to note that the precise solution is anticipated to be determined in the foreseeable future due to current research efforts.

6.1.6 CONCLUSION AND FUTURE SCOPE

The present study investigated the development and advancement of the SSAM technique. The domain of AM is seeing significant growth, with projections indicating its widespread adoption across several industries and a substantial contribution of around 78% to overall production in the foreseeable future. This study aims to conduct an analytical bibliometric assessment of the existing publications on SSAM, marking the first examination of its kind. The aforementioned initial inquiry has shown the worldwide publishing pattern, as well as the most pertinent papers, reports, researchers, and nations in the field of SSAM technology.

- This technology is now being employed for the purpose of creating three-dimensional tangible goods that possess practical utility in many sectors such as automobiles, power electronics, aeronautical, and manufacturing domains.
- The SSAM process is employed for the fabrication of miniature patterns, dies, tool machinery, molds, and commercial products.
- Most medical devices, dental requirements, surgical tools, knee joints, bio implants, and laboratory equipment are currently made using this technology.

Nevertheless, there is a scarcity of academic research about the use of SSAM within the medical domain. Moreover, this approach finds use across several fields of engineering and medical. However, it is anticipated that a precise solution will be determined in the near future as a result of extensive study.

REFERENCES

[1] Mueller, B., AM technologies–Rapid prototyping to direct digital manufacturing. *Assembly Automation*, 2012. **32**(2).

[2] Khodabakhshi, F., et al., Friction stir welding of a P/M Al–Al2O3 nanocomposite: Microstructure and mechanical properties. *Materials Science and Engineering: A*, 2013. **585**: p. 222–232.

[3] Bai, Y. and C.B. Williams, Binder jetting AM with a particle-free metal ink as a binder precursor. *Materials & Design*, 2018. **147**: p. 146–156.

[4] Ngo, T.D., et al., AM (3D printing): A review of materials, methods, applications and challenges. *Composites Part B: Engineering*, 2018. **143**: p. 172–196.

[5] Sasahara, H., M. Tsutsumi, and M. Chino, Development of a layered manufacturing system using sheet metal-polymer lamination for mechanical parts. *The International Journal of Advanced Manufacturing Technology*, 2005. **27**(3): p. 268–273.

[6] Srivastava, M., et al., A review on recent progress in solid state friction based metal AM: Friction stir additive techniques. *Critical Reviews in Solid State and Materials Sciences*, 2019. **44**(5): p. 345–377.

[7] Rivera, O., et al., Influence of texture and grain refinement on the mechanical behavior of AA2219 fabricated by high shear solid state material deposition. *Materials Science and Engineering: A*, 2018. **724**: p. 547–558.

[8] Hehr, A. and M.J. Dapino, Dynamics of ultrasonic AM. *Ultrasonics*, 2017. **73**: p. 49–66.

[9] Palanivel, S. and R.S. Mishra, Building without melting: A short review of friction-based AM techniques. *International Journal of Additive and Subtractive Materials Manufacturing*, 2017. **1**(1): p. 82–103.

[10] Gandra, J., et al., Functionally graded materials produced by friction stir processing. *Journal of Materials Processing Technology*, 2011. **211**(11): p. 1659–1668.

[11] Miranda, R., et al., Reinforcement strategies for producing functionally graded materials by friction stir processing in aluminium alloys. *Journal of Materials Processing Technology*, 2013. **213**(9): p. 1609–1615.

[12] Arora, A., et al., Strains and strain rates during friction stir welding. *Scripta Materialia*, 2009. **61**(9): p. 863–866.

[13] Kumar, R., V. Pancholi, and R.P. Bharti, Material flow visualization and determination of strain rate during friction stir welding. *Journal of Materials Processing Technology*, 2018. **255**: p. 470–476.

[14] Khodabakhshi, F., et al., Effects of stored strain energy on restoration mechanisms and texture components in an aluminum–magnesium alloy prepared by friction stir processing. *Materials Science and Engineering: A*, 2015. **642**: p. 204–214.

[15] Khodabakhshi, F., A. Gerlich, and P. Švec, Reactive friction-stir processing of an Al-Mg alloy with introducing multi-walled carbon nano-tubes (MW-CNTs): Microstructural characteristics and mechanical properties. *Materials Characterization*, 2017. **131**: p. 359–373.

[16] Oh, Y., C. Zhou, and S. Behdad, Part decomposition and assembly-based (Re) design for AM: A review. *AM*, 2018. **22**: p. 230–242.

[17] Miyanaji, H., N. Momenzadeh, and L. Yang, Effect of printing speed on quality of printed parts in Binder Jetting Process. *AM*, 2018. **20**: p. 1–10.

[18] Calvert, J.R., *Microstructure and mechanical properties of WE43 alloy produced via additive friction stir technology*. 2015, Virginia Tech.

[19] Palanivel, S., H. Sidhar, and R. Mishra, Friction stir AM: Route to high structural performance. *Jom*, 2015. **67**(3): p. 616–621.

[20] Hang, Z.Y., et al., Non-beam-based metal AM enabled by additive friction stir deposition. *Scripta Materialia*, 2018. **153**: p. 122–130.

[21] Harun, W., et al., A review of powder AM processes for metallic biomaterials. *Powder Technology*, 2018. **327**: p. 128–151.

[22] Sankaran, K.K. and R.S. Mishra, *Metallurgy and design of alloys with hierarchical microstructures*. 2017: Elsevier.

[23] Palanivel, S., et al., Friction stir AM for high structural performance through microstructural control in an Mg based WE43 alloy. *Materials & Design (1980–2015)*, 2015. 65: p. 934–952.

[24] DebRoy, T., et al., AM of metallic components – Process, structure and properties. *Progress in Materials Science*, 2018. **92**: p. 112–224.

[25] Agrawal, P., et al., Excellent strength-ductility synergy in metastable high entropy alloy by laser powder bed AM. *AM*, 2020. **32**: p. 101098.

[26] Froes, F.H. and R. Boyer, *AM for the aerospace industry*. 2019: Elsevier.

[27] Mazumder, J., et al., Closed loop direct metal deposition: Art to part. *Optics and Lasers in Engineering*, 2000. **34**(4-6): p. 397–414.

[28] Gibert, J.M., G. Fadel, and M.F. Daqaq, On the stick-slip dynamics in ultrasonic AM. *Journal of Sound and Vibration*, 2013. **332**(19): p. 4680–4695.

[29] Lavoie, M. and J.L. Addis, Harnessing the potential of AM technologies: Challenges and opportunities for entrepreneurial strategies. *International Journal of Innovation Studies*, 2018. **2**(4): p. 123–136.

[30] Lewandowski, J.J. and M. Seifi, *Metal AM: A review of mechanical properties (postprint)*. 2016.

[31] Ford, S. and M. Despeisse, AM and sustainability: An exploratory study of the advantages and challenges. *Journal of Cleaner Production*, 2016. **137**: p. 1573–1587.

[32] Li, L., et al., AM of near-net-shape bonded magnets: Prospects and challenges. *Scripta Materialia*, 2017. **135**: p. 100–104.

[33] Ziaee, M. and N.B. Crane, Binder jetting: A review of process, materials, and methods. *AM*, 2019. **28**: p. 781–801.

[34] Stevens, E., et al., Density variation in binder jetting 3D-printed and sintered Ti-6Al-4V. *AM*, 2018. **22**: p. 746–752.

[35] German, R.M., Injection molding of metals and ceramics. *Powder Metallurgy*, 1997. **42**: p. 157–160.

[36] Thompson, Y., et al., Fused filament fabrication, debinding and sintering as a low cost AM method of 316L stainless steel. *AM*, 2019. **30**: p. 100861.

[37] Sriraman, M., et al., Thermal transients during processing of materials by very high power ultrasonic AM. *Journal of Materials Processing Technology*, 2011. **211**(10): p. 1650–1657.

[38] Wolcott, P.J., A. Hehr, and M.J. Dapino, Optimized welding parameters for Al 6061 ultrasonic additive manufactured structures. *Journal of Materials Research*, 2014. **29**(17): p. 2055–2065.

[39] Alkimov, A., Gas-dynamic spray method for applying a coating. *US Pat.* 5302414, 1994.

[40] Schmidt, T., et al., Development of a generalized parameter window for cold spray deposition. *Acta materialia*, 2006. **54**(3): p. 729–742.

[41] Papyrin, A., et al., *Cold spray technology*. 2006: Elsevier.

[42] Yin, S., et al., Cold spray AM and repair: Fundamentals and applications. *AM*, 2018. **21**: p. 628–650.

[43] King, W.E., et al., Laser powder bed fusion AM of metals; physics, computational, and materials challenges. *Applied Physics Reviews*, 2015. **2**(4): p. 041304.

[44] Gong, X., T. Anderson, and K. Chou. Review on powder-based electron beam AM technology. in *International Symposium on Flexible Automation*. 2012. American Society of Mechanical Engineers.

[45] Sundaram, M.M., A.B. Kamaraj, and V.S. Kumar, Mask-less electrochemical AM: a feasibility study. *Journal of Manufacturing Science and Engineering*, 2015. **137**(2).

[46] Wang, H. and Y. Zou, Microscale interaction between laser and metal powder in powder-bed AM: conduction mode versus keyhole mode. *International Journal of Heat and Mass Transfer*, 2019. **142**: p. 118473.

[47] King, W.E., et al., Observation of keyhole-mode laser melting in laser powder-bed fusion AM. *Journal of Materials Processing Technology*, 2014. **214**(12): p. 2915–2925.

[48] Körner, C., AM of metallic components by selective electron beam melting—A review. *International Materials Reviews*, 2016. **61**(5): p. 361–377.

[49] Yap, C.Y., et al., Review of selective laser melting: Materials and applications. *Applied Physics Reviews*, 2015. **2**(4): p. 041101.

[50] Michaelraj, A., *Taxonomy of physical prototypes: structure and validation*. 2009, Clemson University.

[51] Milewski, J.O., AM metal, the art of the possible, in *AM of Metals*. 2017, Springer. p. 7–33.

[52] Panchagnula, J.S. and S. Simhambhatla. AM of complex shapes through weld-deposition and feature based slicing. in *ASME International Mechanical Engineering Congress and Exposition*. 2015. American Society of Mechanical Engineers.

[53] Iebba, M., et al., Influence of powder characteristics on formation of porosity in AM of Ti-6Al-4V components. *Journal of Materials Engineering and Performance*, 2017. **26**(8): p. 4138–4147.

[54] Wang, Y. and Y.F. Zhao, Investigation of sintering shrinkage in binder jetting AM process. *Procedia Manufacturing*, 2017. **10**: p. 779–790.

[55] Marya, M., et al., Microstructural development and technical challenges in laser AM: case study with a 316L industrial part. *Metallurgical and Materials Transactions B*, 2015. **46**(4): p. 1654–1665.

[56] Rathee, S., et al., *Friction based AM technologies: Principles for building in solid state, benefits, limitations, and applications.* 2018. CRC Press.
[57] Dilip, J. and G. Janaki Ram, Friction freeform fabrication of superalloy Inconel 718: prospects and problems. *Metallurgical and Materials Transactions B*, 2014. **45**(1): p. 182–192.
[58] He, C., et al., Investigation on microstructural evolution and property variation along building direction in friction stir additive manufactured Al–Zn–Mg alloy. *Materials Science and Engineering: A*, 2020. **777**: p. 139035.
[59] Habibnejad-korayem, M., J. Zhang, and Y. Zou, Effect of particle size distribution on the flowability of plasma atomized Ti-6Al-4V powders. *Powder Technology*, 2021. **392**: p. 536–543.
[60] White, D.R., Ultrasonic consolidation of aluminum tooling. *Advanced Materials & Processes*, 2003. **161**(1): p. 64–66.
[61] Committee, A.I.H., *Properties and selection: Nonferrous alloys and special-purpose materials.* ASM International, 1992. **2**: p. 1143–1144.
[62] Batchelder, J.S., *Method for controlled porosity three-dimensional modeling.* 1997, Google Patents.
[63] Bose, A., et al., Traditional and AM of a new Tungsten heavy alloy alternative. *International Journal of Refractory Metals and Hard Materials*, 2018. **73**: p. 22–28.
[64] Gussev, M., et al., Effect of post weld heat treatment on the 6061 aluminum alloy produced by ultrasonic AM. *Materials Science and Engineering: A*, 2017. **684**: p. 606–616.
[65] Han, T., et al., Effect of preheat temperature and post-process treatment on the microstructure and mechanical properties of stainless steel 410 made via ultrasonic AM. *Materials Science and Engineering: A*, 2020. **769**: p. 138457.
[66] Griffiths, R.J., et al., A perspective on solid-state AM of aluminum matrix composites using MELD. *Journal of Materials Engineering and Performance*, 2019. **28**(2): p. 648–656.
[67] Attar, H., et al., Effect of powder particle shape on the properties of in situ Ti–TiB composite materials produced by selective laser melting. *Journal of Materials Science & Technology*, 2015. **31**(10): p. 1001–1005.
[68] Huang, C., et al., In-situ formation of Ni-Al intermetallics-coated graphite/Al composite in a cold-sprayed coating and its high temperature tribological behaviors. *Journal of Materials Science & Technology*, 2017. **33**(6): p. 507–515.
[69] Hussein, S.A. and A. Hadzley, Characteristics of aluminum-to-steel joint made by friction stir welding: A review. *Materials Today Communications*, 2015. **5**: p. 32–49.
[70] Fayazfar, H., et al., A critical review of powder-based AM of ferrous alloys: Process parameters, microstructure and mechanical properties. *Materials & Design*, 2018. **144**: p. 98–128.
[71] Khodabakhshi, F. and A. Gerlich, Potentials and strategies of solid-state additive friction-stir manufacturing technology: A critical review. *Journal of Manufacturing Processes*, 2018. **36**: p. 77–92.
[72] Sharma, A., et al., A new process for design and manufacture of tailor-made functionally graded composites through friction stir AM. *Journal of Manufacturing Processes*, 2017. **26**: p. 122–130.
[73] Huang, G., et al., Friction stir brazing of 6061 aluminum alloy and H62 brass: Evaluation of microstructure, mechanical and fracture behavior. *Materials & Design*, 2016. **99**: p. 403–411.

[74] Liu, H., H. Zhang, and L. Yu, Effect of welding speed on microstructures and mechanical properties of underwater friction stir welded 2219 aluminum alloy. *Materials & Design*, 2011. **32**(3): p. 1548–1553.

[75] Padhy, G., C. Wu, and S. Gao, Friction stir based welding and processing technologies-processes, parameters, microstructures and applications: A review. *Journal of Materials Science & Technology*, 2018. **34**(1): p. 1–38.

[76] Dressler, M., T. Studnitzky, and B. Kieback. AM using 3D screen printing. in 2017 International Conference on Electromagnetics in Advanced Applications (ICEAA). 2017. IEEE.

[77] Bonnard, R., P. Mognol, and J.-Y. Hascoët, A new digital chain for AM processes. *Virtual and Physical Prototyping*, 2010. **5**(2): p. 75–88.

[78] Ivanova, O., C. Williams, and T. Campbell, AM (AM) and nanotechnology: promises and challenges. *Rapid prototyping journal*, 2013. **19**(5): p. 353–364.

[79] Ding, D., et al., Wire-feed AM of metal components: Technologies, developments and future interests. *The International Journal of Advanced Manufacturing Technology*, 2015. **81**(1): p. 465–481.

[80] Ligon, S.C., et al., Polymers for 3D printing and customized AM. *Chemical Reviews*, 2017. **117**(15): p. 10212–10290.

[81] de Fuentes García-Romero, J.M., Industry 4.0: Managing the digital transformation. *Economía industrial*, 2018(410): p. 179–181.

[82] da Rosa Righi, R., A.M. Alberti, and M. Singh, *Blockchain technology for Industry 4.0*. 2020: Springer.

[83] Tofail, S.A., et al., AM: Scientific and technological challenges, market uptake and opportunities. *Materials Today*, 2018. **21**(1): p. 22–37.

[84] Nayyar, A. and A. Kumar, *A roadmap to industry 4.0: smart production, sharp business and sustainable development*. 2020: Springer.

[85] Rao, M.S. and N. Ramanaiah, Optimization of Process parameters for FSW of Al-Mg-Mn-Sc-Zr alloy using CCD and RSM. *Strojnícky časopis-Journal of Mechanical Engineering*, 2018. **68**(3): p. 195–224.

[86] Mishra, R.S. and Z. Ma, Friction stir welding and processing. *Materials Science and Engineering: R: Reports*, 2005. **50**(1-2): p. 1–78.

[87] Tuncer, N. and A. Bose, Solid-state metal AM: A review. *Jom*, 2020. **72**(9): p. 3090–3111.

[88] Li, W., et al., Solid-state AM and repairing by cold spraying: A review. *Journal of Materials Science & Technology*, 2018. **34**(3): p. 440–457.

[89] Dosta, S., M. Couto, and J. Guilemany, Cold spray deposition of a WC-25Co cermet onto Al7075-T6 and carbon steel substrates. *Acta Materialia*, 2013. **61**(2): p. 643–652.

[90] Assadi, H., Kreye, H., Gärtner, F., Klassen, T.J. Cold spraying–A materials perspective. *Acta Materialia*. 2016 Sep 1;116:382–407.

[91] Friel, R.J. and R.A. Harris, Ultrasonic AM–a hybrid production process for novel functional products. *Procedia Cirp*, 2013. **6**: p. 35–40.

[92] Suhonen, T., et al., Residual stress development in cold sprayed Al, Cu and Ti coatings. *Acta Materialia*, 2013. **61**(17): p. 6329–6337.

[93] Yuan, W., et al., Material flow and microstructural evolution during friction stir spot welding of AZ31 magnesium alloy. *Materials Science and Engineering: a*, 2012. **543**: p. 200–209.

[94] Yuan, W. and R. Mishra, Grain size and texture effects on deformation behavior of AZ31 magnesium alloy. *Materials Science and Engineering: A*, 2012. **558**: p. 716–724.

[95] Agrawal, P., et al., Processing-structure-property correlation in additive friction stir deposited Ti-6Al-4V alloy from recycled metal chips. *AM*, 2021. **47**: p. 102259.

[96] Russell, M., et al., Recent developments in the friction stir welding of titanium alloys. *Welding in the World*, 2008. **52**(9): p. 12–15.

[97] Jordon, J., et al., Direct recycling of machine chips through a novel solid-state AM process. *Materials & Design*, 2020. **193**: p. 108850.

[98] Avery, D., et al., Influence of grain refinement and microstructure on fatigue behavior for solid-state additively manufactured Al-Zn-Mg-Cu alloy. *Metallurgical and Materials Transactions A*, 2020. **51**(6): p. 2778–2795.

[99] Perry, M.E., et al., Morphological and microstructural investigation of the non-planar interface formed in solid-state metal AM by additive friction stir deposition. *AM*, 2020. **35**: p. 101293.

[100] Lu, I., *Friction Stir AM (FSAM) of 2050 Al-Cu-Li Alloy*. 2019, University of South Carolina.

7 Direct Energy Deposition Process (DED) and an Insight into DED Process Using Flux-cored and Metal-cored Wires

R. Suryanarayanan and Vishvesh Badheka

7.1 INTRODUCTION

In recent years, metal-based additive manufacturing (MAM) has garnered greater interest among the automobile, aerospace, oil and gas, and marine industries due to its ability to produce complex shaped components at shorter lead times and reduced costs. MAM processes are classified based on the energy source and material feed-based. The energy sources used for MAM are laser-based, electron beam, electric arc, and friction (Figure 7.1).

The material feed for MAM is classified as powder-based and wire-based. As mentioned in Figure 7.1, the MAM systems which use laser power source are Selective Laser Sintering (SLS), Selective Laser Melting (SLM), Direct metal laser sintering (DMLS), and Powder bed Laser fusion (PBF-L). The systems which use electron beam is electron beam melting (EBM). The major constraints in the systems that are powered through laser and electron beam are the component size, slower printing speed, increased material costs, and wastes. The wire-based Direct energy deposition (DED) systems work on the principle of gas welding. The DED systems use powder and wire for printing the metallic structures. Moreover, the cost of the metallic wire feedstock is low compared to the cost of the powder feed. The DED systems that use wire feed stocks were introduced in 1925 as a method to produce metallic ornaments [1]. The DED process has an upper hand over the powder-based MAM systems in terms of production of larger components, faster printing speeds and reduced wastes. Moreover, with the help of the automation, DED process's employability across various industries has seen increased growth over the last few years.

DOI: 10.1201/9781003484325-7

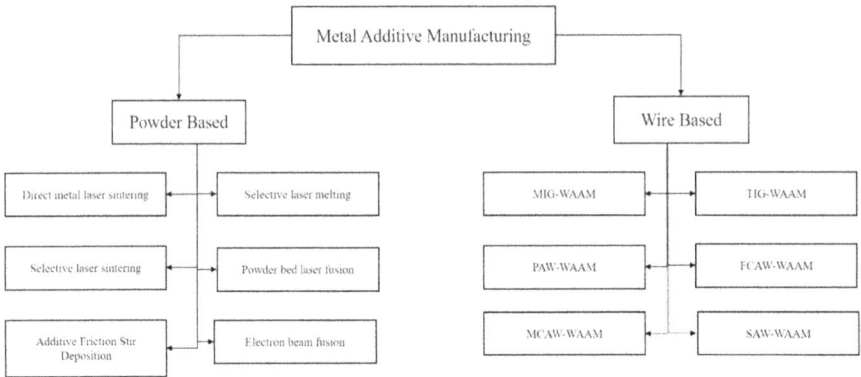

FIGURE 7.1 Metal additive manufacturing classification.

DED process has been reported to print metallic parts of different metals like Al, Ti, steel, and bi-metallic structures. DED processes compared to conventional manufacturing processes (subtractive manufacturing) has reported to reduce the production costs compared to conventional manufacturing processes like forging and casting. Recently, DED was demonstrated to produce Ti-based pressure vessel, which was employed in space exploration. The research team reported that the DED process produced the pressure vessel at reduced the manufacturing costs, time, and waste when compared to forging process [2].

The DED process is classified based on the type of the heat source are Metal inert gas welding (MIG) or Gas metal arc welding (GMAW), Tungsten inert gas welding (TIG) or Gas Tungsten inert gas welding (TIG), plasma arc welding (PAM) and submerged arc welding (SAW). The heat sources have their own unique influence on the deposition of the weld bead and on the quality of the weld. The MIG-based DED is reported to possess higher deposition rates than the TIG-based DED and PAM-based DED, but the MIG-based DED has low stability and produces increased fumes due to the action of the electric current on the wire spool. Such constraints in the MIG-based DED were overcome by the introduction of cold metal transfer system (CMT), which was able to produce quality prints at the faster rates. The print and its performance produced by the DED processes depends upon the process parameters, material, deposition modes, traverse modes, etc. [3,4]. Various researchers have reported the review on the mechanical and metallurgical performance of the DED prints, process control, defects with various metal alloys, however, with the recent variations in the conventional DED prints with flux-cored and metal-cored wires has created the need for systematic analysis of these techniques.

Therefore, in this review article, the influence of DED parameters on the mechanical and microstructural properties of the prints, deposition techniques, post-processing techniques for DED prints, and finally the need for flux-cored and metal-cored wires in DED process are highlighted.

7.2 INFLUENCE OF DED PARAMETERS ON PRINT'S PROPERTIES

The DED parameters can be classified into two types, deposition parameters and spray parameters, as observed in Figure 7.2. The deposition parameters are welding voltage, welding current, deposition rate, welding speed, travel path, and backing plate material, whereas the spray parameters are globular, pulsed, short circuit, and spray.

The DED parameters have a key role in deciding the heat input acting during the process. The heat input during the DED process is presented in equation 7.1 [3]:

$$\text{Heat input,} \quad H_i = \eta \frac{WI}{U} \tag{7.1}$$

W is the welding voltage, V, I is the welding current, A, U is the welding speed, mm/s, and η is the arc thermal efficiency.

Su et al. [5] studied the effects of printing conditions like welding current, welding voltage, wire deposition rate, and welding speed on the print's microstructural, metallurgical, and mechanical properties in DED of Al-Mg alloys. The authors reported that the increase in the ratio of wire deposition rate and welding speed increased the cooling rates during DED, and had a significant influence on the grain size, metallurgy across the print layers. The authors also reported that the DED fabricated Al-Mg alloy possessed high mechanical strength compared to cast Al-Mg alloy. Kumar and Maiji [6] reported that the double-wired feed mechanism (Figure 7.3) enabled the authors to fabricate high strength prints in DED of AISI 304 L austenite steel compared to single feed mechanism. The authors credited the higher deposition rates and formation of completely developed ferrite grains for increased mechanical properties compared to single feed system. Ayed et al. [7] reported that the increase in the head travel speed had a significant influence on the print properties in DED of Ti-6Al-4V. Yilidz et al. [8] identified that the ratio of wire feed rate and travel speed were important conditions to affect the heat input during the DED process. The authors observed that the variation in the heat input during the DED was observed to have a significant effect on the bead characteristics, mechanical and metallurgical properties of the HSLA print fabricated via CMT mode.

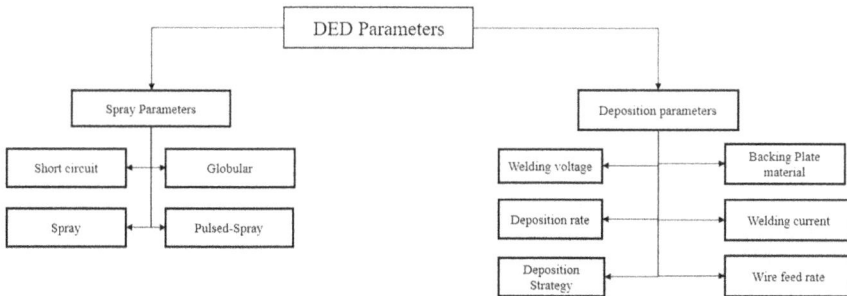

FIGURE 7.2 DED parameters.

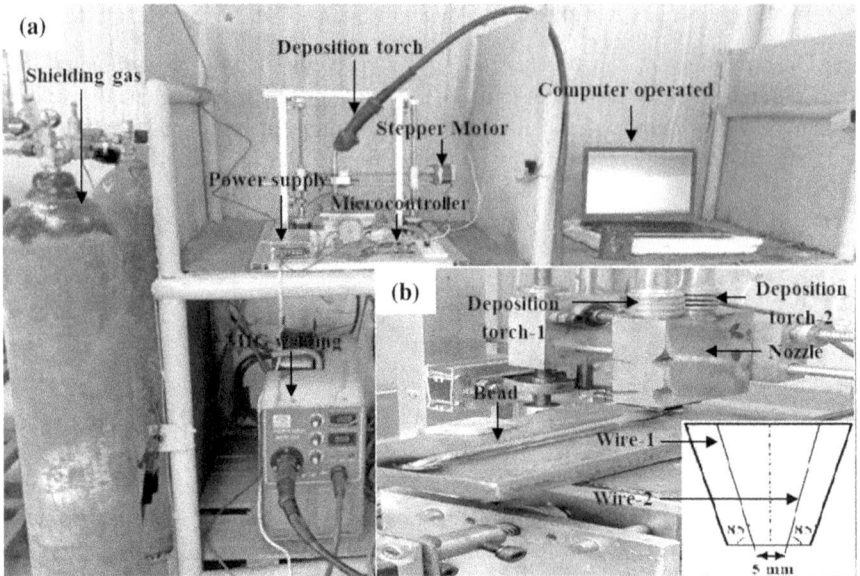

FIGURE 7.3 Double wire feed mechanism in DED [6].

Durate et al. [9] applied forging action on the printed layers in DED of AISI 316 L steel to enhance the mechanical properties of the prints. The authors termed this process as hot wire forging DED (Figure 7.4). The forging action resulted in inducing the plastic deformation on the prints, which ensured the closing of the of the pores to enhance the strength of the prints. Sun et al. [10] reported that the wire feed rate and travel speed were key parameters that determined the bead geometry. The combination of higher wire feed rate and low travel speeds were observed to produce incomplete fills during the DED process.

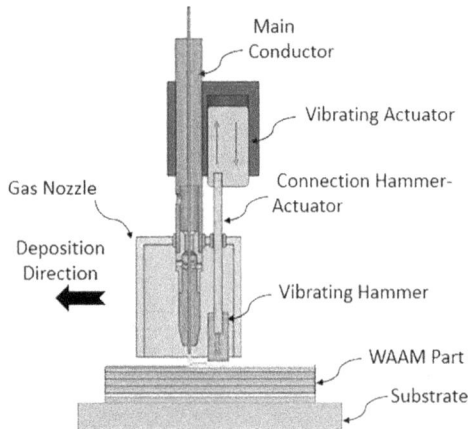

FIGURE 7.4 Hot wire forging DED [9].

Wang et al. [11] introduced two modes of arc transfer, speed arc, and speed pulse, in DED of stainless steel. The authors reported that the variation in the heat input during the DED with respect to the modes of arc transfer and the mechanical strength of the prints were higher than the prints produced by CMT mode. Wu et al. [12] observed that the increase in the powder feed rate during the DED of high nitrogen steel resulted in improving the mechanical and metallurgical properties of the print. The authors credited that the increase in the powder feed rate increased the nitrogen content in the prints compared to the wire fed prints, which initiated the metallurgical changes in the prints to enhance the mechanical strength. The authors used a GTAW-based heat source for the process. Han et al. [13] produced bi-metallic structures by CMT-based DED built on single-use water-cooled aluminum block. The author noted the absence of the defects and intermetallic compounds in the prints. The authors also observed the variations in the hardness across the layers, which was credited to the variation in the temperature across the layers. Fu et al. [14] reported that the introduction of hot wire processing in DED of Al alloys resulted in reducing the net heat input in DED and also reducing the hydrogen contamination in the prints. The authors observed that the welding current played a deciding role in determining the porosity, which in turn affected the print's mechanical strength. Kumar and Maiji [6] in their another work optimized the process parameters in bi-metallic DED of ASS 304 L steel and Inconel 125 alloy by response surface methodology. The authors observed that the prints with intermediate layers possessed higher mechanical strength compared to that of the prints without the intermediate layers. The authors also observed variations in the microhardness and the microstructure of the prints. Prado-Cerqueira et al. [15] reported the effects of different modes of heat source on the mechanical and microstructural properties in DED of copper-coated steel wire. The authors observed variations in the Brinell hardness profiles which was credited to the variations in the heat input with respect to MIG, CMT, and its variations. The authors reported that the prints with mechanical strength > 550 MPa were not recommended as they could not assure the homogeneity in the mechanical behavior of the prints.

Rodriguez et al. [16] reported that the ratio of wire feed rate and travel speed determined the thickness of the deposition in ultra-cold DED of HSLA steel. The authors used TIG-based DED and MIG-based DED heat sources to fabricate the single wall prints and compared their mechanical and microstructural properties. The authors achieved the reduced heat input and higher cooling rates in ultra-cold DED by making the electrical circuit between the tungsten electrode and the wire feedstock, which reduced the chances of the accumulation of the molten metal onto the electrode and transfer less heat onto the substrate. Hauser et al. [17] produced multi-material prints of Al alloys by CMT-based DED. The authors reported that the mechanical properties of the prints were dependent on the individual alloy rather than the transition region between the two alloys. Yangfan et al. [18] studied the effects of travel speeds on the mechanical and microstructural properties of Inconel 625 alloy printed via CMT-based DED. The authors credited the variations in the microstructure at the bottom, middle, and top layer to the variation in the heat dissipation rates.

The authors observed that the hardness distribution was not observed to be affected with the increase in the travel speed. The tensile strength was observed to increase with the increase in the travel speed, which was credited to the fine grains and precipitation. Li et al. [19] produced low-carbon steel prints at high deposition efficiencies with the help of SAW-based DED. The variations in the mechanical strength across the different regions were credited to the variations in the temperature across the layers. The authors observed that the impact toughness of the prints had high enhancements. Such a phenomenon was credited to the characteristic heat-treatment procedures which improved the mechanical properties of the prints. The authors in their recent work reported that the similar results, with the improvement in the impact toughness due to the microstructural enhancement in the SAW-based DED [20]. Wang et al. [21] employed twin wire with PAW-based DED to print Titanium-aluminide prints on a heated platform at 500°C, to prevent the cracking. The authors observed microstructural variations at the top and middle regions, with the equiaxed, dendritic grains in the interdendritic phase at top and lamellar grains in the middle. The authors observed anisotropy in the compression strength of the prints, which was credited to the grain morphology and crystallography. Veiga et al. [22] produced Ti alloy prints with the assistance of PAW-based DED considering low distortion. The authors used the overlap deposition strategy to print the Ti alloy. The authors observed anisotropy in mechanical strength and also stressed the need for heat treatment to enhance the mechanical strength. The authors also reported the effects of up-milling, down-milling, and slot-milling on the DED prints. Lin et al. [23] used pulsed PAW-based DED to print Ti alloy print structure. The authors observed that the grain formation was severely dependent on the heat input, which was observed to be low due to the use of pulsed PAW heat source. The authors credited the increased mechanical strength of the prints to the fine grains. Li et al. [24] reported the print properties and the effect of heat-treatment on the P91 steels printed by PAW-based DED heat source. The heat-treatment procedures were observed to induce homogenization which improved the hardness and influenced the metallurgy of the prints. The homogenization also led to improve the mechanical strength of the prints. Sinhgal et al. [25] reported the effects of printing conditions in SAW-based DED of AISI 1023 steel. The printing conditions were optimized by AHP-TOPSIS technique. The increase in the welding voltage and the wire feed rate was observed to increase the weld bead width and penetration. However, the weld bead width reduced with the increase in the travel speed due to reduced heat input. The provision of preheating current reduced the weld bead width as it reduced the wire feeding rate and burning rate of the wire. Choi et al. [26] studied the effects of microstructure on the low-temperature fracture toughness in DED prints of low carbon and low alloy steel produced by SAW heat source. The grain morphology across the print layers was credited to the different cooling rates. The impact toughness of the prints was observed to reduce with the decrease in the test temperature. Rojas et al. [27] fabricated TiC reinforced NiCrBSi matrix with the help of PAW-based DED and optimized the DED parameters by Taguchi technique. The authors observed defects like end

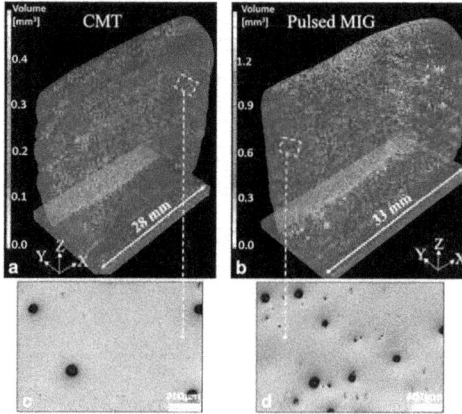

FIGURE 7.5   X-ray CT scans of CMT and pulsed MIG samples showing pore volume [29].

collapsing, uneven height, unstable deposition in the prints. The bead characteristics were dependent by powder flow rate, carrier gas flow rate, and travel speed. Rodriguez et al. [28] compared the properties of the DED prints of stainless steel produced via CMT and TopTIG modes. The authors observed the clear variations in the bead deposition characteristics under varying welding parameters in continuous and pulsed mode in CMT-based DED prints and pulsed mode was observed to provide the best results. Moreover, the results on the provision of the dwell time between the deposition were also highlighted in the study. The bead deposition study was conducted for the TopTIG mode to determine the best deposition condition. On comparing the deposition rates of the CMT and TopTIG, the CMT mode had the higher deposition rate. The mechanical strengths of the prints obtained via CMT and TopTIG modes displayed anisotropy. Derekar et al. [29] reported the effects of deposition modes, heat input, dwell time, interlayer temperature on pore formation, and hydrogen dissolution. The authors used pulsed MIG-based DED and CMT-based DED to print Al-Mg-Mn alloy. The pore volume (Figure 7.5) was observed to be low in the CMT-based DED prints compared to pulsed MIG-based DED prints.

The authors credited the differences in the heat input and deposition style among the different DED heat sources for the variation in the pore volume in the prints. However, the pore volume in the prints was observe to be affected with the changes in the heat input. The similar trend was also observed with the changes in the interlayer temperature and the dwell time. Such a trend was credited to the contribution of these processing conditions to the heat accumulation during the DED process. The increased melt pool temperature in pulsed MIG-based DED prints results in increased in hydrogen content in them compared to CMT-based DED prints which possessed reduced hydrogen content due to reduced melt pool temperature. Stenson et al. [30] compared the effects of DED parameters on the prints properties of mild steel printed via MIG and CMT-based heat source. The microhardness survey was observed to be affected by the printing conditions

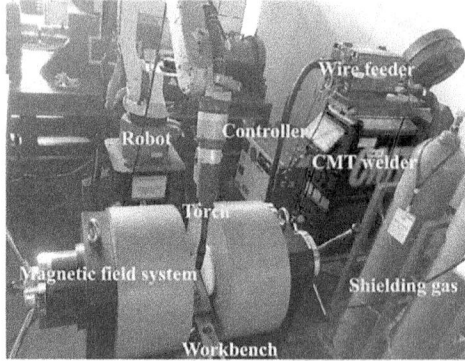

FIGURE 7.6 Magnetic field supported CMT-based DED [33].

irrespective of the changes in the MIG and CMT DED heat source. The microstructural investigations also concurred with the hardness survey indicating that the printing conditions had the significant input, compared to the changes in the heat source. The microstructure of the prints was observed to contain acicular ferrite, and its morphology varied with the increase in the heat input. Tian et al. [31] reported the print properties of dissimilar Ti-Al alloy prints produced using CMT-based DED method. The authors produced two different prints, with one print having Ti alloy as the bottom layer and the second print with the Al alloy as the bottom print. The EDS analysis at the interface showed the formation of the Ti-Al intermetallic compounds. The mechanical strength of the dissimilar prints was observed to be low and showed brittle mode of fracture. The formation of $TiAl_3$ brittle intermetallic compound was credited for the low mechanical strength and cause for the propagation of the failure through micro cracks. Wang et al. [32] reported the print properties of AZ31B alloy printed by CMT-based DED method. The authors varied the heat input during the CMT and observed the variation in the microstructure. The mechanical strength of the prints was observed to show anisotropy. Wang et al. [33] reported that the use of magnetic field in CMT-based DED (Figure 7.6) was observed to produced superior characteristic Inconel 625 prints.

On comparing the microstructures (Figure 7.7) of the prints produced under the influence of magnetic field and without the magnetic field, the samples printed under the influence of magnetic field had a refined dendritic structure compared to coarser dendrites in the prints produced in conventional mode. The formation of the refined microstructure was credited to the magnetic field which caused a stirring effect to break down the coarser dendrites into newer and finer dendrites.

Moreover, the application of the magnetic field was also observed to diffuse the alloying elements in the melt pool and denting the formation of the lave phases in the microstructure. The mechanical strength of the print produced under the influence of magnetic field was also enhanced compared to the mechanical strength of the print produced by conventional mode. Shen et al. [34] reported the print

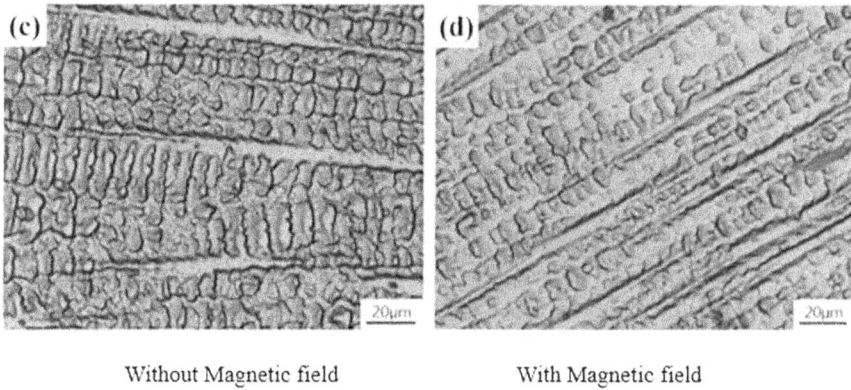

Without Magnetic field	With Magnetic field

FIGURE 7.7   SEM micrographs of CMT DED prints [33].

characteristics of Al–Co–Cr–Fe–Ni high entropy alloy produced by CMT-based DED technique. The authors observed the variations in the print height with the changes in the travel speed, which was credited to the influence on the heat input. The microstructural observations of the print structure and the cast structure showed significant difference. The AM-manufactured high entropy alloy was observed to possess high compression strength compared to the cast structure.

7.2.1 DEPOSITION STRATEGIES IN DED

The deposition patterns have a key role in DED as they contribute to residual stresses or distortions and deposition efficiency or in other words, reducing or avoiding depressions and producing better travel paths to obtain desired layer thickness [35]. Liu et al. [36] classified the deposition patterns into two types, linear pattern and contour pattern and their combination (Figure 7.8).

The linear pattern consisted of raster and zig-zag. The raster pattern is the simplest technique, as it has been reported to implement at shorter computational times; however, it also induced higher residual stresses on the layers due to uniform deposition [35,36]. The zig-zag was most suited for the complex shaped prints were observed with difficult implementation and for printing

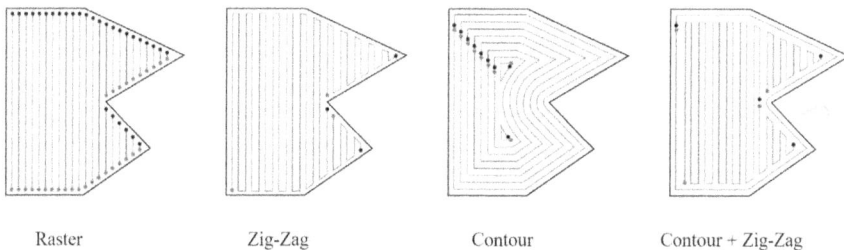

Raster	Zig-Zag	Contour	Contour + Zig-Zag

FIGURE 7.8   Deposition strategies in DED [36].

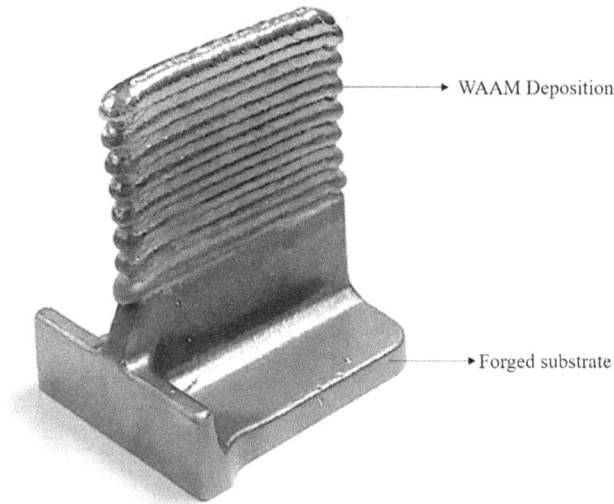

FIGURE 7.9   DED on forged substrate [42].

larger dimension components resulted in increased heat input. The problem with the zig-zag pattern for larger dimensions was overcome with shorter segmentations or sub-geometries [36]. Zhang et al. [37] used a combination of zig-zag and contour offset pattern to produce prints with reduced residual stresses compared to the conventional zig-zag pattern. Ogino et al. [38] reported that the deposition shape was dependent on temperature control and deposition direction. The authors verified the earlier statement numerically and experimentally in MIG-based DED of mild steel. The numerical simulations showed that the application of inter pass cooling resulted in thinning of the deposition shape thinner and thicker without cooling. The authors recommended to reverse the direction of printing in order to prevent the differences in height of the prints, which was credited to the variation in the temperature. Kohler et al. [39] observed larger pores at the regions which experienced higher temperatures and at arc start and stop point. The authors also found the lack of fusion at the first layer deposition region in spiral and S-type patterns, which were mainly found at right-angle turns. Li et al. [40] reported the effects of four types of deposition patterns on the temperature history, distortion during and post-processing in DED of H-13 steel. Nguyen et al. [41] studied deposition strategies based upon the continuous Euler tool path in DED of rib-web structures. The authors used the machine learning technique to avoid the formation of the void defect at the junction and the printing time. Mishurova et al. [42] used the forged substrate to build the DED prints (Figure 7.9). The authors used the alternating deposition patterns and varied the welding current to reduce the heat input generated during the process. The authors observed the clear variations in the grain orientations at the substrate, transition zone, and the print region.

Gu et al. [43] compared the effects of arc deposition modes in DED of Al 2319 alloy over the Al 2219-T87 substrate plate. The authors used pulsed CMT and pulsed MIG heat sources to print the Al alloy. The authors observed the formation of fine equiaxed columnar grains at the interlayer boundary and the walls for pulsed CMT, whereas for pulsed MIG was found with mixed miscellaneous and larger thinner dendritic columnar grains at the walls. The authors credited the variation in the heat transfer rates of melt pool shape generated by two different heat sources for producing varying microstructures. The research team found that the mechanical strength of the prints and the substrate irrespective of the deposition strategy was observed to improve. Aldalur et al. [44] used oscillating and overlapping deposition strategy to print mild steel structures by MIG-based DED. The use of different deposition strategies for the different walls, which resulted in varying cooling effects and microstructures. The wall printed with the overlapping strategy was produced with finer grains, which produced higher mechanical strength compared to oscillating strategy.

Su et al. 2021 [45] used different deposition strategies based upon their orientation to print Al–Si alloys by CMT-based DED. The 90o deposition had a reversing strategy, while the 45o deposition strategy followed deposition along 45o line. The clockwise 90 o deposition strategy followed a clockwise direction path. The authors observed that the three deposition strategies had their own cooling rates, which resulted in varying microstructures and grain sizes. Xu et al. [46] used oscillating, parallel, and weaving strategies to print maraging steel by CMT-based DED. The authors observed the formation of holes on the prints in oscillating modes due to accumulation of high heat and low cooling which resulted in shrinking, formation of holes due to exit of gas. The parallel mode prints were observed with glossy prints and without holes, however, lack of fusion defects was observed. The print with weaving mode was also observed without holes and was observed with waviness. Wang et al. [47] reported that the provision of dwell time between the printing of the interlayer and the change in the direction of the print resulted in variation in the microstructure of the prints in DED of Ti-6Al-4V. The variation in the microstructures was caused due to the differences in the heat accumulation and heat dissipation with respect to the deposition direction and the dwell time. Wu et al. [48] printed bi-metallic prints of Steel–Nickel material with the help of TIG-based DED heat source and Ar gas shielding. The authors used an overlapping strategy with the twin wire feed system for printing the print with one-half side of steel and the other half side of nickel (Figure 7.10).

The microstructural and metallurgical analysis at the interface indicated the melting of Ni and its fusion with the Fe particles. The microhardness survey indicated the loss of hardness at the top layer on the nickel side due to the thermal cycle. The steel side indicated higher hardness, due to fine grains and precipitation of carbides. The interfacial hardness was observed to increase and fluctuate. The mechanical strength of the bi-metallic print structure was observed to be greater than that of the base metals. The authors credited the fusion of Fe and Ni atoms for enhanced strength of the b-metallic prints. Zhang et al. [49] reported the effects of printing conditions on fracture toughness and fatigue propagation in DED of Ti-6Al-4V alloy. The authors found that the prints made with oscillation deposition path had a coarser microstructure, which enhanced the fracture

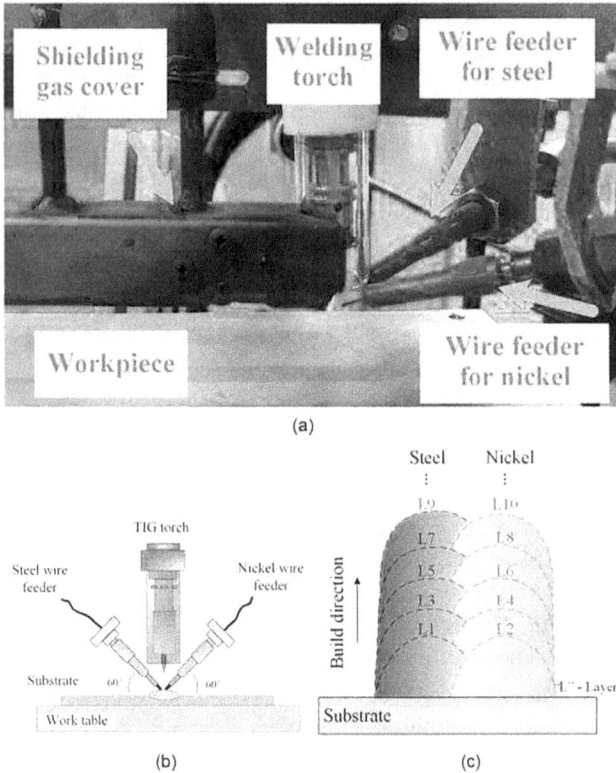

(a)

(b)   (c)

FIGURE 7.10 Double wire feed mechanism of steel–nickel by DED [48].

toughness of the prints. Campatelli et al. [50] used two different printing strategies in printing of Al alloy via DED. The authors employed CMT and CMT mixed mode of strategies and varied the idle time during these techniques to maintain the cooling rates. The bead produced by the CMT mixed mode was observed to be larger in dimension compared to the bead produced by conventional CMT. The microstructural variations produced by the two different deposition modes were observed to have a significant impact on the mechanical strength.

7.2.2 POST-PROCESSING OF DED PRINTS

Tanvir et al. [51] reported the effects of post-print heat-treatment in Inconel 625 prints produced by CMT-based DED. The prints were heat treated to 980°C for varying times of 30 minutes, 60 minutes, and 120 minutes followed by water quenching. The authors observed no significant changes in the grain size and orientation with the heat treatment. The brittle laves in the as deposited samples were observed to disappear with the heat treatment. However, the heat treatment also induced the formation of delta phases earlier than expected due to the localization of niobium content at interdendritic zones. The increase in heat-treatment period also increased the size of the carbide particles and delta phases. Caballero et al. [52] deposited 17-4 PH stainless steel wire with three different deposition strategies on the 304-austenite stainless steel substrate by CMT-based DED. The authors observed the variation in the hardness with respect to the deposition strategies, shielding gas ratios and the post-print heat-treatment procedure on the print properties. The authors stress the requirement of solution heat treatment and direct aging to achieve tensile strength of the print's closer to the required standard.

Qiu et al. [53] reported that the heat-treatment temperature had a significant effect on the metallurgy, which in turn affected the mechanical strength and ductility of haste alloy prints produced by TIG-based DED. The prints were heat treated under two different temperature and time, with one condition being 871°C for 6 hrs, followed by quenching and 1177°C for 30 min and water quenching. The microstructure of the printed sample was found with columnar grains aligned along the direction of the deposition. The sample heat treated to 871°C was observed with similar features, however, the sample heat treated to 1177°C showed a homogeneous microstructure. The metallurgical analysis showed that the heat-treating temperature of 1177°C resulted in dissolving of p phase and the release of molybdenum atoms into the matrix, which enhanced the ductility and the mechanical strength. Tanvir et al. [54] in their continuation of their previous study reported that the two-hour heat-treatment improved the mechanical properties of the Inconel 625 prints produced by CMT-based DED. The lave phases in the printed samples hampered the dislocation movement, to dent the mechanical strength. The rise in the temperature was observed to dissolve these phases to improve the mechanical strength. However, the authors also credited the Ti-based carbides to reshape the grain boundaries to enhance the mechanical strength.

Alonso et al. [55] reported the effects of milling parameters on the Inconel 718 alloy printed by plasma-based DED. The prints showed higher mechanical strength compared to the cast product. The microstructural composition of the prints consisted of fine dendrites, columnar grains distribution in the vertical direction, and also presence of lave and carbide particles (Figure 7.11). The authors observed typical milling defects like burrs and adhesion of Inconel on the milling tool (Figure 7.12). The surface roughness was observed to improve with the increase in the cutting speed.

The increased plastic deformation caused due to the compression action during the tool and the print interaction led to the formation of the burrs, whereas the adhesion of the Inconel alloy onto the tool was credited to the print's high ductility. Schneider [56] reported that the heat-treated Inconel 718 samples 3d printed via laser powder bed fusion, blown powder diffusion and DED were observed with similar mechanical strength even with the variation in the microstructure. The samples printed with the laser powder bed fusion were observed with fine grains, whereas the WAAM samples had coarser grains. Seow et al. [57] also reported the disappearance of the lave particle in the heat-treated Inconel 718 alloy to produce a homogenous microstructure. The authors used a plasma-based DED to print the Inconel alloy. Vazquez et al. [58] used two different deposition strategies and heat-treatment procedures in printing of

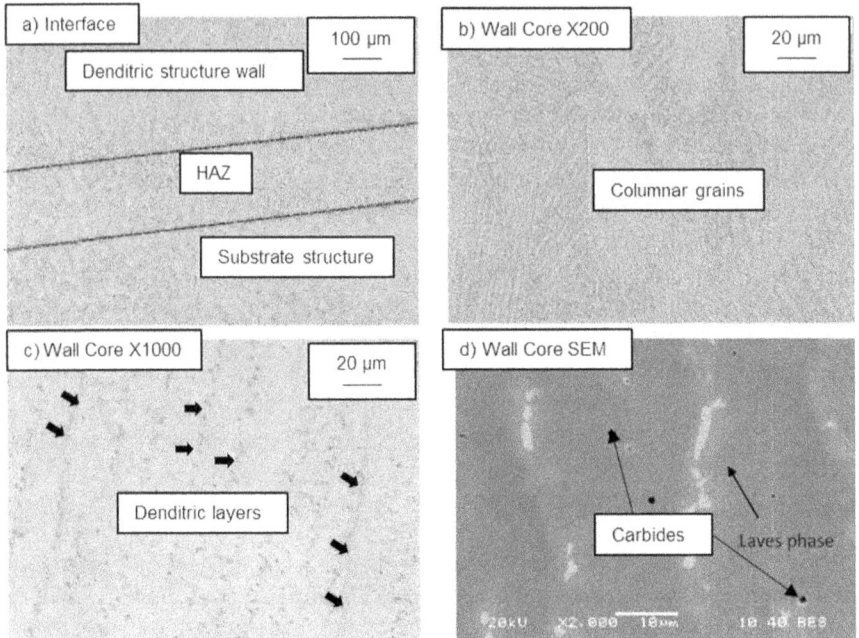

FIGURE 7.11 Optical microstructure of Inconel 718 alloy printed by PAW-based DED [55].

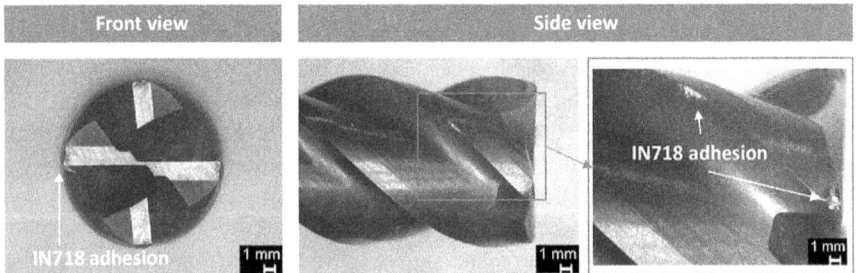

FIGURE 7.12 Adhesion of Inconel on milling tool [55].

Ti-6Al-4V alloy by CMT-based DED. The authors used a 3-bead technique (Figure 7.13) to print thicker walls and a single-bead technique for thinner walls.

The authors observed that the mechanical strength of the prints post-heat-treatment procedures reduced with the increase in the elongation and also the anisotropy was observed, irrespective of the deposition strategy. The micro-structural observations revealed that the increase in the heat-treatment tempera-tures resulted in coarsening of the α phases. The print's mechanical properties were high compared to cast product, irrespective of with or without heat treatment. Bermingham et al. [59] reported the effects of stress-reliving treatment, heat-treatment procedures, and hot isostatic pressing processes on the mechanical

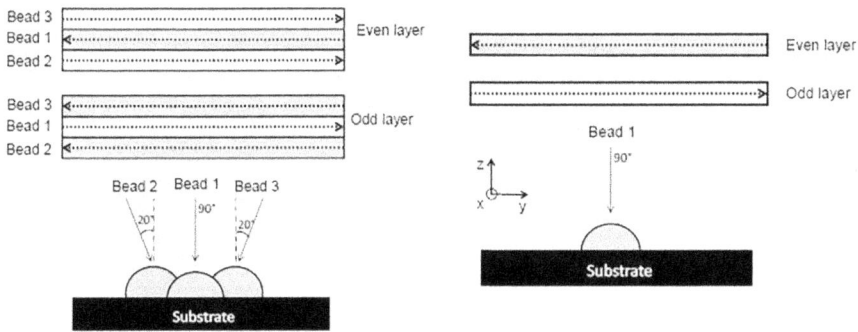

FIGURE 7.13   Three-point deposition and single-point deposition [58].

properties of Ti-6Al-4V alloy produced by DED. The authors observed variations in the microstructure with the changes in the heat-treatment procedures, with no significant changes in the microstructure with respect to the print and the stress-relived sample; however, the mechanical strength of the print was lower than that of the stress-relived sample due to the presence of residual stress. The vacuum annealing and hot isostatic pressing process did not significantly affect the mechanical strength of the prints compared to solution treatment process. Dharmendra et al. [60] studied the effects of different heat-treatment procedures on nickel aluminum bronze, NAB prints fabricated via MIG-based DED on a stainless steel substrate. The coarsening of the precipitates was observed with the increase in the heat-treatment temperature. The mechanical strength of the prints was observed to improve at the cost of ductility with the increase in the heat-treatment temperatures. Wang et al. 2021 [61] employed hybrid post-processing technique, a combination of cavitation peening and electrochemical polishing to process 316 L stainless steel produced via SLM. The hybrid post-processing technique increased the surface hardness. Qi et al. [62] produced Al 2024 prints on Al 2024 substrate through TIG-based DED by using double wire feed mechanism of ER2319 alloy and ER5087 alloy and studied the influence of heat-treatment procedure on the print's properties. The microstructure evaluation of the prints was observed with dendritic features, which were observed to disappear with the heat-treatment procedure. The SEM micrographs showed the presence of pores, which was formed due to the escaping of the hydrogen gas. The heat-treatment procedures were observed to enhance the print strength. The authors observed isotropy in mechanical strength for the printed samples; however, the heat-treated samples showed anisotropy. The reason for such a behavior was credited to the pores in the prints. Elmer et al. [63] reported anisotropy in the mechanical strength of the stainless steel prints produced by MIG-based DED. The authors employed rolling and heat-treatment procedures to eliminate the anisotropy and texture in the prints. The recrystallization phenomenon induced by the rolling was observed to improve the mechanical strength of the prints compared to mechanical strength of as printed sample and cast product. Chi et al. [64] reported the influence of combined processing procedure consisting

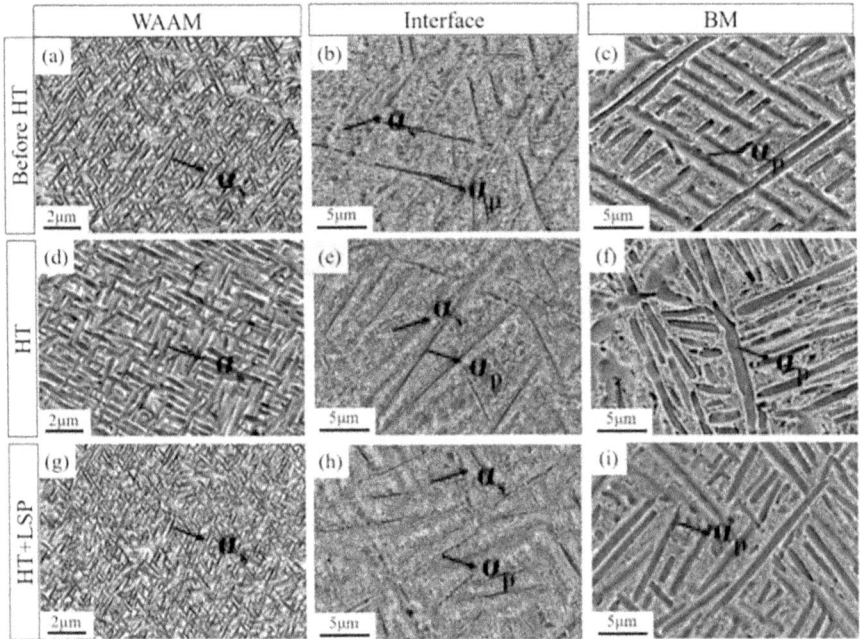

FIGURE 7.14   Influence of combined post-processing treatments on the microstructure [64].

of heat treatment and laser shock peening on Ti alloy print fabricated by TIG-based DED. The combination of the post-processing techniques was observed to induce the grain refinement, across the print, the substrate, and the interface (Figure 7.14).

The residual stresses across the different regions were observed to be influenced with the inducement of combined post-processing techniques. The microhardness distribution (Figure 7.15) was observed to decrease with the heat treatment; however, it increased with the laser shock peening treatment, due to the inducement of plastic deformation and grain refinement produced by laser treatment.

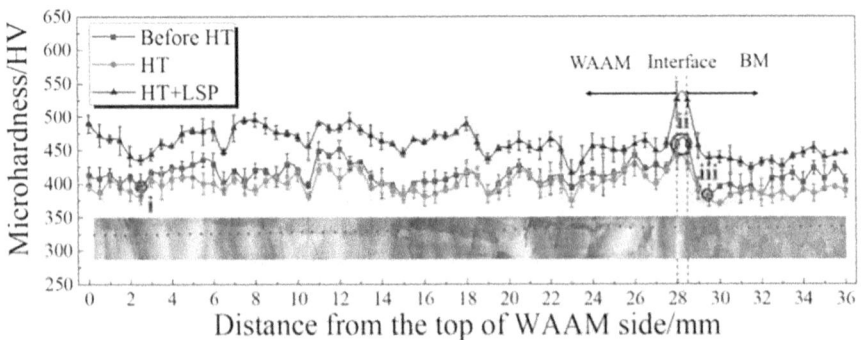

FIGURE 7.15   Microhardness distribution on DED prints w.r.t post-processing techniques [64].

The tensile strengths of the prints showed that the combined post-processing techniques were observed to improve due to the refined grains, plastic deformation, and positive effects of the residual stress. Sun et al. [65] reported that the use of laser shock treatment resulted in grain refinement and enhanced mechanical properties of the Al 2319 prints produced via TIG-based DED. The microhardness was observed to improve post the laser shock treatment, as the laser modified the surface to produce low distortions and mechanical twins. Moreover, the laser treatments produced beneficial residual stresses and enhanced the mechanical strength. Li et al. [66] reported that the formability of the Al-Si-Mg alloy print was observed to get enhanced with the use of heat treatment. The mechanical properties of the prints showed low anisotropy and were observed to improve with the increase in the heat input during CMT-based DED process. Zhang et al. [67] reported the print properties of high nitrogen Cr-Mn steel alloys produced via CMT-based DED. The authors used a custom wire for the CMT process. The authors recommended the use of low periods of high temperature heat-treatment to enhance the mechanical properties of the prints. Such a phenomenon in the mechanical strength was credited to the changes in the metallurgy of the prints. Kindermann et al. [68] reported the effects of processing conditions and heat-treatment procedure on the Inconel 718 prints obtained via CMT-based DED. The authors stressed that the wire feed rate and the travel speed were key to control the heat input during CMT and also avoid the formation of the humping defect, which was identified by the breaking up of the weld bead into droplets at higher heat input. The metallurgy and the microstructure of the prints was significantly improved by precipitation of the finer Nb rich precipitates and its increased dispersion, reduced dendrite arm spacing, and reduced lave phases, respectively, at higher travel speeds. Such a phenomenon was explained by the increasing cooling and faster solidification at higher travel speeds. The heat-treatment procedures were observed to have a positive influence on the microstructure and the metallurgy of the prints.

7.2.3   THE NEED FOR FLUX-CORED AND METAL-CORED WIRE IN DED

From the previous sections, it is observed that the printing conditions, deposition strategies, and post-processing techniques were key to obtain the desired microstructure and mechanical properties. The use of flux-cored wires was reported to have a better adjustability and flexibility to have on the wire composition, which enabled the control over the microstructure and mechanical properties. The DED with the flux-cored wires works on the principle of flux-cored arc welding (FCAW) [69]. Zhang et al. [69] produced duplex steel prints through FCAW-based DED. The authors observed the increased percentage of austenite in the prints and the mechanical properties of the prints showed anisotropy. The authors developed novel flux-cored wires which were analyzed to show increased chromium content and reduced nickel content, and more vitally, the austenite percentage in the print was reduced. One of the advantages of DED process is that the components can be made from the

Metal-cored wire   Solid wire

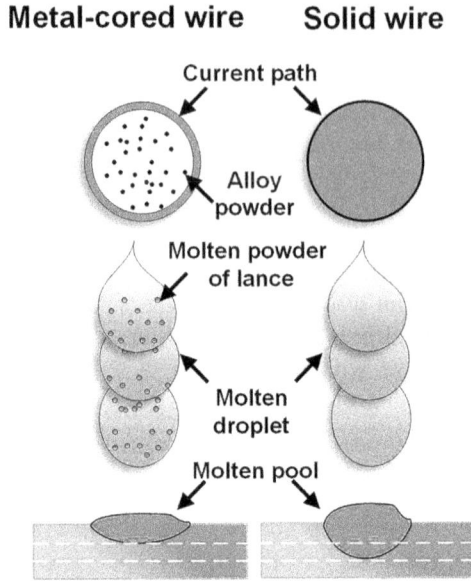

FIGURE 7.16 Comparison of metal transfer mode in metal core wire and solid wire [70].

tailormade wires. Metal-cored wires have made the possibility of producing tailormade wires, as these wires are hollow sheath filled with different powders. Panickiewicz 2021 [70] reported the mechanical and microstructural character-istics of maraging steel prints produced by flux-cored wire-based DED. The authors cited the reduced solidification rate provided by the flux and the shielding gas during the printing stage as the beneficial characteristic in using flux-cored wire-based DED. The chemical composition analysis showed the increased presence of carbon, which led to increased brittleness and the formation of titanium and molybdenum carbides, which also reduced the formation of the inter metallics post-aging. This phenomenon was also well supported with a reduced percentage of nickel. The hardness across the print was credited to the thermal cycles and precipitation. On comparing with the flux-cored wires, the metal-cored wires possess less flux agents to reduce the formation of the slag. Moreover, the fillings can be optimized to provide custom solutions and reduce defects. The other advantage of the metal-cored wire in DED process (Figure 7.16) is that the heat input during the process is significantly less compared to conventional DED technique; this is due to the generated arc melting the powder, which is deposited onto the substrate.

During this stage, the heat transfer does not happen through the powder, which results in reduced energy to melt the metal-cored wire compared to the solid wire [71]. Lin et al. [71] used metal-cored wires to print mild steel with the help of MIG-based DED. The hardness variations across the layers were credited to the precipitate size, and grain size. The precipitate distribution and the grain size across the layers were credited to the different thermal

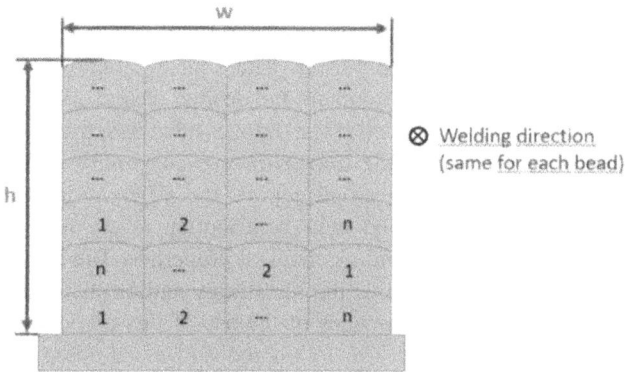

FIGURE 7.17   Alternating Deposition Strategy [73].

profiles. The print's mechanical strength was observed to fall from the top to bottom layer and also showed anisotropy. Lin [72] also printed stellite 6, a cobalt and chromium-based alloy by metal-cored DED. The authors observed that the use of high heat input resulted in grain growth and dilution and reduced the hardness. The authors developed a process window to obtain defect-free stellite prints. From the literatures, it is observed that the researchers were able to produce superior prints in terms of microstructure, mechanical properties by the use of flux-cored and metal-cored wires. Since the information available on such studies are relatively new makes the possibility of exploring newer concepts in these techniques is stressed. The parameters in flux-cored wire and metal-cored wire DED can be optimized, the effects of different deposition patterns can also be explored. Fuchs et al. [73] reported the use of uncoated metal-cored wire G4Si1 material to print its layer with the assistance of electron beam heating. The authors determined to find the feasible wire feed rate, electron beam current, and amplitude of deflected electron beam. The authors used an alternating deposition strategy (Figure 7.17) to print the layers and varied the heat input during the printing phase to maintain the thermal cycle.

The research team found that the use of higher values of electron beam current resulted in increased heat input and increased dilution, whereas, the use of high wire feed rates resulted in low dilution. The microstructural observations indicated the formation of the pores across the weld bead and the metallurgical variations, explained by the variations in the thermal history. Fard et al. [74] printed stellite 6 alloy with the assistance of TIG-based DED on a stainless steel substrate. The authors varied the welding current during the printing of each layer, with the goal to reduce the temperature during the printing. The authors used co-ordinated heating and wire feeding strategy to reduce the distortion during the printing and moreover, this system gave a control over the bead deposition size. The microhardness analysis showed that the sample printed with controlled heating and feeding strategy had a higher hardness, which could be due to the finer grains.

7.3 SUMMARY

The literature survey identifies the various heat sources employed in DED. The control of the heat input was key to obtain the desired characteristics of the prints and avoid the formation of defects. The article identifies the various process parameters that were to control the microstructure and improve the mechanical properties of the DED prints. The article reports the different deposition patterns were used in DED and highlighted their contribution to the heat input, microstructural evolution, metallurgy, and mechanical properties. The importance of the heat-treatment procedures to improve the metallurgy and mechanical properties of the prints were stressed. The advantages of the use of flux-cored and metal-cored wires in DED were highlighted. The flux-cored and metal-cored wires in DED present a chance to the researcher studies to explore the further into this space is highlighted.

REFERENCES

[1] Ralph, B. (1925). U.S. Patent No. 1,533,300. Washington, DC: U.S. Patent and Trademark Office.

[2] https://www.metal-am.com/wire-arc-additive-manufacturing-builds-titanium-pressure-vessel-for-space-exploration/

[3] Jafari, D., Vaneker, T. H., & Gibson, I. (2021). Wire and arc additive manufacturing: Opportunities and challenges to control the quality and accuracy of manufactured parts. Materials & Design, 202, 109471.

[4] Jin, W., Zhang, C., Jin, S., Tian, Y., Wellmann, D., & Liu, W. (2020). Wire arc additive manufacturing of stainless steels: A review. Applied sciences, 10(5), 1563.

[5] Su, C., Chen, X., Gao, C., & Wang, Y. (2019). Effect of heat input on microstructure and mechanical properties of Al-Mg alloys fabricated by WAAM. Applied Surface Science, 486, 431–440.

[6] Kumar, A., & Maji, K. (2021). Selection of process parameters for near-net shape deposition in wire arc additive manufacturing by genetic algorithm. Journal of Materials Engineering and Performance, 29, 3334–3352.

[7] Ayed, A., Valencia, A., Bras, G., Bernard, H., Michaud, P., Balcaen, Y., & Alexis, J. (2020). Effects of WAAM process parameters on metallurgical and mechanical properties of Ti-6Al-4V deposits. In Advances in Materials, Mechanics and Manufacturing (pp. 26–35). Springer, Cham.

[8] Yildiz, A. S., Davut, K., Koc, B., & Yilmaz, O. (2020). Wire arc additive manufacturing of high-strength low alloy steels: Study of process parameters and their influence on the bead geometry and mechanical characteristics. The International Journal of Advanced Manufacturing Technology, 108(11), 3391–3404.

[9] Duarte, V. R., Rodrigues, T. A., Schell, N., Miranda, R. M., Oliveira, J. P., & Santos, T. G. (2020). Hot forging wire and arc additive manufacturing (HF-WAAM). Additive manufacturing, 35, 101193.

[10] Sun, L., Jiang, F., Huang, R., Yuan, D., Su, Y., Guo, C., & Wang, J. (2020). Investigation on the process window with liner energy density for single-layer parts fabricated by wire and arc additive manufacturing. Journal of Manufacturing Processes, 56, 898–907.

[11] Wang, L., Xue, J., & Wang, Q. (2019). Correlation between arc mode, microstructure, and mechanical properties during wire arc additive manufacturing of 316L stainless steel. Materials Science and Engineering: A, 751, 183–190.

[12] Wu, T., Liu, J., Wang, K., Wang, L., & Zhang, X. (2021). Microstructure and mechanical properties of wire-powder hybrid additive manufacturing for high nitrogen steel. Journal of Manufacturing Processes, 70, 248–258.

[13] Han, S., Zhang, Z., Liu, Z., Zhang, H., & Xue, D. (2020). Investigation of the microstructure and mechanical performance of bimetal components fabricated using CMT-based wire arc additive manufacturing. Materials Research Express, 7(11), 116525.

[14] Fu, R., Tang, S., Lu, J., Cui, Y., Li, Z., Zhang, H., ... & Liu, C. (2021). Hot-wire arc additive manufacturing of aluminum alloy with reduced porosity and high deposition rate. Materials & Design, 199, 109370.

[15] Prado-Cerqueira, J. L., Camacho, A. M., Diéguez, J. L., Rodríguez-Prieto, Á., Aragón, A. M., Lorenzo-Martín, C., & Yanguas-Gil, Á. (2018). Analysis of favorable process conditions for the manufacturing of thin-wall pieces of mild steel obtained by wire and arc additive manufacturing (WAAM). Materials, 11(8), 1449.

[16] Rodrigues, T. A., Duarte, V. R., Miranda, R. M., Santos, T. G., & Oliveira, J. P. (2021). Ultracold-Wire and arc additive manufacturing (UC-WAAM). Journal of Materials Processing Technology, 296, 117196.

[17] Hauser, T., Reisch, R. T., Seebauer, S., Parasar, A., Kamps, T., Casati, R., ... & Kaplan, A. F. (2021). Multi-Material Wire Arc Additive Manufacturing of low and high alloyed aluminium alloys with in-situ material analysis. Journal of Manufacturing Processes, 69, 378–390.

[18] Yangfan, W., Xizhang, C., & Chuanchu, S. (2019). Microstructure and mechanical properties of Inconel 625 fabricated by wire-arc additive manufacturing. Surface and Coatings Technology, 374, 116–123.

[19] Li, Y., Wu, S., Li, H., & Cheng, F. (2021). Dramatic improvement of impact toughness for the fabricating of low-carbon steel components via submerged arc additive manufacturing. Materials Letters, 283, 128780.

[20] Li, Y., Wu, S., Li, H., Dong, Y., & Cheng, F. (2021). Submerged arc additive manufacturing (SAAM) of low-carbon steel: Effect of in-situ intrinsic heat treatment (IHT) on microstructure and mechanical properties. Additive Manufacturing, 46, 102124.

[21] Wang, L., Zhang, Y., Hua, X., Shen, C., Li, F., Huang, Y., & Ding, Y. (2021). Fabrication of γ-TiAl intermetallic alloy using the twin-wire plasma arc additive manufacturing process: Microstructure evolution and mechanical properties. Materials Science and Engineering: A, 812, 141056.

[22] Veiga, F., Gil Del Val, A., Suárez, A., & Alonso, U. (2020). Analysis of the machining process of titanium Ti6Al-4V parts manufactured by wire arc additive manufacturing (WAAM). Materials, 13(3), 766.

[23] Lin, J. J., Lv, Y. H., Liu, Y. X., Xu, B. S., Sun, Z., Li, Z. G., & Wu, Y. X. (2016). Microstructural evolution and mechanical properties of Ti-6Al-4V wall deposited by pulsed plasma arc additive manufacturing. Materials & Design, 102, 30–40.

[24] Li, K., Klecka, M. A., Chen, S., & Xiong, W. (2021). Wire-arc additive manufacturing and post-heat treatment optimization on microstructure and mechanical properties of Grade 91 steel. Additive Manufacturing, 37, 101734.

[25] Singhal, T. S., Jain, J. K., Kumar, M., & Saxena, K. K. (2021). Effect of filler wire preheating and nozzle cooling with advanced submerged arc welding process on bead geometry and microstructure. Advances in Materials and Processing Technologies, 1–15.

[26] Choi, B. C., Kim, B., Kim, B. J., Choi, Y. W., Lee, S. J., Jeon, J. B., ... & Kim, H. C. (2021). Effect of microstructure on low-temperature fracture toughness of a submerged-arc-welded low-carbon and low-alloy steel plate. Metals, 11(11), 1839.

[27] Rojas, J. G. M., Ghasri-Khouzani, M., Wolfe, T., Fleck, B., Henein, H., & Qureshi, A. J. (2021). Preliminary geometrical and microstructural characterization of WC-reinforced NiCrBSi matrix composites fabricated by plasma transferred arc additive manufacturing through Taguchi-based experimentation. The International Journal of Advanced Manufacturing Technology, 113(5), 1451–1468.

[28] Rodriguez, N., Vázquez, L., Huarte, I., Arruti, E., Tabernero, I., & Alvarez, P. (2018). Wire and arc additive manufacturing: a comparison between CMT and TopTIG processes applied to stainless steel. Welding in the World, 62(5), 1083–1096.

[29] Derekar, K. S., Addison, A., Joshi, S. S., Zhang, X., Lawrence, J., Xu, L., … & Griffiths, D. (2020). Effect of pulsed metal inert gas (pulsed-MIG) and cold metal transfer (CMT) techniques on hydrogen dissolution in wire arc additive manufacturing (WAAM) of aluminium. The International Journal of Advanced Manufacturing Technology, 107(1), 311–331.

[30] Stinson, H., Ward, R., Quinn, J., & McGarrigle, C. (2021). Comparison of properties and bead geometry in MIG and CMT single layer samples for WAAM applications. Metals, 11(10), 1530.

[31] Tian, Y., Shen, J., Hu, S., Gou, J., & Kannatey-Asibu, E. (2019). Wire and arc additive manufactured Ti–6Al–4V/Al–6.25 Cu dissimilar alloys by CMT-welding: Effect of deposition order on reaction layer. Science and Technology of Welding and Joining, 11, 1–18.

[32] Wang, P., Zhang, H., Zhu, H., Li, Q., & Feng, M. (2021). Wire-arc additive manufacturing of AZ31 magnesium alloy fabricated by cold metal transfer heat source: Processing, microstructure, and mechanical behavior. Journal of Materials Processing Technology, 288, 116895.

[33] Wang, Y., Chen, X., Shen, Q., Su, C., Zhang, Y., Jayalakshmi, S., & Singh, R. A. (2021). Effect of magnetic Field on the microstructure and mechanical properties of inconel 625 superalloy fabricated by wire arc additive manufacturing. Journal of Manufacturing Processes, 64, 10–19.

[34] Shen, Q., Kong, X., & Chen, X. (2021). Fabrication of bulk Al-Co-Cr-Fe-Ni high-entropy alloy using combined cable wire arc additive manufacturing (CCW-AAM): Microstructure and mechanical properties. Journal of Materials Science & Technology, 74, 136–142.

[35] Venturini, G., Montevecchi, F., Scippa, A., & Campatelli, G. (2016). Optimization of WAAM deposition patterns for T-crossing features. Procedia Cirp, 55, 95–100.

[36] Liu, J., Xu, Y., Ge, Y., Hou, Z., & Chen, S. (2020). Wire and arc additive manufacturing of metal components: A review of recent research developments. The International Journal of Advanced Manufacturing Technology, 55, 1–50.

[37] Zhang, C., Shen, C., Hua, X., Li, F., Zhang, Y., & Zhu, Y. (2020). Influence of wire-arc additive manufacturing path planning strategy on the residual stress status in one single buildup layer. The International Journal of Advanced Manufacturing Technology, 111(3), 797–806.

[38] Ogino, Y., Asai, S., & Hirata, Y. (2018). Numerical simulation of WAAM process by a GMAW weld pool model. Welding in the World, 62(2), 393–401.

[39] Köhler, M., Sun, L., Hensel, J., Pallaspuro, S., Kömi, J., Dilger, K., & Zhang, Z. (2021). Comparative study of deposition patterns for DED-Arc additive manufacturing of Al-4046. Materials & Design, 210, 110122.

[40] Li, X., Lin, J., Xia, Z., Zhang, Y., & Fu, H. (2021). Influence of deposition patterns on distortion of H13 steel by wire-arc additive manufacturing. Metals, 11(3), 485.

[41] Nguyen, L., Buhl, J., & Bambach, M. (2020). Continuous Eulerian tool path strategies for wire-arc additive manufacturing of rib-web structures with machine-learning-based adaptive void filling. Additive Manufacturing, 35, 101265.

[42] Mishurova, T., Sydow, B., Thiede, T., Sizova, I., Ulbricht, A., Bambach, M., & Bruno, G. (2020). Residual stress and microstructure of a Ti-6Al-4V wire arc additive manufacturing hybrid demonstrator. Metals, 10(6), 701.

[43] Gu, J., Yang, S., Gao, M., Bai, J., & Liu, K. (2020). Influence of deposition strategy of structural interface on microstructures and mechanical properties of additively manufactured Al alloy. Additive Manufacturing, 34, 101370.

[44] Aldalur, E., Veiga, F., Suárez, A., Bilbao, J., & Lamikiz, A. (2020). High deposition wire arc additive manufacturing of mild steel: Strategies and heat input effect on microstructure and mechanical properties. Journal of Manufacturing Processes, 58, 615–626.

[45] Su, C., Chen, X., Konovalov, S., Singh, R. A., Jayalakshmi, S., & Huang, L. (2021). Effect of deposition strategies on the microstructure and tensile properties of wire arc additive manufactured Al-5Si alloys. Journal of Materials Engineering and Performance, 30(3), 2136–2146.

[46] Xu, X., Ding, J., Ganguly, S., Diao, C., & Williams, S. (2019). Preliminary investigation of building strategies of maraging steel bulk material using wire+ arc additive manufacture. Journal of Materials Engineering and Performance, 28(2), 594–600.

[47] Wang, J., Lin, X., Li, J., Hu, Y., Zhou, Y., Wang, C., ... & Huang, W. (2019). Effects of deposition strategies on macro/microstructure and mechanical properties of wire and arc additive manufactured Ti6Al4V. Materials Science and Engineering: A, 754, 735–749.

[48] Wu, B., Qiu, Z., Pan, Z., Carpenter, K., Wang, T., Ding, D., ... & Li, H. (2020). Enhanced interface strength in steel-nickel bimetallic component fabricated using wire arc additive manufacturing with interweaving deposition strategy. Journal of Materials Science & Technology, 52, 226–234.

[49] Zhang, X., Martina, F., Ding, J., Wang, X., & Williams, S. W. (2017). Fracture toughness and fatigue crack growth rate properties in wire+ arc additive manufactured Ti-6Al-4V. Fatigue & Fracture of Engineering Materials & Structures, 40(5), 790–803.

[50] Campatelli, G., Campanella, D., Barcellona, A., Fratini, L., Grossi, N., & Ingarao, G. (2020). Microstructural, mechanical and energy demand characterization of alternative WAAM techniques for Al-alloy parts production. CIRP Journal of Manufacturing Science and Technology, 31, 492–499.

[51] Tanvir, A. N. M., Ahsan, M. R., Ji, C., Hawkins, W., Bates, B., & Kim, D. B. (2019). Heat treatment effects on Inconel 625 components fabricated by wire+ arc additive manufacturing (WAAM)—part 1: Microstructural characterization. The International Journal of Advanced Manufacturing Technology, 103(9), 3785–3798.

[52] Caballero, A., Ding, J., Ganguly, S., & Williams, S. (2019). Wire+ Arc Additive Manufacture of 17-4 PH stainless steel: Effect of different processing conditions on microstructure, hardness, and tensile strength. Journal of Materials Processing Technology, 268, 54–62.

[53] Qiu, Z., Wu, B., Wang, Z., Wexler, D., Carpenter, K., Zhu, H., ... & Li, H. (2021). Effects of post heat treatment on the microstructure and mechanical properties of wire arc additively manufactured Hastelloy C276 alloy. Materials Characterization, 177, 111158.

[54] Tanvir, A. N. M., Ahsan, M. R., Seo, G., Kim, J. D., Ji, C., Bates, B., ... & Kim, D. B. (2020). Heat treatment effects on Inconel 625 components fabricated by wire+ arc additively manufacturing (WAAM)—part 2: Mechanical properties. The International Journal of Advanced Manufacturing Technology, 110(7), 1709–1721.

[55] Alonso, U., Veiga, F., Suárez, A., & Del Val, A. G. (2021). Characterization of Inconel 718® superalloy fabricated by wire Arc Additive Manufacturing: Effect on mechanical properties and machinability. Journal of Materials Research and Technology, 14, 2665–2676.

[56] Schneider, J. (2020). Comparison of microstructural response to heat treatment of Inconel 718 prepared by three different metal additive manufacturing processes. JOM, 72(3), 1085–1091.

[57] Seow, C. E., Coules, H. E., Wu, G., Khan, R. H., Xu, X., & Williams, S. (2019). Wire+ Arc additively manufactured Inconel 718: Effect of post-deposition heat treatments on microstructure and tensile properties. Materials & Design, 183, 108157.

[58] Vazquez, L., Rodriguez, M. N., Rodriguez, I., & Alvarez, P. (2021). Influence of post-deposition heat treatments on the microstructure and tensile properties of Ti-6Al-4V parts manufactured by CMT-WAAM. Metals, 11(8), 1161.

[59] Bermingham, M. J., Nicastro, L., Kent, D., Chen, Y., & Dargusch, M. S. (2018). Optimising the mechanical properties of Ti-6Al-4V components produced by wire+ arc additive manufacturing with post-process heat treatments. Journal of Alloys and Compounds, 753, 247–255.

[60] Dharmendra, C., Amirkhiz, B. S., Lloyd, A., Ram, G. J., & Mohammadi, M. (2020). Wire-arc additive manufactured nickel aluminum bronze with enhanced mechanical properties using heat treatments cycles. Additive Manufacturing, 36, 101510.

[61] Wang, B., Castellana, J., & Melkote, S. N. (2021). A hybrid post-processing method for improving the surface quality of additively manufactured metal parts. CIRP Annals.

[62] Qi, Z., Qi, B., Cong, B., Sun, H., Zhao, G., & Ding, J. (2019). Microstructure and mechanical properties of wire+ arc additively manufactured 2024 aluminum alloy components: As-deposited and post heat-treated. Journal of Manufacturing Processes, 40, 27–36.

[63] Elmer, J. W., Fisher, K., Gibbs, G., Sengthay, J., & Urabe, D. (2021). Post-build thermomechanical processing of wire arc additively manufactured stainless steel for improved mechanical properties and reduction of crystallographic texture. Additive Manufacturing, 102573.

[64] Chi, J., Cai, Z., Wan, Z., Zhang, H., Chen, Z., Li, L., ... & Guo, W. (2020). Effects of heat treatment combined with laser shock peening on wire and arc additive manufactured Ti17 titanium alloy: Microstructures, residual stress and mechanical properties. Surface and Coatings Technology, 396, 125908.

[65] Sun, R., Li, L., Zhu, Y., Guo, W., Peng, P., Cong, B., ... & Liu, L. (2018). Microstructure, residual stress and tensile properties control of wire-arc additive manufactured 2319 aluminum alloy with laser shock peening. Journal of Alloys and Compounds, 747, 255–265.

[66] Li, C., Gu, H., Wang, W., Wang, S., Ren, L., Wang, Z., ... & Zhai, Y. (2020). Effect of heat input on formability, microstructure, and properties of Al–7Si–0.6 Mg alloys deposited by CMT-WAAM process. Applied Sciences, 10(1), 70.

[67] Zhang, X., Zhou, Q., Wang, K., Peng, Y., Ding, J., Kong, J., & Williams, S. (2019). Study on microstructure and tensile properties of high nitrogen Cr-Mn steel processed by CMT wire and arc additive manufacturing. Materials & Design, 166, 107611.

[68] Kindermann, R. M., Roy, M. J., Morana, R., & Prangnell, P. B. (2020). Process response of Inconel 718 to wire+ arc additive manufacturing with cold metal transfer. Materials & Design, 195, 109031.

[69] Zhang, Y., Cheng, F., & Wu, S. (2021). The microstructure and mechanical properties of duplex stainless steel components fabricated via flux-cored wire arc-additive manufacturing. Journal of Manufacturing Processes, 69, 204–214.

[70] Pańcikiewicz, K. (2021). Preliminary process and microstructure examination of flux-cored wire arc additive manufactured 18Ni-12Co-4Mo-Ti maraging steel. Materials, 14(21), 6725.

[71] Lin, Z., Goulas, C., Ya, W., & Hermans, M. J. (2019). Microstructure and mechanical properties of medium carbon steel deposits obtained via wire and arc additive manufacturing using metal-cored wire. Metals, 9(6), 673.

[72] Lin, Z. (2019). Wire and arc additive manufacturing of thin structures using metal-cored wire consumables: Microstructure, mechanical properties, and experiment-based thermal model.

[73] Fuchs, J., Schneider, C., & Enzinger, N. (2018). Wire-based additive manufacturing using an electron beam as heat source. Welding in the World, 62(2), 267–275.

[74] Shahryari Fard, S., Frei, H., Huang, X., & Yao, M. (2021). Coordinated heat and feed printing strategy for wire and arc additive manufacturing of metal-cored wires. *Journal of Materials Engineering and Performance*, 30, 8841–8888.

8 Multimaterial 3D Printing of Metamaterials
Design, Properties, Applications, and Advancement

Debashish Gogoi, Tanyu Donarld Kongnyui, Manjesh Kumar, and Jasvinder Singh

8.1 INTRODUCTION

The functional as well as compositional gradation of the materials has been demonstrated by the nature of several optimized structural designs for specific kinds of applications. Several multimaterials (MMs) have been identified for 3D printing (3DPg), which are inspired by the nature developed with different kinds of structures as shown in Figure 8.1.

The functional gradation of the material has demonstrated excellent performance with the combination of the different kinds of materials as in Figure 8.1. However, the natural components have more complexity than the human-designed components. Therefore, they are not meeting the requirements of some specific applications due to their limitation of design and methods of processing. As the design freedom in 3DPg provides unlimited freedom; these techniques are very promising for the manufacture of functionally graded material (FGM) in comparison to the conventional methods of fabrication. Also, the latest developments in 3DPg have led to the fabrication of MM for the fabrication of a single component. Due to these advancements, researchers can manufacture nature-inspired components with a variety of design complexity. Yasuga et al. fabricated a scaffold using multiple materials using their own developed experimental setup of FLUID3EAMS. The developed process was validated by fabrication the of complex structures such as soft tissues. Mirzaali et al. [1] also developed bio-inspired components. The fabricated components were further investigated for the determination of fracture as well as deformation behavior composite materials which had applicability in advanced industries like soft robots, aerospace, and automobiles. Numerous studies

DOI: 10.1201/9781003484325-8

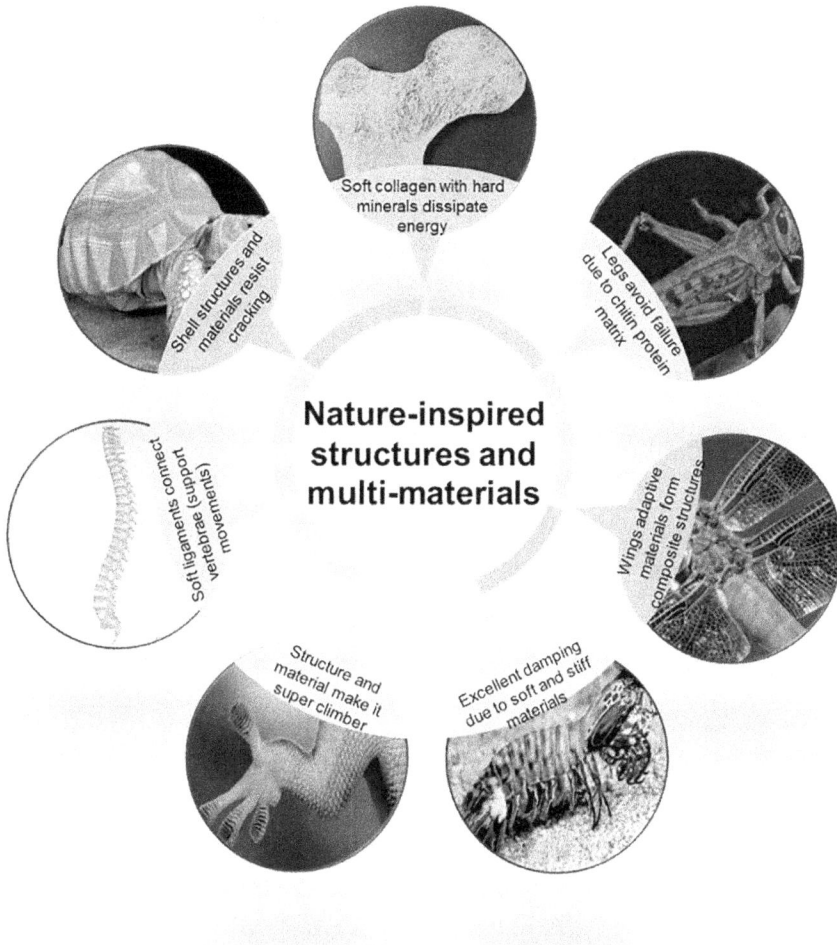

FIGURE 8.1 Demonstration of nature-inspired structures and MM.

have been reported for the designing of multiple nature-inspired devices using 3DPg for several applications in automotive parts with lightweight [2], the sensor in soft tissues [3], energy harvest projects [4], 3D-printed dentures, etc. [5].

Following the progress made in the improvement of MM parts via 3DPg, multiple review articles have been reported on different aspects. However, in the past five years, a strategical and systematic review is still to be prepared which encompasses the advancements in this domain of research. Wang et al. [6] have reported a review article on MM fabrication using laser powder bed fusion (LPBF) process in terms of their characteristics, their challenges in manufacturing, and their scope in the recent world. Goh et al. [7] reviewed the 3DPg process of electronic parts in terms of techniques for fabrication, design concepts, and challenges in micro-sized electronics. A review has also been prepared concerning technologies and their advancements. A systematic review of MM 3DPg has been conducted by Collado et al. [8] and Zheng et al. [9] by considering the multiple factors. MM powder deposition mechanism of LPBF has also been reviewed by Wei and Li [10] along with the survey of the molten metal pool environment and the effect of process parameters on it. A report has also been furnished by Zhang et al. [11] and Loh et al. [12] in connection with progress in 3DPg technologies in designing, modeling, and fabrication of FGMs. The applications of 3DPg in aeronautical and automotive industries using MM fabrications have been reviewed by Blanco et al. [13]. A comprehensive assessment has also been done by Ravanbakhsh et al. [14] the reveal emerging technologies that are employing the MMs in biomedical fabrications. In a similar direction, Viola et al. [15] reported the utilization of MM in artificial tissue construction. Researchers have also reviewed the articles of MM 4D printing having application in bio-inspired parts and energy absorption [16–18]. The analysis of the several review articles revealed about the lack of comprehensive assessment of 3DPg techniques coupled with MM fabrications and their progress in the recent phase. The lack in assessment includes designing of MM with applicable analysis, technological enhancements, software developments for finite element analysis (FEA), post-processing, and properties customization in MM 3DPg. Therefore, for the future possibilities of MM 3DPg, technical concerns, restrictions, and potential applications should be discussed comprehensively.

The present chapter discusses various MM 3DPg techniques. It provides an overview of various types of MMs (MM), encompassing combinations like metal–metal (M–M), metal–ceramics (M–C), polymer–polymer (P–P), and others. There is an elaborate explanation of MM modeling and the numerical simulations applied to objects and structures using these MMs. The discussion shifts to AM processes capable of producing MMs, along with an exploration of research gaps and the limitations tied to these processes. Moreover, the post-processing methods for objects manufactured using MMs are the focal point of this chapter. The review presents an illustration of MM applications and their potential across various industries. Additionally, the chapter discusses the limitations and crucial challenges associated with adopting MM Additive Manufacturing (MMAMg). Finally, it encapsulates the conclusion of this review, summarizing the overall future trends within the domain of MMAMg research.

8.2 MATERIALS FOR 3D PRINTING

Nature shows outstanding functional abilities and performance by making use of the combination of excepsion. Additionally, thermaltional structures and compositions of materials [19–23]. MMAMg method is one of the growing AM technologies that can improve the functioning of 3D-printed products by blending MM or variable compositions of the same material in a particular unit and it may achieve such excellent customized performance for engineered designs that are inspired by nature. As shown in Figure 8.2, MMAMg provides several chances for building highly customized, complicated, useful, and expensive products with enhanced

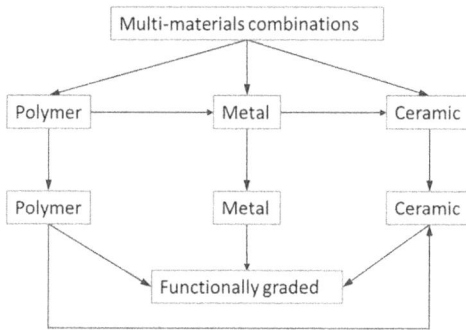

FIGURE 8.2 Classification of MMs.

qualities. Based on material types and potential property enhancements, elementary and common combinations of the materials developed via an appropriate AM technique are shown in Figure 8.2. The features of the products, including their electrical and multifunctional qualities, may be tailored by integrating materials at different sizes [24,25].

Almost every kind of material has been used in MMAMg. The material combinations under investigation either fall into one category (P–P, for example) or two different categories (P–M, for example). The two most extensively researched material groups are metals and polymers. To increase strength, stability, functionality, etc., the most often studied MM polymers are PLA [25–38], ABS [31,38–45], PEEK [29,46,47], and PET [35]. Elastomers, such as hydrogels and TPU [31,36,37,44], are the least researched MMs. TPU is a challenging elastomer to print because of partial melting, material feeding problems brought on by buckling, increased viscosity and melt strength, and incomplete layer and raster integration [48]. Hydrogels' wide temperature ranges for gelation and low viscosity make accurate AM of them challenging as well [49]. Therefore, further research is required to understand most of these material combinations.

The choice of an appropriate MMAMg technique for producing multifunctional parts demands a thorough understanding of the benefits as well as drawbacks of each AM technique. The proper distribution of numerous materials and solid bonding of materials with varied kinds and compositions are the basic problems for MMAMg.

8.2.1 Metal–Metal

Although there are numerous AM techniques for printing metals, most of them can only produce a particular metal combination at a time due to machine-specified constraints like a PF unit. Therefore, this constraint may be overcome by a novel PF design with a few materials feeding sources. Alternatively, multifunctional parts can be printed using premixed powders (Figure 8.3) [50].

Dimensional accuracy and postprocess requirements are two additional material- and process-related constraints. The bonding of different materials is the most difficult part. The compatibility of the MMs is therefore a critical factor

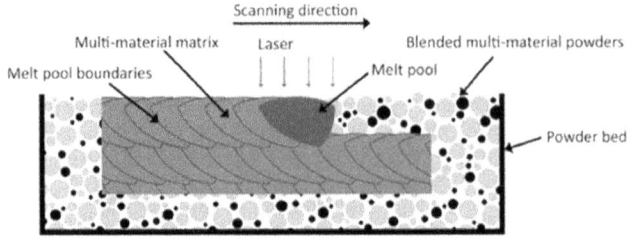

FIGURE 8.3 Illustration of laser powder-based fusion processing premixed dissimilar powders [50].

for MMAMg. For use in the automotive, rail, and aviation sectors, the powder bed fusion (PBF) technique was used to have MM elements and improve its physical and operating qualities [51]. Demir et al. [52] studied in-situ powder blending units for developing pure Fe, Fe/Al-12Si, and Al-12Si MM parts employing LPBF, which showed sufficiently strong bonding [52]. Despite the positive outcomes Demir et al. had, the manufacturability of the Fe/Al- 12Si MM showed a lower dimensional precision. With the use of LPBF, Wei et al. [53] investigated two metallurgically suitable Ti5Al2.5- Sn and Ti6Al4V production methods. An interface with no flaws and an elemental diffusion zone showed strong bonding. Despite this, Scaramuccia et al. [54] found significant cracking between Ti6Al4V and IN718, showing metallurgical incompatibility that leads to the production of a brittle intermetallic phase. The DED process, one of the MMAMg methods, has drawn a lot of research interest because it can produce large-scale components at a fast construction rate without the need for powder recycling. It is also possible to use numerous or mixed powder-feeding systems [55]. The research has shown that IN718, which was developed and helpfully made crack formation easier, was brittle. The SS316L/IN718 MM component's multifunctionality was demonstrated by the other half of the product, which was made of SS316L had ductile characteristics, and showed plastic strain while tensile deformation [56].

8.2.2 METAL–CERAMIC

The MMAMg composed of ceramics and metals is currently in the research and developmental phase. It is very difficult to create strong bonds between metals and ceramics. When ceramic melt has a flowability that makes metals evaporate, it becomes difficult to control the intended MM composition. Additionally, thermal stress and a wide range of thermal expansion coefficients, which uphold fissures between metals and ceramics, weaken the tensile strength of M-C MMAMg products. The products can only be used in certain structural applications as a result LOM, a type of lamination technique, effectively produces M-C structures by processing at lower temperatures and avoiding the constraints brought on by thermal stress (Figure 8.4) [50].

The Al_2O_3/Cu–O composite structure was described using LOM and heat treatment. The findings demonstrated the manufacture of geometrically complicated

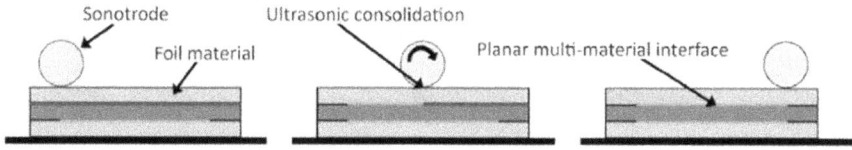

FIGURE 8.4 Demonstration of MM sheet lamination process [50].

elements like gears from thick materials with sufficient mechanical and electrical characteristics while keeping adequate bond properties [57]. Both a combination of Ti_6Al_4V powders and 3.1 wt% TiB_2 powders treated employing PBF [58] and Titanium and 1.5 wt% Bromide (B) powder have been the subject of recent investigations. The titanium's hardness and wear characteristics were enhanced by both MMAMg techniques. Also investigated for biological applications were LENS fabrications of Ti_6Al_4V with 6 wt% hydroxyapatite [59] and Cp-Ti with 9% CaP [60]. Results indicated improved wear characteristics when the matched $CaTiO_3$ and $Ca(PO_4)_3$ phases were produced and applied as a tribological defensive layer to a surface of biomaterial. For mechanical working of biomedical elements, however, managing the reaction phases, such as Ti_5P_3, $CaTiO_3$, and $Ca(PO_4)_3$, is essential since these delicate phases create a cracking susceptibility owing to the heat variations in metal and ceramic. Lattice bone implants made of $Fe30CaSiO_3$ (in wt %) were also created using the material extrusion MMAMg process [61]. As compared to pure-Fe lattice structure, the Fe30CaSiO3 biomaterial dramatically enhanced in vivo osteosynthesis and showed promising outcomes for the treatment of bone cancer.

8.2.3 METAL–POLYMER

Polymer AM processing has received substantial attention due to its lower temperature and added controlled processing conditions than metals, that are used in medical and aerospace sectors etc. [62]. PLA has been widely investigated and used as a filament material by AM technologies among other polymers. The FDM technique's low working temperature allows for the application of materials with lower melting temperatures [63]. M–P MMAMg has been utilized for nature-inspired soft robot production [64], combining the benefits of PLA with the mechanical robustness of metals. The insufficiently low process temperature, however, precludes metals and polymers from producing a solid bond. The utilization of M–P hybrid filaments to improve the characteristics of polymer materials (PMs) has therefore been the main focus of M–P MMAMg. Metals are added to polymers as fillers to improve their performance. Concerningly, there hasn't been adequate research done on the mechanical properties of polymer filaments incorporating a variety of filler metals. When the metal ratio in M–P MMAMg rises, pore formation seems to be the major challenge. Using the FDM approach, researchers investigated copper-reinforced PLA [65]. The most effective mechanical properties were demonstrated by Cu-PLA composites with a 25 wt% Cu addition, suggesting that the properties may be changed by varying the metal filler

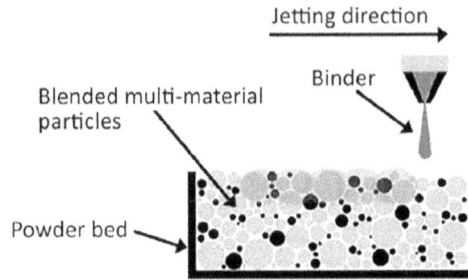

FIGURE 8.5   Illustration of premixed MM binder jetting [70].

content. As a PLA filler, Diaz-Garcia [66] used maraging steel to provide magnetic functionality to the MM made via the extrusion AM technique [66]. After sintering, the items manufactured with bronze-PLA filament displayed roughly 21% shrinkage and substantial porosity after raising the metal ratio to 86 wt% with 16 wt% of the polymer [67]. Although there is not much study on metal-polymer MMAMg, an M–P MM structure has been effectively shown utilizing a grouping of FFF, and extrusion-based additive manufacturing (AM) methods [68]. As previously stated, realizing the M–P bond with a unit AM technique is difficult. As a result, it is proposed to use two separate AM techniques to create metal-polymer structures [69]. The M–P interface, however, displayed consistent strength with an interlocking pattern, suggesting suitable polymer penetration in both macro and micro-level mechanical structures. In IN625 alloy produced employing BJ technique with anticipated reaction during polymer bond breaking, a strengthening Cr_3C_2 phase (Figure 8.5) was shown polymer binder-metal powder interface [70].

   This demonstrated the utility of the polymer-metal response, which may alter the properties of the finished goods. Future studies should focus on improving the M–P ratio, decreasing variability in M–P MM melting points, and improving the sintering parameters to decrease the porosity of the printed components to attain the objective of creating higher-quality metal elements [71].

8.2.4   POLYMER–POLYMER

A variety of materials have been employed in material extrusion AM to build MM yet monolithic elements. MMAMg may be used to produce functionally gradient structures with multicolor with changes in mechanical, and thermal characteristics. Various polymer combinations have been found in the examined pieces of literature based on the usages and requirements of the manufacturing process. Thermoplastic polymer chains provide significant obstacles in mobilizing by chemical reaction, which may readily produce thermoset plastics. In the MMAMg system, the addition of thermoset and thermoplastic polymers is hardly employed. The widening application of the MMAMg system in several research fields, such as the investigation of a connection of various materials and their mechanical characteristics on the yield of monolithic buildings, was made possible by the dominant role of thermoplastic in hot-melt extrusion-based systems. In MM 3D manufactured elements, the edges created between discrete materials on their geometrical border are considered a crucial factor, which varies based on the characteristics of the materials and printing factors [72]. There hasn't been any significant study on MM fused filament fabrication (FFF)-based AM that has looked at the relationship between printing parameters and diverse surface properties, in contrast to unit material extrusion 3DPg [35]. On MM extrusion-based 3DPg, there continues to be a lack of information, especially about the hardware, element design, and printing material limitations. The following are the main process constraints caused by the equipment: extrusion heads must first be calibrated, and the printing area must be reduced. Then, during the deposition process, the two (or more) extrusion heads must be precisely aligned. MM printing for component design permits the creation of numerous geometrical interactions in the component. However, compared to the strength of bulk material, the discrete material characteristics at the interface result in a weaker bonding strength. Another part of the design-related problem is the filament's residence time in extruder. The extruded material's thermophysical and rheological properties are frequently impacted by this residence period, which leaves little space for flow rate modification in a dual extrusion system. As a result, while employing traditional extrusion methods, the plastic drools more often [73]. The challenges with materials are caused by the composition and preparation of the thermoplastics used in the extrusion process. Discrete PMs have distinct chemical structures in contrast to thermosets, which perform the crosslinking of open monomers and regulate the curing procedure. For instance, a developed polymer chain is typically present in thermoplastic materials. Various PMs may cause problems due to variations in the thermal expansion coefficients. Due to uneven shrinkage during cooling, the components eventually become deformed and exhibit low dimensional stability. Moreover, the mechanical integrity predicted at boundaries of different printed materials lowers in the event of chemical incompatibility or inadequate affinity [35].

8.2.5 FUNCTIONALLY GRADED MM

Functionally graded MMs (FGMMs) provide amazing applications when a material gradient is needed, and the working conditions and material needs change across the component [74]. When looking for variations in temperature in elements for aerospace and aviation applications, engineers and designers make use of how the material properties interact. To achieve desirable physical or chemical outcomes, the center component of the blade of a turbine, for instance, must have adequate tensile strength and fatigue resistance due to the relatively low temperature in the blade's core [75]. As a result, functional grading of at least two materials is required [74]. There are two types of FGMMs: (1) those made by construction, such as stacking to create an element, and (2) those made by mass transit to get material gradient [74]. Current AM techniques are suitable for the productive way of producing FGMMs employing endless composites with complex geometries. The primary objective of employing AM in the fabrication of FGMMs is to create performance-based freeform elements directed by their gradient in material characteristics [76]. In the DED method, the powder is put into a melt pool when a dynamic laser is utilized. This technique may be used to produce FGMMs by altering the composition of powder between layers. Several metal alloys have had graded structures created using this method. Traditional or robotic DED instrumentations configured with more than one powder feed may handle it with ease [74]. DED provides many advantages, especially when compared to customary production methods like casting or powder metallurgy. To replicate the design of interior features and passages, the components are constructed layer by layer. In addition, it doesn't need specialized equipment like those used in casting or powder metallurgy. As a result, it can produce intricate shapes and coatings effectively [77]. By combining DED with multi-nozzles, FGMMs, which are traditional MM structures incorporating compositional gradient, may be produced quickly. Multi-nozzles can meet this need during DED by altering the types of powder supplied and their respective ratios. This is a strong benefit of this approach vs. other AM techniques, including PBF [77]. Researchers have recently tried to create FGMMs using PBF-AM methods [78]. Using complex morphology or dynamically created gradients, functionally graded MMAMg handles the MM issue. The last element, which tries to strengthen interfacial bonding between different materials, is controlled by the geometry and material arrangements [77].

This is shown in Figure 8.6(b). To eliminate the harsh properties of two different materials, a compositional alteration from diffused to a unified 2nd phase structure; layered employing disconnected compositional factors can be achieved. Because of discrete differences in material characteristics, this approach can thus avoid the usual failures of traditional MMAMg (Figure 8.6(a) [12]. The component created after the three-dimensional union of two materials using a dynamic gradient demonstrates the best qualities of both materials. It can have a range of weights while still maintaining its mechanical or chemical, or physical, qualities, such as toughness, impact resistance, wear resistance, or wear resistance. To attain

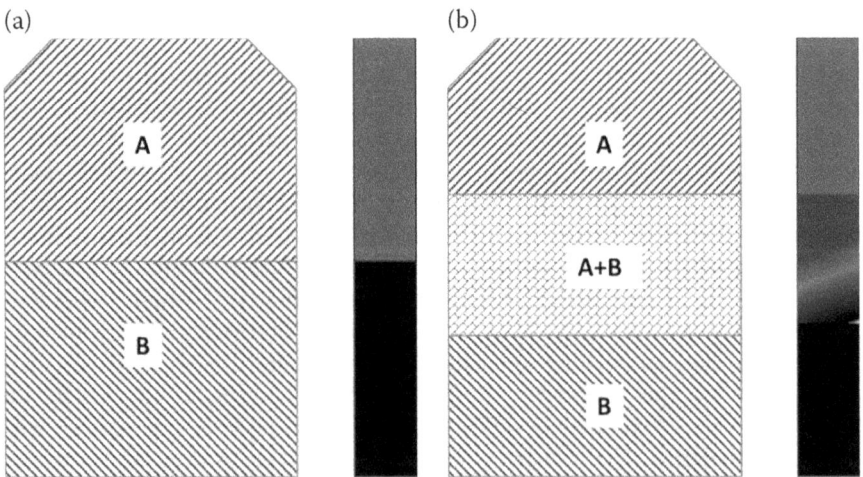

FIGURE 8.6   (a) Traditional MMAMg vs. (b) functionally graded MMAMg.

the desired attributes of components, diverse combinations of materials no longer need to compromise their inherent features [12]. MM, M–C, C–C, and polymer-based material combinations are among the possible MMAMgs for FGMs [73,75,77,79,80], as well as metal-ceramic, ceramic-ceramic, and ceramic-ceramic. The aerospace, military, biomedical, and automotive sectors are only a few of the fields where FGMMs may find use. In addition to these uses, MMAMg may generate specific components for die-mold applications [80]. FG materials may be used to regulate deformation, pressure, wear, and corrosion, even though they were first developed for heat-resistant applications. Their use has traditionally been restrained to the design of composite materials and ceramic coatings, and the grading of metallic materials hasn't gotten much attention. However, the emergence of FGMs with complicated metallic gradients of different alloys has been facilitated by the advent of AM methods like the aforementioned L-DED [80]. It takes work to create strong metallurgical connections between different alloys. In ternary or multicomponent alloys, complete solubility throughout all compositions and temperatures is substantially more complicated than it is in binary alloys [80]. Compositional gradients might be used to solve this problem. AM was employed by Aremu et al. [81] to create FG lattice structures with varying cell characteristics. However, there were numerous difficulties, such as adjusting various variables, unpredictable temperature fields, and tuning the AM process for the production of FGMs. There continue to be a number of challenges to be resolved even if a few of these issues have been solved by advancing artificial intelligence approaches, the design and development of AM process diagnostic tools, and the development of thermodynamic databases [81].

8.3 MM COMPONENTS DESIGN AND ANALYSIS

8.3.1 Design and Modeling

MM additive manufacturing (MMFGAM), which takes into account the element of MM gradients and/or complicated morphological characteristics, is also known as FGMM AM. The qualities and functions of MM elements are controlled by configurations of material and/or morphological gradience. By adopting a diverse compositional transition or moderate concentration gradients, distinct boundaries in MMAMg components may be removed. The use of various materials in integrated structures has enhanced functionality, decreased weight, and enhanced the production processes by fusing assembly and production into an individual processing step [82]. If we say in short, the desired characteristics may be constructed by fusing several materials into a unit integrated element. The design of MM elements may be categorized depending on the way different materials are incorporated into a unit object. A MM elements material can be dispersed in several ways, based on the required features and usefulness. To produce high-quality tool components for the MM AM process, Qiu and Langrana [83] set up a computer-aided design (CAD) system. Again, Bhashyam et al. [84] designed a CAD program containing libraries of planning algorithms and materials configurations that may be used to design the compositions and geometries of MM elements. The AM workflow for a unit material is comparable to the MM approach. However, the main distinction is that the MM process prioritizes the specification and portion of the materials inside the designated MM component [12]. Yao et al. [85] developed a comprehensive framework in recent research for developing MM components that may be produced utilizing the AM method. The four interconnected modules that make up their design framework are the choice of primary material, the AM technique, the MM constitution, and the finding of element geometry. They also created a database with information on the restrictions, capabilities, norms, and guidelines of AM techniques, that aided in decision-making during various stages of MM product creation. For designers who are in the field of MMAMg research, their methodology might be an essential reference. Shin et al. [86] suggested two techniques for creating MM elements. The first technique uses reproductive design, where the object's shape and material distribution were optimized utilizing homogenization design methods [87]. The designers were able to produce an MM item utilizing their own experiences and unique ideas using the second technique, which involved applying a knowledge-based variation method. Wargnier et al. [88] established a conceptual method for creating MM items using the material-searching technique based on the functional requirements of the suggested components. For each AM technique and different materials, MM design guidelines still need to be researched. In-depth research should also be done on how each AM processes' MM component design. It is also necessary to conduct additional research into CAD software programs that can design, optimize across several disciplines, and simulate MM components.

8.3.2 ADVANCES IN NUMERICAL SIMULATION

Numerical studies are crucial to the MMAMg design process because they help with product reliability, stress estimation, and parameter processing optimization. Study, observation, and modeling of MMAMg processes are challenging, time-consuming, and computationally costly because fusing MMs is a complicated process [89]. This is especially true for the fusion of MM including metals. The use of simulations to forecast and resolve MM-related issues, whether dependent on material fusion mechanics or manufacturing techniques, is now possible because of the advent of high-performing computers and workstations [90]. In order to understand and predict melt-pool events like convection, surface tension gradients, recoil pressure, vaporization, and momentum losses in mushy zones, numerical simulations using the finite element method (FEM), discrete element method (DEM), computational fluid dynamics (CFD), and molecular dynamics (MD) were implemented for MMAMg designs in the DED and PBF processes [91]. The MMAMg design simulation simulates the unequal temperature distribution induced by thermophysical parameter differences prior to testing. Moreover, instead of repeating trials, a simulation may simply modify the powder fraction for MMAMg design. However, most simulation methodologies face various hurdles, such as high processing power requirements and difficulties in gathering MM data; just a few software products on the market today provide tools for constructing and enhancing MMs. Furthermore, the present FEM packages only give a limited degree of functionality for MMs. There isn't yet a software program made exclusively for designing, optimizing, and simulating MM-related investigations. For design, optimization, and FEA, the majority of researchers in the reviewed literature used generic software, such as ABAQUS [30], ANSYS [92], level set model [93], COMSOL Multiphysics [94], Rhino [95], and MATLAB [96]. It should be noted that these programs weren't created with research and development on MM in mind. Therefore, substantial work is needed in order to use these programs for MM-related research.

8.3.3 MM Lattice Structures

AM can only be used to realize an optimal and application-specific structure when their 3D models can be developed digitally utilizing software [97]. Nonetheless, most commercially existing software can create models with little material data, which is crucial if the component is meant to be manufactured using MMs. It is a difficult process if the models must be constructed using lattice architecture [98]. Although their technique needs a large amount of manual labor, Wei et al. [99] provided a manual data preparation method to develop MM structures. As a result, their method cannot be used for mass manufacturing or industrial purposes. Therefore, creating a superior data interface and 3D file is critical for combining geometric and MM data while concurrently linking to the AM system for MM structure manufacturing [12]. Few studies have investigated the 3D-printed lattice architectures of MMs made from a variety of materials. Zhang et al. [100] investigated MM porous structures made by SLM of four different materials in a unit specimen. They determined that mixing high-strength brittle material with low-strength PM can improve the compression behavior of lattice constructions. Boley et al. [101] investigated design and manufacturing challenges using 4D printing of varied morphable lattices. These MM 3D-printed structures have the potential to change a wide range of applications. Recently, photosensitive materials were used to create strut-based lattice structures [102] in order to achieve elastomeric and stiff material qualities in a unit component. Nevertheless, the design techniques for MM structures are quite restricted. As a result, they necessitate innovative computer modeling approaches that include extensive geometry and material data while regulating material distribution at the voxel stage. In general, voxel-stage material distribution and structural modeling are costly. Richards [103] suggested a modeling technique for describing material geometry-topology using volumetric texture maps in order to lessen the amount of computation. The designer may be able to make changes at the voxel level thanks to this approach. As a result, the functional representation-based modeling techniques may encompass intricate lattice structures and successfully characterize several materials.

8.4 MM ADDITIVE MANUFACTURING PROCESSES

Conventional manufacturing procedures cannot be used to fabricate MM components because of the high degree of design and material distribution complexity. In the late 1990s, this constraint led several researchers to investigate MMAMg technologies [8]. As a result, practically all AM techniques can now create components using several materials in a single component because of ongoing advances in AM technologies. Nearly all AM techniques can produce MMs. The most often used techniques are material extrusion, PBF, and material jetting. Additionally, the vat-photopolymerization method, often known as SLA, was the first AM technique and has been extensively employed to produce MMs. When printing numerous materials simultaneously, an AM technique faces a number of difficulties. As a result, hybrid AM technologies [104] have attracted a lot of scientific interest. Four AM technologies—FFF, direct ink writing (DIW), aerosol jetting (AJ), and inkjet (IJ)—were combined in a recent study [105] to create a hybrid 3D printer that can quickly fabricate complex multiple materials with a variety of functionalities. An MMAMg system often comprises layer bonding systems and the delivery of several materials [106]. It is essential to use an AM production technique to enable strong bonding between the various types and compositions of materials while building multifunctional MMAMg components. Thus, the use of AM technology and the adjustment of AM process parameters are principally responsible for the success of creating such components. Each AM technique has specific benefits and drawbacks for processing multiple materials. With a recently developed powder-feeding system or powder mixing process, each AM technique may fabricate MMs. The development of MMAMg procedures is still ongoing, nevertheless, in order to address the problems brought on by the use of unusual materials. The MMAMg technique, in theory, joins disparate materials to improve the performance of the construction while taking location-specific features into account. The bonding quality, however, continues to be a significant obstacle to the success of MMAMg procedures. The choice of MM for MMAMg becomes crucial because Lumpe et al.'s investigation of MM interfaces revealed that bonding quality depends on the material combination [107]. If not, it is possible to find the failure in the interface zones [108]. Recent research [109] has concentrated on the gradient change of the material composition for MMAMg procedures to improve the interface bonding. The impact of the material transition zone and gradient composition, however, is yet unclear. As a result, more study is needed to understand the transition zone's features with regard to merging several materials [110].

8.4.1 Extrusion-based Systems

Extrusion-based AM displaces the material feedstock via one or more nozzles to allow the production of a 3D structure using a variety of materials [111]. Thermoplastics [112], metal-filled thermoplastics, composites [113], flexible elastomers [40], and mixed MM powder-based feedstock [114] are just a few of the materials that may be printed using this method. Similar to the BJ technique, the process's success depends on the suitability of the binder with the feedstock to account for unwanted interactions or reactions during extrusion or sintering. The components created by the extrusion process need post-AM de-binding and sintering to maintain structural reliability. Additionally, the powder recycling process is not necessary since the binder is added to the material instead of being extruded on a powder bed. It saves both time and money. However, this technology's main drawback is the printing outcome, which depends on the nozzle diameter and powder size. Additionally, manufacturing items with wide aspect ratios and overhanging shapes is difficult and necessitates a steady flow of material and quick solidification. An MM feedstock design is essential in order to achieve this. The novel hybrid material extrusion and bonding technique has been considered to create different MM systems in which Al-based elements were extensively studied, allowing continuous extrusion and high-quality bonding by applying friction stir welding [115]. The widely used extrusion-based AM method, however, is FFF. The FFF system is the most widely used AM system. It specifically employs thermoplastic-based feedstock filaments that are extruded layer by layer via a nozzle on the build platform after going through a number of thermal processes, such as heating and melting in a hot extruder. The FFF system frequently employs numerous extrusion heads to create monolithic pieces and deposit separate thermoplastic materials. A model material is often used when the main material is extruded through one extrusion head, and a support material is typically used when secondary material is extruded through a different head. However, an MM structure may be made using two extrusion heads, and composite parts can be made using a secondary extrusion head for another thermoplastic or other materials (such as carbon fiber). The fabrication of multifunctional parts, such as embedded sensors [116], electrical circuits [117], reinforced composite parts [118], heating elements [119], and functionally gradient structures [75], is greatly facilitated by the FFF system's ability to produce continuous fiber, filament, and wires. No of the kind of thermoplastic, the FFF MM system faces a number of difficulties. The FFF system has limits in terms of component design, process development, and material selection, much like the polymer–polymer MMAMg system. For instance, a certain volume of the thermoplastic material is where fiber material can be deposited. The matrix material inside the component is reduced and the voids are increased with a greater volume percentage of fiber content, which affects the mechanical characteristics and increases anisotropy [120]. The accessibility of the produced component is another illustration of a design restriction on a part. The manufacture of monolithic, multifunctional devices is made possible by embedded electronics fabrication using an FFF system, which is a special field of application. The process of assembly and disassembly for service has not yet been studied in detail. In order to boost the end-users trust in quickly replacing the conventional components and deploying them for multifunctional usage, the dependability of the produced components should be sufficiently good. A number of techniques, including wire coextrusion systems [121], wire embedding, fiber embedding, and fiber impregnation have been developed and are preserved under the patent for the FFF MMAMg system. Although each technique increased MMAMg's capacity, they all had limitations regarding fiber insertion, control, volume fraction maintenance, or initial deployment of the continuous fiber/wire. Moreover, the process growth of MMAMg systems is constrained by a mismatch in the CTE of the discrete material. Typically, the continuous fiber, including carbon fiber, glass fiber, and any conductive and resistance wires, is significantly reduced by the thermal conductivity and CTE of the thermoplastic-based matrix materials.

8.4.2 Powder Bed Fusion

The basic idea behind the PBF technique is to apply a layer of metal powder to a substrate employing a powder-feeding system and then use a laser PBF (LPBF) as a heat source to selectively melt or sinter the area selected by the CAD of the desired element [122]. This process continues for each subsequent layer by lowering the build stage, depositing powder on top of the layer before it, and melting or sintering the present layer in addition to the layer(s) before it [111]. However, the production of sole-metal powder utilizing a single-powder bed dispensing mechanism is the most common use of this approach [123]. A combination of M–M or M–C powders can be printed even with a single-powder delivery system to overcome this constraint as single-powder delivery technology specifically restricts the manufacturing of MM components. MM PBF, which uses blended materials, has been accomplished for a variety of material combinations [124]. A number of elements need to be taken into account in order to create a solid and enduring relationship between various different materials in order to fully fulfill the planned MM system's potential [125]. The main issues for the fabrication's success are the adjustment of the PBF process parameters to allow for adequate remelting into the preceding layers and strengthening the link between layers [126], which may be impacted by the powder properties [127]. The element is prone to showing a lack of fusion during process variables optimization to prevent undesirable defects because of the low-energy density of the heat source, wherein high-melting-point alloying elements are located, or the applied energy density of porosity remains constant because of existing low-melting-point elements. If not, variations in thermal characteristics are what induce fissures across unrelated materials [128]. To improve their surface area and quantity of energy absorbed during the PBF process, it has been suggested to use smaller powder sizes for materials with high-melting points [129]. Thus, thorough consideration of both powder characteristics and process conditions optimization is required when producing MM elements utilizing PBF systems. Additionally, even after an effective fabrication, postprocess heat treatment may still be required to provide the proper compositional homogeneity. Also, being a developing technology for MMAMg, PBF methods must provide an improved mechanism for recycling and reusing the powder throughout the building phase in comparison to DED systems [129].

8.4.3 Jetting-based System

Binder jetting (BJ), which is like PBF in that it needs a powder bed, uses a different type of bonding. Although the PBF uses thermal energy to solidify the powder, the BJ employs an adhesive to bind the powders together to build the structure at low temperatures [111]. A suitable binder with metal powder is required for this method, such as polymer binders, metal inks, and metallic slat composites [130]. Bonding may be applied on a powder bed using the same method as on a PBF. In order to produce MM elements, the PBF process has the same challenges as the BJ process, including multiple powder feeding and efficient powder recycling systems. In contrast to the PBF process, components that go through the BJ processing step require post-AM heat treatment to maintain structural integrity. However, as various materials have varying thermal properties, heat treatment after AM causes structural shrinkage, which might be detrimental to the structural application [131]. Because it may be used with numerous jetting heads, material jetting (MJ) or inkjet 3DPg systems are suitable for MMAMg. Polyjet (PJ) [132] is a superb illustration of MJ technology that has been widely applied for multicolor AM. The problems of the MJ technique, however, include the restricted availability of materials [133] and the assessment of material rheology [134] during printing.

8.4.4 Directed Energy Deposition

According to component design, the material is fed via a nozzle which can conduct multiaxial displacement during the DED process and is then melted using the heat source [135]. When employing wire and arc additive manufacturing (WAAM), the capabilities of complicated designs and dimensional precision are reduced [136]. Additionally, the WAAM technique is widely used for fabricating MM components due to its powder-feeding architecture, which makes the MMAMg approach simple [137]. However, there are issues with how well WAAM's final products function, including excessive grain growth brought on by heat buildup and a reduction in mechanical characteristics [138]. Powder-based DED techniques, such as WAAM, are known as the best way to satisfy these requirements even though near-net shape AM items need to have high accuracy and lower surface roughness since they allow the delivery of multiple materials to the heat source in order to create multi-functional components [139]. The most challenging aspect for DED and other analogous AM technologies is forging a strong connection between diverse materials. Although powder feeding is a more suited technique than PBF, material misalliances such as thermal expansion coefficients, thermal conductivity, laser absorption, and melting temperatures are material-specific challenges that result in residual stress and metallurgical flaws during DED fabrications [139].

8.4.5 Hybrid MMAMg System

A hybrid manufacturing technique that combines additive and subtractive manufacturing is able to preserve dimensional precision and surface polish while boosting efficiency through the use of deep layers. By utilizing each process's strengths while avoiding its weaknesses, hybrid manufacturing may be able to produce a system that performs better than each process on its own [140]. The benefits of the hybrid technique have been widely applied in the manufacture of metal elements, with parts requiring precise dimensions or surface polish machining and complex geometry being manufactured using AM [140]. Furthermore, hybrid methodologies have been employed to create polymer and MM elements [141]. Recently, a method that combines metal AM and metal forming has been developed to enhance the stiffness and wear resistance of made elements while also enhancing the geometry of the material layers that have been deposited. New ideas from sheet metal forming and bulk metal forming methods are being included in this technology, which is now evolving and increasing quickly. Early hybrid metal

AM systems sought to boost productivity and component quality by combining metal AM and metal cutting with a variety of thermal energy sources. The very first hybrid AM systems were introduced in the middle of the 2010s as a result of these developments [140]. To produce M–P elements, Weflen and Frank recommended a hybrid manufacturing system technique. First, the execution of the polymer extrusion AM tool is evaluated, and process variables are put in the machining center. The relationship between element cooling and form is looked at to assess the thermal characteristics of polymer extrusion in the manufacturing center. In order to compare the structural characteristics of extruded polymer and M–P interface, the component's strength is then baselined on different cooling intervals and material flow rates [140]. Recently, hybrid MMAMg machines and metal AM configurations have been improved [142]. Muguruza et al. [143] created a hybrid method for producing ceramic particle-reinforced photosensitive resins, which consists of DLP technology and a 2D drop-on-demand inkjet printing system. The DLP method photopolymerizes photosensitive resins to produce 3D objects, whereas an inkjet printing technology deposits tiny droplets of conductive ink. Similar techniques have been used to produce functional materials (like printed circuits) for electrical devices. Ma [129] created a unique technology called "hybrid deposition manufacturing" by merging the FDM and MJ AM techniques to produce MM components for robotics and mechatronics-related applications in another study. When compared to other techniques, their solution significantly reduced production time and needed manual labor, waste material, and assembly complexity [144]. It is possible to produce novel items with MM forms and surface treatments using the DED process in conjunction with a multi-axis machining center that are not feasible with only AM methods. The successful implantation of a Bluetooth active sensor into a high-strength steel component made with pH 13-8 stainless steel and Invar 36 alloy MM deposition has been shown using hybrid DED systems as a consequence [145]. Despite the benefits of enhanced productivity, surface quality, and dimensional precision [11], hybrid MMAMg systems provide a number of obstacles and constraints. For example, because DED is a fusion-based technique, there may be structural issues with dissimilar metal welding, such as the development of undesired intermetallic phases [18]. The adoption of the hybrid technique still faces several technical challenges (such as process planning, decision planning, the use of cutting fluids, and the requirement for post-processing). As new hardware advancements happen, corresponding software tools keep falling short of hybrid systems' promise. Further research is required in order to create full hybrid AM software solutions because of the gap in the literature [146]. Dilberoglu et al. [146] suggested a hybrid manufacturing simulator that allows users to read G-codes, allowing them to visualize tool tracks and morphologies of created things. As a result, overcoming the aforementioned restrictions will ultimately improve hybrid system capabilities, particularly for MM production.

8.5 POST-PROCESSING OF MMAMG

The post-processing of MMAMg elements varies from that of unit material elements. Temperature and other parameters should be changed in case of thermal post-processing to account for the thermal qualities of both materials. As a result, rather than accounting for only one substance, a new post-processing procedure should be devised that is compatible with all components in combination. Post-processing validates that characteristic of AM elements, including mechanical characteristics, surface qualities, geometric correctness, tolerance, and aesthetics, meets standards and design criteria. Post-processing methods for AM elements in general include but are not constrained to, chemical processing, laser microma-chining, electro-polishing, etc. [147]. For instance, Goh et al. [7] recently reviewed a number of post-processing methods for 3D-printed MM electronics. Certain technologies may manufacture a variety of materials in an arrangement of an unremitting and progressive process using one of the AM processes. In contrast to PBF-based methodologies [148], which can create distinct material gradients, DED and LENS AM techniques may produce MMs with unremitting and/or distinct material gradients within and across the layers. The great capability of the aforementioned MMAMg methods does, however, entail significant post-processing demands to guarantee dimensional correctness and accurate form. Components that are printed using a homogenous material, have simple designs, and were made using only one type of material may all be handled by commercially accessible post-processing processes. Therefore, a unique approach must be created for the post-processing of MMAMg or functionally graded AM components in order to advance toward the achievement of completely functioning components for a number of applications. To reduce the need for this null-value-addition procedure, the post-processing aspect of MM components must also be taken into account during the design stage [85]. Additionally, efficient post-processing will increase production efficiency and reduce component costs overall by taking into account post-heat treatment, support material removal, and considerations during part design.

8.6   APPLICATIONS AND OPPORTUNITIES

Recent research and technological advances focused on the increasing multi-functionality in many applications brought on by MM goods, which significantly benefited the global industrial and research communities. When compared to traditional structures, the ability to create customized MM structures using 3DPg technology allowed for specific material selections and enhanced a number of desired component characteristics [149]. Numerous high-tech engineering disci-plines, particularly the aerospace industry, have benefited greatly from the development of MMAMg. The value of developing and utilizing MM components has also been acknowledged due to their contribution to the development of lightweight designs and their ability to support tooling testing in space. For the production of MM components utilized in any aircraft or space mission, the choice of material, such as metal powder, ceramics, polymer, and reinforced composites, as well as its performance, is essential [31]. Additionally, MMAMg-related tech-nology developments have led to significant improvements in medicine. In tissue engineering designs for sensitive human body parts, significant advancements have been made, including the use of biodegradable polymers for cell encapsulation and drug delivery systems [150]. Additionally, crucial accomplishments of MMAMg in the medical area include shape-memory polymers and 4D printing of diverse components [151]. Thus, the subsequent subsections have provided further detail on the key uses of MMAMg in various sectors.

8.6.1 MMAMg of Electronic Component

MMAMg has permitted the integration of electrically different materials such as conductors, semiconductors, and dielectrics, all of which are required in the direct manufacturing of three-dimensional electronic devices. MMAMg may be used to create a variety of electrical components, including pressure sensors, smart sensor integration, and microelectromechanical systems. Hainsworth et al. [152] developed a malleable soft robot actuator with embedded sensors and incorporated it into a robotic arm's end effector gripping mechanism to assist human-occupied activities that would be impossible to undertake without 3DPg technology due to assembly labor. Several MMAMg processes, such as FDM, which can construct embedded sensors and MM extrusion technology, may use several materials to fabricate electrical devices or components [153]. Electronic components manufactured using single-layer deposition (conventional approach) have various drawbacks that make them unsuitable for high-performance electronic applications. However, multilayered capacitors and inductors may be manufactured, removing the prior design constraint and opening new options for high-performance circuits. Although multilayered MM removes numerous usually necessary steps in traditional manufacturing [154], multilayered circuit creation faces various constraints due to the availability of restricted alternatives in printing methods. Specific research revealed the successful application of AM techniques for creating several types of totally printed multilayered circuits [155–157]. The functional components of the OLED display were made up of six layers that were 3D-printed using MMs. The capacity of MMAMg 3DPg to include multiple materials in the detailed design of LEDs is promising for generating creative electronics, and it is now being explored to replace the traditional technique for electronic component manufacture. 3D-printed energy devices are relatively new in comparison to other electrical parts and gadgets, and they represent an emerging study topic. Energy devices are classified as either energy harvesting or energy storage devices.

8.6.2 MMAMg Sustainability

The use of MMAMg will mark a turning point for quick manufacturing, adaptable design, and structural uses. By utilizing the combined or hybridized qualities of many materials, MM 3DPg has the ability to be used in structural engineering applications. Additionally, it enables the speedy construction of strong, high-quality buildings that have the characteristics of all blended materials. [143]. MM printing is more environmentally friendly than single-material AM and is able to create new, distinctive 4D structures with particular shapes and features. Additionally, MMAMg-fabricated components have better mechanical characteristics, durability, and tensile strength [158]. Future space missions will rely heavily on manufacturing to lower launch costs and supply equipment on demand for protracted operations that don't need to be refueled from Earth. ABS was recently printed aboard the International Space Station (ISS) in microgravity using the FDM process; however, the performance of ABS in microgravity is yet unclear. The development of novel materials and more intricate metal structures in space has reached this point. In order to ensure self-sufficiency, the main objective of in-space manufacturing is to develop considerable space exploration that is less dependent on Earth. MMAMg has decreased waste materials, which is crucial for in-space manufacturing applications since self-sustenance in space is essential. Only plastic parts could previously be recycled and used again. However, researchers plan to create a completely integrated manufacturing facility that will be a metal and MM recycler. As was already mentioned, MMAMg has the potential to offer several sustainability advantages. A wide number of industries will be impacted by advances in 4D printing and smart MMAMg technology, especially for the continuation of deep space exploration and self-sufficiency [158].

8.6.3   4-Dimensional (4D) Printing and MMAMg for Soft Robotics

In contrast to the inflexible, hard joints seen in conventional robots, soft robotics makes use of flexible, pliable materials for robotic applications. They have a large number of usages in bio-robotics, medicine, and aviation because of their increased degrees of freedom. The primary objective of soft robotics is to produce flexible, human-like robots [159–163]. The advancement in soft robotics and actuators in a multitude of sectors holds great promise for MMAMg technology. Widespread application of AM methodologies for the development of soft robots is now viable due to the latest developments on 3DPg of soft materials. FFF, direct ink writing, DLP, and SLA are the primary AM processes used in the production of soft robots. A soft robot's actuation, or the process of converting energy into mechanical work, is its essential component [160]. As a result, it has an impact on a variety of elements, including the motions produced by soft robots and the required manufacturing process. A few of the actuation methods that have been developed include hybrid actuation, shape-memory materials (SMMs), electroactive polymers (EAP), flexible fluidic actuation (FFA), and cable-driven actuation [161–163]. The area of soft robotics was transformed by an AM-based pneumatic network (PneuNet) actuator developed by a Harvard team of researchers to increase the wall thickness and the number of chambers of PneuNets based on an FFA actuation system [164]. Shintake et al. [165] created a novel soft actuator based on a hybrid actuation system by employing a 3D-printed PLA mold in addition to FFA soft robots. Soft robots are gaining popularity because of their versatility in a number of applications, notably for jobs requiring safety, dexterity, and conformal deformation. Applications for soft robotics include the study of as-yet-unexplored fields including biology, aviation, electronics, biomimetics, food and agriculture, e-textiles, object manipulation, and automation. These robots, known as MMAMg, may soon be developed as a result of the current rate of research in the field of soft robotics, which will help to greatly speed up this development. Despite all the developments in system design and materials science, creating soft robots still faces considerable challenges in terms of connecting the various components and fabricating different materials. However, MMAMg technologies are quickly gaining importance in soft robotics. Researchers should concentrate on improving the use of MMAMg in soft robotics, process variables of unique materials, and methods for printing to overcome a number of printing issues that adversely affect 3D-printed objects. These printing issues include the lack of soft materials on the marketplace and the accessibility of commercial 3D printing equipment that can print soft materials [165].

8.6.4 MM Bio-Printing

The term "bio-printing" describes a "material transfer process" that entails the utilization of cells, living substances, and chemicals to produce biomaterials. Due to the rising number of organ transplants but the inadequate supply of organs for transplantation, the development of 3D or 4D bio-printing is particularly beneficial [166]. The physical and chemical properties of things created by 3D bio-printing using a single material are frequently not particularly different. In order to make a 3D printable composite with exceptional qualities like ideal mechanical properties, needed biological compatibility, enhanced biomimicking of tissue buildings, and suitable fidelity, there is a swiftly growing interest in altering and diversifying generic materials for printing by integrating them with other materials demonstrating special characteristics [167]. With MM bio-printing, the issues with single-material fabrication are eliminated, and sophisticated design of biological tissue and organs is reproduced in scaffolds [168]. With the integration of many features into a single project, MM 3D bio-printing creates practical products that are essential in terms of strength and performance [169]. Medical science has greatly advanced as a result of MMAMg technological advancements. Both the utilization of decomposable polymers for cell encapsulation and drug distribution methods, as well as the manufacture of tissue engineering [170] constructs for sensitive human body parts, have made significant strides. This has made it feasible to develop functional, diverse, multicellular, and multiscale hepatic constructs that can operate swiftly and effectively. Co-printing of hydrogel and decomposable materials, which is made possible by the MM approach, also allows for the use of MM bio-printing for producing physically suitable vasculature. In bioresearch, human organ replication has been achieved using both tissue modeling and drug testing specimens. Additionally, the ability of MM bio-printing to reproduce complicated 3D microenvironments of tumors holds the promise of producing more accurate biomimetic tumor models. Depending on their dynamics, bio-printed diverse tumor models have the ability to accelerate the growth of a tumor [171]. MM bio-printing presents the prospective to produce precise biomimetic tumor models due to its capability of reproducing the complex 3D microenvironments of tumors. Dynamically, bio-printed diverse tumor models allow the growth of tumors [172].

8.6.5 MMAMG FOR ARCHITECTURE AND CONSTRUCTION

The difficulties the building sector faces lead to excessive material and energy use and almost no increase in productivity. The present building methods have a number of problems, such as strict construction constraints, the labor-intensive task of aligning the components to fit inside them, and a significant risk of construction mistakes brought on by poor coordination between design and implementation. A sizable amount of building activity is carried out inefficiently and laboriously [173]. The present manufacturing and building processes may be considerably changed by integrating AM with diverse materials to address these concerns. Without extra assembly or MM architectural design, MMAMg may produce items with different material characteristics. By minimizing the number of necessary production processes and addressing problems related to the connection of certain materials or pieces, this change can minimize inefficiencies in manufacturing and construction [174]. A limited amount of research has been done to adapt these techniques to the architecture and construction industry, despite the fact that the majority of AM procedures may utilize a variety of materials. Concrete material offers reasonable building quality at a fair price. Mixed concrete offers a wider variety of qualities and uses than conventional applications since it may be produced depending on the various features of the mixture and the individual additions or fillers in concrete. Contextually, a variety of strategies may be used to change these characteristics by modifying the composition of the concrete mixture, making concrete an appropriate medium for MMAMg. In MMAMg of concrete, changing the material density is the most common technique. The MIT Mediated Matter group investigated the use of aluminum powder, often known as variable density concrete, as a foaming agent for concrete. When uncured concrete and aluminum powder are present, the generation of hydrogen gas causes the concrete mix to foam, reducing the overall weight and consumption [175]. Furthermore, a thorough analysis of lightweight concrete based on additives revealed that additives had a substantial impact on the overall performance of concrete mixtures [176]. Alternatively, fibers can be added to MMAMg and used with concrete to increase the material's tensile strength and ductility. The construction industry faces a number of more pervasive issues that MMAMg seeks to address, including high material consumption, a lack of component adaptability to changing structural requirements and environmental conditions, and an excess of difficulty in the assembly and creation processes presently in use.

8.7 CONSTRAINTS AND CHALLENGES

The use of any material's advantages for favorable purposes is not implied by MMAMg. Researchers sometimes take advantage of the shortcomings of linking MMs, such as weak bi-material bonding and residual stress at the border. Controlling the AM process is challenging because various materials have varied properties and have variable miscibility and wetting constraints [177]. In practice, flaws like cracks, pores, and residual stresses that affect the integrity of the components' dimensional stability, crack development resistance, and mechanical qualities can be produced by bi-material systems like M–P, M–C, and P–C [18].

Future research should thus focus on the material choosing, part design, and manufacturing characteristics of MMAMg elements, especially the residual and surface stresses generated during additive printing of metallic components [178]. Understanding the bonding and optimal sharing of appropriate materials, as well as the mechanisms of reaction kinetics, bonding, residual stresses, and cracking mechanics, is essential for building MMs. Before MMAMg production, efforts in material designs, features, chemical compositions, and qualities were necessary because of the intricacy of the planned MM components' nonuniform traits [179]. Additionally, for the particular procedure, alloy production at the in-situ interface of the two materials is challenging to define and distant from an equilibrium condition [177]. Microcracks and other flaws at the grain boundaries are common in the PBF process for metal materials. It's also important to address other outstanding problems including low production throughput, poor scalability, rough surfaces, weak interfacial bonding, and high cross-contamination [105]. Technical advances, such as the creation of CAD programs, design protocols, and procedures, should be established in addition to the fundamental knowledge [179]. Numerous material changes may be required for the AM of a complex MM element. Therefore, MMAMg processes are slower than single-material AM methods. Additionally, since extrusion-based printing processes need distinct printing procedures that change with the jetting nozzles, replacing the ink cartridge or spool is not required when employing various materials. As a result, when employed to produce MM parts, the majority of AM techniques that have been developed for manufacturing unit material components face enormous challenges. Additionally, a full grasp of 3D printable materials, chemical compositions, optimal printing settings, and manufacturing constraints is required when defining and distributing several elements in a unit component [180]. Due to various constraints, such as a restricted number of 3D printable materials, there are few design standards on material compatibility and MM 3D printability, which is a key obstacle in implementing MMAMg part manufacturing for the end user. A large number of previously evaluated works investigated the properties of MMAMg elements using empirical methods. To validate experimental data, however, many of them did not employ FEM. A few researchers have also improved their production processes or MM designs by using optimization methods. In general, there isn't a lot of literature about MMAMg software development. As a result, in order to fully realize the potential of MM research, MMAMg software requires further research and development. The slicing of MM components and the choice of an appropriate AM method are yet another crucial aspect of the program to take into account [180]. Although Hascoet [103] attempted to define MM gradient slicing, unique strategies for MM and/or functionally graded component preparation, analysis, and slicing are still necessary. Few research teams have focused on complex interior structures that need powerful computers and optimization tools, as well as AM approaches that integrate MM production capabilities. Flexible materials that incorporate metamaterials and cutting-edge lattice structures may be applied in a variety of industries, such as tissue engineering, civil engineering, textiles, and aerospace. By merging micro/nano MM AM, metamaterials, and lattice structures, new functional components may be made [180].

8.8 CONCLUSION AND FUTURE PERSPECTIVE

The development of different MM study areas, including diverse MM composition types, component design, modeling, and analytical methodologies, is summarized in this article. A significant amount of discussion and assessment is also given to applications, post-processing, technological hurdles, and possible research needs.

MMAMg is a rapidly developing field with a wide range of applications. Nevertheless, there are still several challenges that need to be addressed before MMAMg can be broadly accepted. One challenge is the lack of research on alternative materials to polymers and metals. Elastomers and hydrogels, for example, have not been studied as extensively as other materials, but they could offer unique properties for certain applications. Another challenge is the lack of research on the link between printing parameters and the quality of the interaction between dissimilar materials. This is particularly important for metal MMAMg, where dimensional accuracies and post-processing needs are critical. In addition, most of the research on MMAMg has focused on the compressive and tensile properties of components. Other properties, such as chemical, fatigue, and impact properties, also need to be investigated. The design of MMAMg components is also a challenging area. There are no standardized design guidelines for each AM technique and material, and most available research has concentrated on basic properties. More in-depth studies are required on the design of MMAMg components for each AM technique. Finally, there are limited software packages available for designing and optimizing MMAMg components. The available packages do not offer full-range FEA of MMAMg, and there is no software bundle particularly developed for the design, optimization, and simulations of MMAMg studies. This is a notable research gap that needs to be addressed. Overall, MMAMg is a promising technology with a wide range of prospective applications. MMAMg is poised for a revolutionary change in the future. This cutting-edge manufacturing technique is expected to transform several sectors because of continual improvements in technology and materials. The creation of extremely complex and customized components with specialized material qualities will be made possible by the growth of materials that are already accessible, including metals, polymers, ceramics, and composites. This degree of personalization will help sectors like healthcare, where tailored implants and prostheses are becoming a reality, as well as automotive and aerospace, where weight reduction and functional integration are important objectives. The inclusion of electronics and sensors in 3D-printed products, in addition to the advantages of sustainability and educational possibilities, all highlight the enormous potential of MMAMg. However, in order to assure safety and quality, this future also involves regulatory difficulties and the requirement for industry standards. We can foresee a dynamic environment where MMAMg plays a key role in transforming the way we design and create items as the market expands and competition heats up. This review highlights the key challenges and research gaps in MMAMg, and it provides insights for researchers and engineers who are working in this field.

REFERENCES

[1] M. J. Mirzaali *et al.*, "Mechanics of bioinspired functionally graded soft-hard composites made by MM 3D printing," *Compos. Struct.*, vol. 237, p. 111867, 2020, doi: 10.1016/j.compstruct.2020.111867.

[2] D. Wang and S. Li, "Material selection decision-making method for MM lightweight automotive body driven by performance," *Proc. Inst. Mech. Eng. Part L J. Mater. Des. Appl.*, vol. 236, no. 4, pp. 730–746, Nov. 2021, doi: 10.1177/14644207211055661.

[3] M. O. F. Emon, F. Alkadi, D. G. Philip, D.-H. Kim, K.-C. Lee, and J.-W. Choi, "MM 3D printing of a soft pressure sensor," *Addit. Manuf.*, vol. 28, pp. 629–638, 2019, doi: 10.1016/j.addma.2019.06.001.

[4] M. He, X. Zhang, L. dos Santos Fernandez, A. Molter, L. Xia, and T. Shi, "MM topology optimization of piezoelectric composite structures for energy harvesting," *Compos. Struct.*, vol. 265, p. 113783, 2021, doi: 10.1016/j.compstruct.2021.113783.

[5] C.-P. Jiang, M. F. R. Hentihu, S.-Y. Lee, and R. Lin, "Multiresin additive manufacturing process for printing a complete denture and an analysis of accuracy.," *3D Print. Addit. Manuf.*, vol. 9, no. 6, pp. 511–519, Dec. 2022, doi: 10.1089/3dp.2021.0007.

[6] D. Wang *et al.*, "Recent progress on additive manufacturing of MM structures with laser powder bed fusion," *Virtual Phys. Prototyp.*, vol. 17, no. 2, pp. 329–365, 2022, doi: 10.1080/17452759.2022.2028343.

[7] G. L. Goh, H. Zhang, T. H. Chong, and W. Y. Yeong, "3D printing of multilayered and multimaterial electronics: A review," *Adv. Electron. Mater.*, vol. 7, no. 10, p. 2100445, Oct. 2021, doi: 10.1002/aelm.202100445.

[8] A. García-Collado, J. M. Blanco, M. K. Gupta, and R. Dorado-Vicente, "Advances in polymers based MM Additive-Manufacturing Techniques: State-of-art review on properties and applications," *Addit. Manuf.*, vol. 50, p. 102577, 2022, doi: 10.1016/j.addma.2021.102577.

[9] Y. Zheng, W. Zhang, D. M. B. Lopez, and R. Ahmad, "Scientometric analysis and systematic review of MM additive manufacturing of polymers," *Polymers.*, vol. 13, no. 12, 2021, doi: 10.3390/polym13121957.

[10] C. Wei and L. Li, "Recent progress and scientific challenges in MM additive manufacturing via laser-based powder bed fusion," *Virtual Phys. Prototyp.*, vol. 16, no. 3, pp. 347–371, 2021, doi: 10.1080/17452759.2021.1928520.

[11] C. Zhang *et al.*, "Additive manufacturing of functionally graded materials: A review," *Mater. Sci. Eng. A*, vol. 764, p. 138209, 2019, doi: 10.1016/j.msea.2019.138209.

[12] G. H. Loh, E. Pei, D. Harrison, and M. D. Monzón, "An overview of functionally graded additive manufacturing," *Addit. Manuf.*, vol. 23, pp. 34–44, 2018, doi: 10.1016/j.addma.2018.06.023.

[13] D. Blanco, E. M. Rubio, M. M. Marín, and J. P. Davim, "Advanced materials and MMs applied in aeronautical and automotive fields: A systematic review approach," *Procedia CIRP*, vol. 99, pp. 196–201, 2021, doi: 10.1016/j.procir.2021.03.027.

[14] H. Ravanbakhsh, V. Karamzadeh, G. Bao, L. Mongeau, D. Juncker, and Y. S. Zhang, "Emerging technologies in MM bioprinting," *Adv. Mater.*, vol. 33, no. 49, p. 2104730, Dec. 2021, doi: 10.1002/adma.202104730.

[15] M. Viola, S. Piluso, J. Groll, T. Vermonden, J. Malda, and M. Castilho, "The importance of interfaces in MM biofabricated tissue structures," *Adv. Healthc. Mater.*, vol. 10, no. 21, p. 2101021, Nov. 2021, doi: 10.1002/adhm.202101021.

[16] A. Mitchell, U. Lafont, M. Hołyńska, and C. Semprimoschnig, "Additive manufacturing — A review of 4D printing and future applications," *Addit. Manuf.*, vol. 24, pp. 606–626, 2018, doi: 10.1016/j.addma.2018.10.038.

[17] N. S. Ha and G. Lu, "A review of recent research on bio-inspired structures and materials for energy absorption applications," *Compos. Part B Eng.*, vol. 181, p. 107496, 2020, doi: 10.1016/j.compositesb.2019.107496.

[18] A. Bandyopadhyay and B. Heer, "Additive manufacturing of MM structures," *Mater. Sci. Eng. R Rep.*, vol. 129, pp. 1–16, 2018, doi: 10.1016/j.mser.2018.04.001.

[19] A. Bandyopadhyay, K. D. Traxel, and S. Bose, "Nature-inspired materials and structures using 3D Printing," *Mater. Sci. Eng. R Rep.*, vol. 145, pp. 1–51, 2021, doi: 10.1016/j.mser.2021.100609.

[20] M. Dinar and D. W. Rosen, "A Design for Additive Manufacturing Ontology," *J. Comput. Inf. Sci. Eng.*, vol. 17, no. 2, Feb. 2017, doi: 10.1115/1.4035787.

[21] J. Jiang, Y. Xiong, Z. Zhang, and D. W. Rosen, "Machine learning integrated design for additive manufacturing," *J. Intell. Manuf.*, vol. 33, no. 4, pp. 1073–1086, 2022, doi: 10.1007/s10845-020-01715-6.

[22] D. Rosen and S. Kim, "Design and manufacturing implications of additive manufacturing," *J. Mater. Eng. Perform.*, vol. 30, no. 9, pp. 6426–6438, 2021, doi: 10.1007/s11665-021-06030-6.

[23] H. Yasuga *et al.*, "Fluid interfacial energy drives the emergence of three-dimensional periodic structures in micropillar scaffolds," *Nat. Phys.*, vol. 17, no. 7, pp. 794–800, 2021, doi: 10.1038/s41567-021-01204-4.

[24] X. Zheng, C. Williams, C. M. Spadaccini, and K. Shea, "Perspectives on MM additive manufacturing," *J. Mater. Res.*, vol. 36, no. 18, pp. 3549–3557, 2021, doi: 10.1557/s43578-021-00388-y.

[25] C. Yuan, F. Wang, B. Qi, Z. Ding, D. W. Rosen, and Q. Ge, "3D printing of MM composites with tunable shape memory behavior," *Mater. Des.*, vol. 193, p. 108785, 2020, doi: 10.1016/j.matdes.2020.108785.

[26] S. Kumar, R. Singh, T. P. Singh, and A. Batish, "Flexural, pull-out, and fractured surface characterization for MM 3D printed functionally graded prototype," *J. Compos. Mater.*, vol. 54, no. 16, pp. 2087–2099, Dec. 2019, doi: 10.1177/0021998319892067.

[27] S. Kumar, R. Singh, T. P. Singh, and A. Batish, "Comparison of mechanical and morphological properties of 3-D printed functional prototypes: Multi and hybrid blended thermoplastic matrix," *J. Thermoplast. Compos. Mater.*, vol. 35, no. 5, pp. 692–707, May 2020, doi: 10.1177/0892705720925136.

[28] I. Mustafa and T.-H. Kwok, "Development of Intertwined Infills to Improve MM Interfacial Bond Strength," *J. Manuf. Sci. Eng.*, vol. 144, no. 3, Aug. 2021, doi: 10.1115/1.4051884.

[29] H. Yazdani Sarvestani, A. H. Akbarzadeh, D. Therriault, and M. Lévesque, "Engineered bi-material lattices with thermo-mechanical programmability," *Compos. Struct.*, vol. 263, p. 113705, 2021, doi: 10.1016/j.compstruct.2021.113705.

[30] Y. Wang and X. Li, "4D-printed bi-material composite laminate for manufacturing reversible shape-change structures," *Compos. Part B Eng.*, vol. 219, p. 108918, 2021, doi: 10.1016/j.compositesb.2021.108918.

[31] D. Baca and R. Ahmad, "The impact on the mechanical properties of MM polymers fabricated with a single mixing nozzle and multi-nozzle systems via fused deposition modeling," *Int. J. Adv. Manuf. Technol.*, vol. 106, no. 9–10, pp. 4509–4520, 2020, doi: 10.1007/s00170-020-04937-3.

[32] T. Cadiou, F. Demoly, and S. Gomes, "A hybrid additive manufacturing platform based on fused filament fabrication and direct ink writing techniques for MM 3D

printing," *Int. J. Adv. Manuf. Technol.*, vol. 114, no. 11–12, pp. 3551–3562, 2021, doi: 10.1007/s00170-021-06891-0.

[33] G. Peralta Marino, S. De la Pierre, M. Salvo, A. Díaz Lantada, and M. Ferraris, "Modelling, additive layer manufacturing and testing of interlocking structures for joined components," *Sci. Rep.*, vol. 12, no. 1, pp. 1–11, 2022, doi: 10.1038/s41598-022-06521-z.

[34] R. Johnston and Z. Kazancı, "Analysis of additively manufactured (3D printed) dual-material auxetic structures under compression," *Addit. Manuf.*, vol. 38, p. 101783, 2021, doi: 10.1016/j.addma.2020.101783.

[35] L. R. Lopes, A. F. Silva, and O. S. Carneiro, "MM 3D printing: The relevance of materials affinity on the boundary interface performance," *Addit. Manuf.*, vol. 23, pp. 45–52, 2018, doi: 10.1016/j.addma.2018.06.027.

[36] D. Yavas, Q. Liu, Z. Zhang, and D. Wu, "Design and fabrication of architected MM lattices with tunable stiffness, strength, and energy absorption," *Mater. Des.*, vol. 217, p. 110613, 2022, doi: 10.1016/j.matdes.2022.110613.

[37] K. Y. Shah, A. F. Mohamed, and I. N. Tansel, "Additively manufactured MM parts with defect detection capabilities," *Procedia Manuf.*, vol. 39, pp. 493–501, 2019, doi: 10.1016/j.promfg.2020.01.406.

[38] J. Kluczyński *et al.*, "The examination of restrained joints created in the process of MM FFF additive manufacturing technology," *Materials* ., vol.13, no. 4, 2020, doi: 10.3390/ma13040903.

[39] J. Yin, M. Li, G. Dai, H. Zhou, L. Ma, and Y. Zheng, "3D printed MM medical phantoms for needle-tissue interaction modelling of heterogeneous structures," *J. Bionic Eng.*, vol. 18, no. 2, pp. 346–360, 2021, doi: 10.1007/s42235-021-0031-1.

[40] J. Yin, C. Lu, J. Fu, Y. Huang, and Y. Zheng, "Interfacial bonding during MM fused deposition modeling (FDM) process due to inter-molecular diffusion," *Mater. Des.*, vol. 150, pp. 104–112, 2018, doi: 10.1016/j.matdes.2018.04.029.

[41] N. R. Khatri and P. F. Egan, "Tailored energy absorption for 3D printed MM cellular structures using ABS and TPU." Nov. 01, 2021. doi: 10.1115/IMECE2021-73699.

[42] D. J. Roach, C. M. Hamel, C. K. Dunn, M. V. Johnson, X. Kuang, and H. J. Qi, "The m4 3D printer: A MM multi-method additive manufacturing platform for future 3D printed structures," *Addit. Manuf.*, vol. 29, p. 100819, 2019, doi: 10.1016/j.addma.2019.100819.

[43] D. L. Edelen and H. A. Bruck, "Predicting failure modes of 3D-printed MM polymer sandwich structures from process parameters," *J. Sandw. Struct. Mater.*, vol. 24, no. 2, pp. 1049–1075, Jul. 2021, doi: 10.1177/1099636221102 0445.

[44] S. Hasanov, A. Gupta, F. Alifui-Segbaya, and I. Fidan, "Hierarchical homogenization and experimental evaluation of functionally graded materials manufactured by the fused filament fabrication process," *Compos. Struct.*, vol. 275, p. 114488, 2021, doi: 10.1016/j.compstruct.2021.114488.

[45] G. L. Goh *et al.*, "Additively manufactured MM free-form structure with printed electronics," *Int. J. Adv. Manuf. Technol.*, vol. 94, no. 1–4, pp. 1309–1316, 2018, doi: 10.1007/s00170-017-0972-z.

[46] H. Jiang, P. Aihemaiti, W. Aiyiti, and A. Kasimu, "Study Of the compression behaviours of 3D-printed PEEK/CFR-PEEK sandwich composite structures," *Virtual Phys. Prototyp.*, vol. 17, no. 2, pp. 138–155, Apr. 2022, doi: 10.1080/17452759.2021.2014636.

[47] P. Feng *et al.*, "A Multimaterial scaffold with tunable properties: Toward bone tissue repair," *Adv. Sci.*, vol. 5, no. 6, p. 1700817, Jun. 2018, doi: 10.1002/advs. 201700817.

[48] D. Rahmatabadi, I. Ghasemi, M. Baniassadi, K. Abrinia, and M. Baghani, "3D printing of PLA-TPU with different component ratios: Fracture toughness, mechanical properties, and morphology," *J. Mater. Res. Technol.*, vol. 21, pp. 3970–3981, 2022, doi: 10.1016/j.jmrt.2022.11.024.

[49] X. N. Zhang, Q. Zheng, and Z. L. Wu, "Recent advances in 3D printing of tough hydrogels: A review," *Compos. Part B Eng.*, vol. 238, p. 109895, 2022, doi: 10.1016/j.compositesb.2022.109895.

[50] N. E. Putra, M. J. Mirzaali, I. Apachitei, J. Zhou, and A. A. Zadpoor, "MM additive manufacturing technologies for Ti-, Mg-, and Fe-based biomaterials for bone substitution," *Acta Biomater.*, vol. 109, pp. 1–20, 2020, doi: 10.1016/j.actbio. 2020.03.037.

[51] K. Chen *et al.*, "Selective laser melting 316L/CuSn10 MMs: Processing optimization, interfacial characterization and mechanical property," *J. Mater. Process. Technol.*, vol. 283, p. 116701, 2020, doi: 10.1016/j.jmatprotec.2020.116701.

[52] A. G. Demir and B. Previtali, "MM selective laser melting of Fe/Al-12Si components," *Manuf. Lett.*, vol. 11, pp. 8–11, 2017, doi: 10.1016/j.mfglet.2017. 01.002.

[53] K. Wei, X. Zeng, F. Li, M. Liu, and J. Deng, "Microstructure and mechanical property of Ti-5Al-2.5Sn/Ti-6Al-4V dissimilar titanium alloys integrally fabricated by selective laser melting," *Jom*, vol. 72, no. 3, pp. 1031–1038, 2020, doi: 10.1007/ s11837-019-03988-6.

[54] M. G. Scaramuccia, A. G. Demir, L. Caprio, O. Tassa, and B. Previtali, "Development of processing strategies for multigraded selective laser melting of Ti6Al4V and IN718," *Powder Technol.*, vol. 367, pp. 376–389, 2020, doi: 10.1016/ j.powtec.2020.04.010.

[55] M. J. Sagong *et al.*, "Interface characteristics and mechanical behavior of additively manufactured MM of stainless steel and Inconel," *Mater. Sci. Eng. A*, vol. 847, p. 143318, 2022, doi: 10.1016/j.msea.2022.143318.

[56] B. Heer and A. Bandyopadhyay, "Compositionally graded magnetic-nonmagnetic bimetallic structure using laser engineered net shaping," *Mater. Lett.*, vol. 216, pp. 16–19, 2018, doi: 10.1016/j.matlet.2017.12.129.

[57] S. Pfeiffer, H. Lorenz, Z. Fu, T. Fey, P. Greil, and N. Travitzky, "Al2O3/Cu-O composites fabricated by pressureless infiltration of paper-derived Al2O3 porous preforms," *Ceram. Int.*, vol. 44, no. 17, pp. 20835–20840, 2018, doi: 10.1016/ j.ceramint.2018.08.087.

[58] C. Cai *et al.*, "In-situ preparation and formation of TiB/Ti-6Al-4V nanocomposite via laser additive manufacturing: Microstructure evolution and tribological behavior," *Powder Technol.*, vol. 342, pp. 73–84, 2019, doi: 10.1016/j.powtec. 2018.09.088.

[59] H. Sahasrabudhe and A. Bandyopadhyay, "In situ reactive MM Ti6Al4V-calcium phosphate-nitride coatings for bio-tribological applications," *J. Mech. Behav. Biomed. Mater.*, vol. 85, pp. 1–11, 2018, doi: 10.1016/j.jmbbm.2018.05.020.

[60] A. Bandyopadhyay, S. Dittrick, T. Gualtieri, J. Wu, and S. Bose, "Calcium phosphate–titanium composites for articulating surfaces of load-bearing implants," *J. Mech. Behav. Biomed. Mater.*, vol. 57, pp. 280–288, 2016, doi: 10.1016/ j.jmbbm.2015.11.022.

[61] H. Ma *et al.*, "3D printing of high-strength bioscaffolds for the synergistic treatment of bone cancer," *NPG Asia Mater.*, vol. 10, no. 4, pp. 31–44, 2018, doi: 10.1038/ s41427-018-0015-8.

[62] G. D. Goh, Y. L. Yap, H. K. J. Tan, S. L. Sing, G. L. Goh, and W. Y. Yeong, "Process–structure–properties in polymer additive manufacturing via material extrusion: A review," *Crit. Rev. Solid State Mater. Sci.*, vol. 45, no. 2, pp. 113–133, Mar. 2020, doi: 10.1080/10408436.2018.1549977.

[63] E. Brancewicz-Steinmetz and J. Sawicki, "Bonding and strengthening the PLA biopolymer in MM additive manufacturing," *Materials .*, vol. 15, Aug. 2022, doi: 10.3390/ma15165563.

[64] E. Natarajan, K. Y. Chia, A. A. M. Faudzi, W. H. Lim, C. K. Ang, and A. Jafaari, "Bio inspired salamander robot with Pneu-Net Soft actuators – design and walking gait analysis," *Bull. Polish Acad. Sci. Tech. Sci.*, vol. 69, no. 3, pp. 1–11, 2021, doi: 10.24425/bpasts.2021.137055.

[65] A. Kottasamy, M. Samykano, K. Kadirgama, M. Rahman, and M. M. Noor, "Experimental investigation and prediction model for mechanical properties of copper-reinforced polylactic acid composites (Cu-PLA) using FDM-based 3D printing technique," *Int. J. Adv. Manuf. Technol.*, vol. 119, no. 7–8, pp. 5211–5232, 2022, doi: 10.1007/s00170-021-08289-4.

[66] Á. Díaz-García, J. Y. Law, M. Felix, A. Guerrero, and V. Franco, "Functional, thermal and rheological properties of polymer-based magnetic composite filaments for additive manufacturing," *Mater. Des.*, vol. 219, p. 110806, 2022, doi: 10.1016/j.matdes.2022.110806.

[67] X. Wei, I. Behm, T. Winkler, S. Scharf, X. Li, and R. Bähr, "Experimental study on metal parts under variable 3D printing and sintering orientations using bronze/PLA hybrid filament coupled with fused filament fabrication," *Materials*, vol. 15, no. 15. 2022. doi: 10.3390/ma15155333.

[68] R. Matsuzaki, T. Kanatani, and A. Todoroki, "MM additive manufacturing of polymers and metals using fused filament fabrication and electroforming," *Addit. Manuf.*, vol. 29, p. 100812, 2019, doi: 10.1016/j.addma.2019.100812.

[69] Y.-H. Chueh, C. Wei, X. Zhang, and L. Li, "Integrated laser-based powder bed fusion and fused filament fabrication for three-dimensional printing of hybrid metal/polymer objects," *Addit. Manuf.*, vol. 31, p. 100928, 2020, doi: 10.1016/j.addma.2019.100928.

[70] P. D. Enrique *et al.*, "Design of binder jet additive manufactured co-continuous ceramic-reinforced metal matrix composites," *J. Mater. Sci. Technol.*, vol. 49, pp. 81–90, 2020, doi: 10.1016/j.jmst.2020.01.053.

[71] B. G. Thiam, A. El Magri, H. R. Vanaei, and S. Vaudreuil, "3D printed and conventional membranes—A review," *Polymers*, vol. 14, no. 5. 2022. doi: 10.3390/polym14051023.

[72] D. Espalin, J. Alberto Ramirez, F. Medina, and R. Wicker, "MM, multi-technology FDM: Exploring build process variations," *Rapid Prototyp. J.*, vol. 20, no. 3, pp. 236–244, Jan. 2014, doi: 10.1108/RPJ-12-2012-0112.

[73] J. D. Gander and A. J. Giacomin, "Review of die lip buildup in plastics extrusion," *Polym. Eng. Sci.*, vol. 37, no. 7, pp. 1113–1126, 1997, doi: 10.1002/pen.11756.

[74] B. E. Carroll *et al.*, "Functionally graded material of 304L stainless steel and inconel 625 fabricated by directed energy deposition: Characterization and thermodynamic modeling," *Acta Mater.*, vol. 108, pp. 46–54, 2016, doi: 10.1016/j.actamat.2016.02.019.

[75] H. S. Ren, D. Liu, H. B. Tang, X. J. Tian, Y. Y. Zhu, and H. M. Wang, "Microstructure and mechanical properties of a graded structural material," *Mater. Sci. Eng. A*, vol. 611, pp. 362–369, 2014, doi: 10.1016/j.msea.2014.06.016.

[76] S. Hasanov *et al.*, "Review on additive manufacturing of MM parts: Progress and challenges," *J. Manuf. Mater. Process.*, vol. 6, no. 1. 2022. doi: 10.3390/jmmp6010004.

[77] W. Li *et al.*, "Fabrication and characterization of a functionally graded material from Ti-6Al-4V to SS316 by laser metal deposition," *Addit. Manuf.*, vol. 14, pp. 95–104, 2017, doi: 10.1016/j.addma.2016.12.006.

[78] F. Hengsbach *et al.*, "Inline additively manufactured functionally graded MMs: microstructural and mechanical characterization of 316L parts with H13 layers," *Prog. Addit. Manuf.*, vol. 3, no. 4, pp. 221–231, 2018, doi: 10.1007/s40964-018-0044-4.

[79] G. Xu, L. Wu, Y. Su, Z. Wang, K. Luo, and J. Lu, "Microstructure and mechanical properties of directed energy deposited 316L/Ti6Al4V functionally graded materials via constant/gradient power," *Mater. Sci. Eng. A*, vol. 839, p. 142870, 2022, doi: 10.1016/j.msea.2022.142870.

[80] M. Ostolaza, J. I. Arrizubieta, A. Lamikiz, and M. Cortina, "Functionally graded AISI 316L and AISI H13 manufactured by L-DED for die and mould applications," *Appl. Sci.*, vol. 11, no. 2. 2021. doi: 10.3390/app11020771.

[81] A. O. Aremu *et al.*, "A voxel-based method of constructing and skinning conformal and functionally graded lattice structures suitable for additive manufacturing," *Addit. Manuf.*, vol. 13, pp. 1–13, 2017, doi: 10.1016/j.addma.2016.10.006.

[82] M. Toursangsaraki, "A review of MM and composite parts production by modified additive manufacturing methods," 2018, [Online]. Available: http://arxiv.org/abs/1808.01861

[83] D. Qiu and N. A. Langrana, "Void eliminating toolpath for extrusion-based multi-material layered manufacturing," *Rapid Prototyp. J.*, vol. 8, no. 1, pp. 38–45, Jan. 2002, doi: 10.1108/13552540210413293.

[84] S. Bhashyam, K. Hoon Shin, and D. Dutta, "An integrated CAD system for design of heterogeneous objects," *Rapid Prototyp. J.*, vol. 6, no. 2, pp. 119–135, Jan. 2000, doi: 10.1108/13552540010323547.

[85] X. Yao, S. K. Moon, G. Bi, and J. Wei, "A MM part design framework in additive manufacturing," *Int. J. Adv. Manuf. Technol.*, vol. 99, no. 9–12, pp. 2111–2119, 2018, doi: 10.1007/s00170-018-2025-7.

[86] K.-H. Shin, H. Natu, D. Dutta, and J. Mazumder, "A method for the design and fabrication of heterogeneous objects," *Mater. Des.*, vol. 24, no. 5, pp. 339–353, 2003, doi: 10.1016/S0261-3069(03)00060-8.

[87] M. P. Bendsøe and N. Kikuchi, "Generating optimal topologies in structural design using a homogenization method," *Comput. Methods Appl. Mech. Eng.*, vol. 71, no. 2, pp. 197–224, 1988, doi: 10.1016/0045-7825(88)90086-2.

[88] H. Wargnier, F. X. Kromm, M. Danis, and Y. Brechet, "Proposal for a MM design procedure," *Mater. Des.*, vol. 56, pp. 44–49, 2014, doi: 10.1016/j.matdes.2013.11.004.

[89] J. Yang *et al.*, "Laser techniques for dissimilar joining of aluminum alloys to steels: A critical review," *J. Mater. Process. Technol.*, vol. 301, p. 117443, 2022, doi: 10.1016/j.jmatprotec.2021.117443.

[90] D.-S. Nguyen, H.-S. Park, and C.-M. Lee, "Applying selective laser melting to join Al and Fe: An investigation of dissimilar materials," *Appl. Sci.*, vol. 9, no. 15. 2019. doi: 10.3390/app9153031.

[91] M. Mehrpouya, D. Tuma, T. Vaneker, M. Afrasiabi, M. Bambach, and I. Gibson, "Multimaterial powder bed fusion techniques," *Rapid Prototyp. J.*, vol. 28, no. 11, pp. 1–19, 2022, doi: 10.1108/RPJ-01-2022-0014.

[92] R. B. Tipton, D. Hou, Z. Shi, T. M. Weller, and V. R. Bhethanabotla, "Optical interconnects on a flexible substrate by MM hybrid additive and subtractive manufacturing," *Addit. Manuf.*, vol. 48, p. 102409, 2021, doi: 10.1016/j.addma.2021.102409.

[93] Z. Kang, C. Wu, Y. Luo, and M. Li, "Robust topology optimization of MM structures considering uncertain graded interface," *Compos. Struct.*, vol. 208, pp. 395–406, 2019, doi: 10.1016/j.compstruct.2018.10.034.

[94] J. Y. Kim, D. Garcia, Y. Zhu, D. M. Higdon, and H. Z. Yu, "A Bayesian learning framework for fast prediction and uncertainty quantification of additively manufactured MM components," *J. Mater. Process. Technol.*, vol. 303, p. 117528, 2022, doi: 10.1016/j.jmatprotec.2022.117528.

[95] A. Cazón-Martín, M. Iturrizaga-Campelo, L. Matey-Muñoz, M. I. Rodríguez-Ferradas, P. Morer-Camo, and S. Ausejo-Muñoz, "Design and manufacturing of shin pads with MM additive manufactured features for football players: A comparison with commercial shin pads," *Proc. Inst. Mech. Eng. Part P J. Sport. Eng. Technol.*, vol. 233, no. 1, pp. 160–169, Nov. 2018, doi: 10.1177/175433711 8811266.

[96] S. Burggraeve, C. Lopez, and J. Stroobants, "Quantifying compactness and connectivity in multimaterial designs," *Int. Math. Forum*, vol. 15, no. 1, pp. 1–23, 2020, doi: 10.12988/imf.2020.91142.

[97] C. Qi, F. Jiang, and S. Yang, "Advanced honeycomb designs for improving mechanical properties: A review," *Compos. Part B Eng.*, vol. 227, p. 109393, 2021, doi: 10.1016/j.compositesb.2021.109393.

[98] A. Nazir, K. M. Abate, A. Kumar, and J. Y. Jeng, "A state-of-the-art review on types, design, optimization, and additive manufacturing of cellular structures," *Int. J. Adv. Manuf. Technol.*, vol. 104, no. 9–12, pp. 3489–3510, 2019, doi: 10.1007/ s00170-019-04085-3.

[99] Z. H. Liu, D. Q. Zhang, S. L. Sing, C. K. Chua, and L. E. Loh, "Interfacial characterization of SLM parts in MM processing: Metallurgical diffusion between 316L stainless steel and C18400 copper alloy," *Mater. Charact.*, vol. 94, pp. 116–125, 2014, doi: 10.1016/j.matchar.2014.05.001.

[100] M. Zhang, Y. Yang, M. Xu, J. Chen, and D. Wang, "Mechanical properties of MMs porous structures based on triply periodic minimal surface fabricated by additive manufacturing," *Rapid Prototyp. J.*, vol. 27, no. 9, pp. 1681–1692, 2021, doi: 10.1108/RPJ-10-2020-0254.

[101] J. W. Boley *et al.*, "Shape-shifting structured lattices via multimaterial 4D printing," *Proc. Natl. Acad. Sci. USA.*, vol. 116, no. 42, pp. 20856–20862, 2019, doi: 10.1073/ pnas.1908806116.

[102] Q. Ge *et al.*, "3D printing of highly stretchable hydrogel with diverse UV curable polymers," *Sci. Adv.*, vol. 7, no. 2, pp. 1–10, 2021, doi: 10.1126/sciadv.aba4261.

[103] J. Hascoet, P. Muller, and P. Mognol, "Manufacturing of complex parts with continuous functionally graded materials (FGM)," *22nd Annu. Int. Solid Free. Fabr. Symp. – An Addit. Manuf. Conf. SFF 2011*, Jan. 2011.

[104] K. Ramaswamy, R. M. O'Higgins, M. C. Corbett, M. A. McCarthy, and C. T. McCarthy, "Quasi-static and dynamic performance of novel interlocked hybrid metal-composite joints," *Compos. Struct.*, vol. 253, p. 112769, 2020, doi: 10.1016/ j.compstruct.2020.112769.

[105] D. Han and H. Lee, "Recent advances in MM additive manufacturing: Methods and applications," *Curr. Opin. Chem. Eng.*, vol. 28, pp. 158–166, 2020, doi: 10.1016/ j.coche.2020.03.004.

[106] M. Vaezi, S. Chianrabutra, B. Mellor, and S. Yang, "Multiple material additive manufacturing – Part 1: a review," *Virtual Phys. Prototyp.*, vol. 8, no. 1, pp. 19–50, Mar. 2013, doi: 10.1080/17452759.2013.778175.

[107] T. S. Lumpe, J. Mueller, and K. Shea, "Tensile properties of MM interfaces in 3D printed parts," *Mater. Des.*, vol. 162, pp. 1–9, 2019, doi: 10.1016/j.matdes. 2018.11.024.

[108] I. Q. Vu, L. B. Bass, C. B. Williams, and D. A. Dillard, "Characterizing the effect of print orientation on interface integrity of MM jetting additive manufacturing," *Addit. Manuf.*, vol. 22, pp. 447–461, 2018, doi: 10.1016/j.addma.2018.05.036.

[109] J. Brackett *et al.*, "Characterizing material transitions in large-scale Additive Manufacturing," *Addit. Manuf.*, vol. 38, p. 101750, 2021, doi: 10.1016/j.addma.2020.101750.

[110] B. Saleh *et al.*, "30 Years of functionally graded materials: An overview of manufacturing methods, applications and future challenges," *Compos. Part B Eng.*, vol. 201, p. 108376, 2020, doi: 10.1016/j.compositesb.2020.108376.

[111] A. S. K. Kiran *et al.*, "Additive manufacturing technologies: An overview of challenges and perspective of using electrospraying," *Nanocomposites*, vol. 4, no. 4, pp. 190–214, Oct. 2018, doi: 10.1080/20550324.2018.1558499.

[112] S. Wickramasinghe, T. Do, and P. Tran, "FDM-based 3D printing of polymer and associated composite: A review on mechanical properties, defects and treatments," *Polymers*, vol. 12, no. 7. 2020. doi: 10.3390/polym12071529.

[113] A. Gupta, S. Hasanov, I. Fidan, and Z. Zhang, "Homogenized modeling approach for effective property prediction of 3D-printed short fibers reinforced polymer matrix composite material," *Int. J. Adv. Manuf. Technol.*, vol. 118, no. 11–12, pp. 4161–4178, 2022, doi: 10.1007/s00170-021-08230-9.

[114] S. Hasanov, A. Gupta, A. Nasirov, and I. Fidan, "Mechanical characterization of functionally graded materials produced by the fused filament fabrication process," *J. Manuf. Process.*, vol. 58, pp. 923–935, 2020, doi: 10.1016/j.jmapro.2020.09.011.

[115] F. Leoni, Ø. Grong, A. Celotto, H. G. Fjær, P. Ferro, and F. Berto, "Process modelling applied to aluminium-steel butt welding by hybrid metal extrusion and bonding (HYB)," *Metals*, vol. 12, no. 10. 2022. doi: 10.3390/met12101656.

[116] M. Saari, B. Xia, B. Cox, P. S. Krueger, A. L. Cohen, and E. Richer, "Fabrication and analysis of a composite 3D printed capacitive force sensor," *3D Print. Addit. Manuf.*, vol. 3, no. 3, pp. 136–141, Sep. 2016, doi: 10.1089/3dp.2016.0021.

[117] D. Espalin, D. W. Muse, E. MacDonald, and R. B. Wicker, "3D printing multifunctionality: Structures with electronics," *Int. J. Adv. Manuf. Technol.*, vol. 72, no. 5–8, pp. 963–978, 2014, doi: 10.1007/s00170-014-5717-7.

[118] K. M. M. Billah, J. L. Coronel, L. Chavez, Y. Lin, and D. Espalin, "Additive manufacturing of multimaterial and multifunctional structures via ultrasonic embedding of continuous carbon fiber," *Compos. Part C Open Access*, vol. 5, p. 100149, 2021, doi: 10.1016/j.jcomc.2021.100149.

[119] K. M. M. Billah *et al.*, "Large-scale additive manufacturing of self-heating molds," *Addit. Manuf.*, vol. 47, p. 102282, 2021, doi: 10.1016/j.addma.2021.102282.

[120] A. N. Dickson, J. N. Barry, K. A. McDonnell, and D. P. Dowling, "Fabrication of continuous carbon, glass and Kevlar fibre reinforced polymer composites using additive manufacturing," *Addit. Manuf.*, vol. 16, pp. 146–152, 2017, doi: 10.1016/j.addma.2017.06.004.

[121] Y. Chen and L. Ye, "Topological design for 3D-printing of carbon fibre reinforced composite structural parts," *Compos. Sci. Technol.*, vol. 204, p. 108644, 2021, doi: 10.1016/j.compscitech.2020.108644.

[122] T. DebRoy *et al.*, "Additive manufacturing of metallic components – Process, structure and properties," *Prog. Mater. Sci.*, vol. 92, pp. 112–224, 2018, doi: 10.1016/j.pmatsci.2017.10.001.

[123] O. Gokcekaya, N. Hayashi, T. Ishimoto, K. Ueda, T. Narushima, and T. Nakano, "Crystallographic orientation control of pure chromium via laser powder bed fusion and improved high temperature oxidation resistance," *Addit. Manuf.*, vol. 36, p. 101624, 2020, doi: 10.1016/j.addma.2020.101624.

[124] C. Wei, Z. Zhang, D. Cheng, Z. Sun, M. Zhu, and L. Li, "An overview of laser-based multiple metallic material additive manufacturing: from macro- to micro-scales," *Int. J. Extrem. Manuf.*, vol. 3, no. 1, p. 12003, 2021, doi: 10.1088/2631-7990/abce04.

[125] S. L. Sing *et al.*, "Emerging metallic systems for additive manufacturing: In-situ alloying and multi-metal processing in laser powder bed fusion," *Prog. Mater. Sci.*, vol. 119, p. 100795, 2021, doi: 10.1016/j.pmatsci.2021.100795.

[126] O. Gokcekaya *et al.*, "Effect of Scan Length on Densification and Crystallographic Texture Formation of Pure Chromium Fabricated by Laser Powder Bed Fusion," *Crystals*, vol. 11, no. 1. 2021. doi: 10.3390/cryst11010009.

[127] O. Gokcekaya, T. Ishimoto, T. Todo, P. Wang, and T. Nakano, "Influence of powder characteristics on densification via crystallographic texture formation: Pure tungsten prepared by laser powder bed fusion," *Addit. Manuf. Lett.*, vol. 1, p. 100016, 2021, doi: 10.1016/j.addlet.2021.100016.

[128] F. Khodabakhshi, M. H. Farshidianfar, S. Bakhshivash, A. P. Gerlich, and A. Khajepour, "Dissimilar metals deposition by directed energy based on powder-fed laser additive manufacturing," *J. Manuf. Process.*, vol. 43, pp. 83–97, 2019, doi: 10.1016/j.jmapro.2019.05.018.

[129] J. A. Glerum *et al.*, "Operando X-ray diffraction study of thermal and phase evolution during laser powder bed fusion of Al-Sc-Zr elemental powder blends," *Addit. Manuf.*, vol. 55, p. 102806, 2022, doi: 10.1016/j.addma.2022.102806.

[130] Y. Bai and C. B. Williams, "Binder jetting additive manufacturing with a particle-free metal ink as a binder precursor," *Mater. Des.*, vol. 147, pp. 146–156, 2018, doi: 10.1016/j.matdes.2018.03.027.

[131] M. Ziaee and N. B. Crane, "Binder jetting: A review of process, materials, and methods," *Addit. Manuf.*, vol. 28, pp. 781–801, 2019, doi: 10.1016/j.addma.2019.05.031.

[132] C. S. Carrillo and M. Sanchez, "Design and 3D printing of four multimaterial mechanical metamaterial using polyjet technology and digital materials for impact injury prevention," *Proc. Annu. Int. Conf. IEEE Eng. Med. Biol. Soc. EMBS*, pp. 4916–4919, 2021, doi: 10.1109/EMBC46164.2021.9630675.

[133] W. Xu *et al.*, "3D printing for polymer/particle-based processing: A review," *Compos. Part B Eng.*, vol. 223, p. 109102, 2021, doi: 10.1016/j.compositesb.2021.109102.

[134] G. Mattana, A. Loi, M. Woytasik, M. Barbaro, V. Noël, and B. Piro, "Inkjet-printing: A new fabrication technology for organic transistors," *Adv. Mater. Technol.*, vol. 2, no. 10, p. 1700063, 2017, doi: 10.1002/admt.201700063.

[135] M. J. Dantin, W. M. Furr, and M. W. Priddy, "Towards an open-source, preprocessing framework for simulating material deposition for a directed energy deposition process," *Solid Free. Fabr. 2018 Proc. 29th Annu. Int. Solid Free. Fabr. Symp. – An Addit. Manuf. Conf. SFF 2018*, no. January 2018, pp. 1903–1912, 2020.

[136] M. Srivastava, S. Rathee, A. Tiwari, and M. Dongre, "Wire arc additive manufacturing of metals: A review on processes, materials and their behaviour," *Mater. Chem. Phys.*, vol. 294, p. 126988, 2023, doi: 10.1016/j.matchemphys.2022.126988.

[137] A. Raj Paul, A. Mishra, M. Mukherjee, and D. Singh, "Stainless steel to aluminium joining by interfacial doping with Al2O3 powder in wire arc direct energy deposition process," *Mater. Lett.*, vol. 330, p. 133349, 2023, doi: 10.1016/j.matlet.2022.133349.

[138] J. Wang, K. Zhu, W. Zhang, X. Zhu, and X. Lu, "Microstructural and defect evolution during WAAM resulting in mechanical property differences for AA5356 component," *J. Mater. Res. Technol.*, vol. 22, pp. 982–996, 2023, doi: 10.1016/j.jmrt.2022.11.116.

[139] A. Bandyopadhyay, K. D. Traxel, M. Lang, M. Juhasz, N. Eliaz, and S. Bose, "Alloy design via additive manufacturing: Advantages, challenges, applications and perspectives," *Mater. Today*, vol. 52, pp. 207–224, 2022, doi: 10.1016/j.mattod.2021.11.026.

[140] E. Weflen and M. C. Frank, "Hybrid additive and subtractive manufacturing of MM objects," *Rapid Prototyp. J.*, vol. 27, no. 10, pp. 1860–1871, Jan. 2021, doi: 10.1108/RPJ-06-2020-0142.

[141] J. P. M. Pragana, R. F. V Sampaio, I. M. F. Bragança, C. M. A. Silva, and P. A. F. Martins, "Hybrid metal additive manufacturing: A state–of–the-art review," *Adv. Ind. Manuf. Eng.*, vol. 2, p. 100032, 2021, doi: 10.1016/j.aime.2021.100032.

[142] J. M. Flynn, A. Shokrani, S. T. Newman, and V. Dhokia, "Hybrid additive and subtractive machine tools – Research and industrial developments," *Int. J. Mach. Tools Manuf.*, vol. 101, pp. 79–101, 2016, doi: 10.1016/j.ijmachtools.2015.11.007.

[143] M. Schneck, M. Horn, M. Schmitt, C. Seidel, G. Schlick, and G. Reinhart, "Review on additive hybrid- and MM-manufacturing of metals by powder bed fusion: State of technology and development potential," *Prog. Addit. Manuf.*, vol. 6, no. 4, pp. 881–894, 2021, doi: 10.1007/s40964-021-00205-2.

[144] R. R. Ma, J. T. Belter, and A. M. Dollar, "Hybrid deposition manufacturing: Design strategies for multimaterial mechanisms via three-dimensional printing and material deposition," *J. Mech. Robot.*, vol. 7, no. 2, May 2015, doi: 10.1115/1.4029400.

[145] U. M. Dilberoglu, V. Haseltalab, U. Yaman, and M. Dolen, "Simulator of an additive and subtractive type of hybrid manufacturing system," *Procedia Manuf.*, vol. 38, pp. 792–799, 2019, doi: 10.1016/j.promfg.2020.01.110.

[146] P. Didier *et al.*, "Consideration of additive manufacturing supports for post-processing by end milling: a hybrid analytical–numerical model and experimental validation," *Prog. Addit. Manuf.*, vol. 7, no. 1, pp. 15–27, 2022, doi: 10.1007/s40964-021-00211-4.

[147] X. Peng, L. Kong, J. Y. H. Fuh, and H. Wang, "A review of post-processing technologies in additive manufacturing," *J. Manuf. Mater. Process.*, vol. 5, no. 2, 2021, doi: 10.3390/jmmp5020038.

[148] B. Li, B. Qian, Y. Xu, Z. Liu, and F. Xuan, "Fine-structured CoCrFeNiMn high-entropy alloy matrix composite with 12 wt% TiN particle reinforcements via selective laser melting assisted additive manufacturing," *Mater. Lett.*, vol. 252, pp. 88–91, 2019, doi: 10.1016/j.matlet.2019.05.108.

[149] B. Blakey-Milner *et al.*, "Metal additive manufacturing in aerospace: A review," *Mater. Des.*, vol. 209, p. 110008, 2021, doi: 10.1016/j.matdes.2021.110008.

[150] S. D. Nath and S. Nilufar, "An overview of additive manufacturing of polymers and associated composites," *Polymers*, vol. 12, no. 11. 2020. doi: 10.3390/polym12112719.

[151] H. Zhou, A. Mohammadi, D. Oetomo, and G. Alici, "A novel monolithic soft robotic thumb for an anthropomorphic prosthetic hand," *IEEE Robot. Autom. Lett.*, vol. 4, no. 2, pp. 602–609, 2019, doi: 10.1109/LRA.2019.2892203.

[152] T. Hainsworth, L. Smith, S. Alexander, and R. MacCurdy, "A fabrication free, 3D Printed, MM, self-sensing soft actuator," *IEEE Robot. Autom. Lett.*, vol. 5, no. 3, pp. 4118–4125, 2020, doi: 10.1109/LRA.2020.2986760.

[153] J. F. Christ, N. Aliheidari, A. Ameli, and P. Pötschke, "3D printed highly elastic strain sensors of multiwalled carbon nanotube/thermoplastic polyurethane nano-composites," *Mater. Des.*, vol. 131, pp. 394–401, 2017, doi: 10.1016/j.matdes.2017.06.011.

[154] V. Correia *et al.*, "Design and fabrication of multilayer inkjet-printed passive components for printed electronics circuit development," *J. Manuf. Process.*, vol. 31, pp. 364–371, 2018, doi: 10.1016/j.jmapro.2017.11.016.

[155] R. Mikkonen, P. Puistola, I. Jönkkäri, and M. Mäntysalo, "Inkjet printable polydimethylsiloxane for all-inkjet-printed multilayered soft electrical applications," *ACS Appl. Mater. Interfaces*, vol. 12, no. 10, pp. 11990–11997, 2020, doi: 10.1021/acsami.9b19632.

[156] J. O. Hardin, C. A. Grabowski, M. Lucas, M. F. Durstock, and J. D. Berrigan, "All-printed multilayer high voltage capacitors with integrated processing feedback," *Addit. Manuf.*, vol. 27, pp. 327–333, 2019, doi: 10.1016/j.addma.2019.02.011.

[157] M.-L. Seol et al., "All 3D printed energy harvester for autonomous and sustainable resource utilization," *Nano Energy*, vol. 52, pp. 271–278, 2018, doi: 10.1016/j.nanoen.2018.07.061.

[158] J.-Y. Lee, J. An, and C. K. Chua, "Fundamentals and applications of 3D printing for novel materials," *Appl. Mater. Today*, vol. 7, pp. 120–133, 2017, doi: 10.1016/j.apmt.2017.02.004.

[159] G. Andrikopoulos, G. Nikolakopoulos, and S. Manesis, "A survey on applications of pneumatic artificial muscles," *2011 19th Mediterr. Conf. Control Autom. MED 2011*, no. June, pp. 1439–1446, 2011, doi: 10.1109/MED.2011.5982983.

[160] F. Liravi and E. Toyserkani, "Additive manufacturing of silicone structures: A review and prospective," *Addit. Manuf.*, vol. 24, pp. 232–242, 2018, doi: 10.1016/j.addma.2018.10.002.

[161] S. Bose, D. Ke, H. Sahasrabudhe, and A. Bandyopadhyay, "Additive manufacturing of biomaterials," *Prog. Mater. Sci.*, vol. 93, pp. 45–111, 2018, doi: 10.1016/j.pmatsci.2017.08.003.

[162] S. Walker, O. D. Yirmibeşoğlu, U. Daalkhaijav, and Y. Mengüç, "14 – Additive manufacturing of soft robots," in *Woodhead Publishing in Materials*, S. M. Walsh, M. S. B. T.-R. S., and A. P. Strano, Eds., Woodhead Publishing, 2019, pp. 335–359. doi: 10.1016/B978-0-08-102260-3.00014-7.

[163] R. T. Shafranek, S. C. Millik, P. T. Smith, C.-U. Lee, A. J. Boydston, and A. Nelson, "Stimuli-responsive materials in additive manufacturing," *Prog. Polym. Sci.*, vol. 93, pp. 36–67, 2019, doi: 10.1016/j.progpolymsci.2019.03.002.

[164] H. Li, J. Yao, P. Zhou, X. Chen, Y. Xu, and Y. Zhao, "High-force soft pneumatic actuators based on novel casting method for robotic applications," *Sensors Actuators A Phys.*, vol. 306, p. 111957, 2020, doi: 10.1016/j.sna.2020.111957.

[165] J. Shintake, B. Schubert, S. Rosset, H. Shea, and D. Floreano, "Variable stiffness actuator for soft robotics using dielectric elastomer and low-melting-point alloy," in *2015 IEEE/RSJ International Conference on Intelligent Robots and Systems (IROS)*, 2015, pp. 1097–1102. doi: 10.1109/IROS.2015.7353507.

[166] Y.-J. Seol, H.-W. Kang, S. J. Lee, A. Atala, and J. J. Yoo, "Bioprinting technology and its applications," *Eur. J. Cardio-Thoracic Surg.*, vol. 46, no. 3, pp. 342–348, Sep. 2014, doi: 10.1093/ejcts/ezu148.

[167] N. Ashammakhi et al., "Bioinks and bioprinting technologies to make heterogeneous and biomimetic tissue constructs," *Mater. Today Bio*, vol. 1, p. 100008, 2019, doi: 10.1016/j.mtbio.2019.100008.

[168] Ž. P. Kačarević et al., "An introduction to 3D bioprinting: Possibilities, challenges and future aspects," *Materials*, vol. 11, no. 11. 2018. doi: 10.3390/ma11112199.

[169] J. M. Lee and W. Y. Yeong, "Design and printing strategies in 3D bioprinting of cell-hydrogels: A review," *Adv. Healthc. Mater.*, vol. 5, no. 22, pp. 2856–2865, Nov. 2016, doi: 10.1002/adhm.201600435.

[170] S. Giwa et al., "The promise of organ and tissue preservation to transform medicine.," *Nat. Biotechnol.*, vol. 35, no. 6, pp. 530–542, Jun. 2017, doi: 10.1038/nbt.3889.

[171] D. Singh, A. Mathur, S. Arora, S. Roy, and N. Mahindroo, "Journey of organ on a chip technology and its role in future healthcare scenario," *Appl. Surf. Sci. Adv.*, vol. 9, p. 100246, 2022, doi: 10.1016/j.apsadv.2022.100246.

[172] S. H. Jariwala, G. S. Lewis, Z. J. Bushman, J. H. Adair, and H. J. Donahue, "3D printing of personalized artificial bone scaffolds.," *3D Print. Addit. Manuf.*, vol. 2, no. 2, pp. 56–64, Jun. 2015, doi: 10.1089/3dp.2015.0001.

[173] S. Siebelink, H. Voordijk, M. Endedijk, and A. Adriaanse, "Understanding barriers to BIM implementation: Their impact across organizational levels in relation to BIM maturity," *Front. Eng. Manag.*, vol. 8, no. 2, pp. 236–257, 2021, doi: 10.1007/s42524-019-0088-2.

[174] S. Keating, "Beyond 3D printing: The new dimensions of additive fabrication," *Des. Emerg. Technol. UX Genom., Robot. Internet Things*, no. July, pp. 379–405, 2015.

[175] N. Oxman, S. Keating, and E. Tsai, "Functionally graded rapid prototyping," *Innov. Dev. Virtual Phys. Prototyp. – Proc. 5th Int. Conf. Adv. Res. Rapid Prototyp.*, no. November, pp. 483–489, 2012, doi: 10.1201/b11341-78.

[176] F. Craveiro, S. Nazarian, H. Bartolo, P. J. Bartolo, and J. Pinto Duarte, "An automated system for 3D printing functionally graded concrete-based materials," *Addit. Manuf.*, vol. 33, p. 101146, 2020, doi: 10.1016/j.addma.2020.101146.

[177] M. Mehrpouya, D. Tuma, T. Vaneker, M. Afrasiabi, M. Bambach, and I. Gibson, "Multimaterial powder bed fusion techniques," *Rapid Prototyp. J.*, vol. 28, no. 11, pp. 1–19, Jan. 2022, doi: 10.1108/RPJ-01-2022-0014.

[178] C. Gullipalli, N. Thawari, P. Burad, and T. V. K. Gupta, "Residual stresses and distortions in additive manufactured Inconel 718," *Mater. Manuf. Process.*, vol. 38, no. 12, pp. 1549–1560, Sep. 2023, doi: 10.1080/10426914.2023.2165663.

[179] P. Muller, P. Mognol, and J. Hascoet, *Functionally Graded Material (FGM) Parts: From Design to the Manufacturing Simulation*, vol. 4. 2012. doi: 10.1115/ESDA2012-82586.

[180] J. C. Steuben, A. P. Iliopoulos, and J. G. Michopoulos, "Implicit slicing for functionally tailored additive manufacturing," *Comput. Des.*, vol. 77, pp. 107–119, 2016, doi: 10.1016/j.cad.2016.04.003.

9 Application of 3D Printing for Energy Storage Devices

Rudranarayan Kandi, Dipesh Kumar Mishra, and Ankit Tripathi

9.1 INTRODUCTION

In recent days, the demand for electrochemical-based energy storage devices (ESDs) in modern society has increased exponentially. Besides, traditional ESDs are associated with problems of poor functionality, thus it does not meet the energy demands in current situations [1]. Conventional manufacturing techniques were found to be one of the major aspects connected with the poor performance of traditional ESDs. In addition, the other aspects such as life cycle, safety cost, power density, energy density, and consistency of traditional ESDs cannot be balanced with the need of current applications. In this context, academics and industries have done significant work to overcome the previously mentioned problems by exploring new materials of cathode/anode, improving the life cycle, power density, and cost and consistency metrics of ESDs that remain not able to meet the demand in current situations [2]. Moreover, consistency and cost are vital parameters for ESDs, particularly during real-load applications. Therefore, it is necessary to develop a new manufacturing technique that provides better structural ESDs in all aspects and further fills the gap of demand for commercial applications [3,4].

Currently, the manufactured ESDs are mainly shown with different internal structures (such as fiber structure, layered structured, and specific pore types) that are widely used for practical applications. In the last decades, a newly developed advanced technique such as additive manufacturing (popularly known as 3D Printing) is widely used for the fabrication of internal structures of ESDs. The accuracy of 3D printing was inferred as a main advantage to meet the demand of commercial applications [3,5,6]. Thereby, 3D printing is popularly studied in various applications such as in the biomedical field, electronics and micro fluids, etc. Mainly, the process using 3D printing for the manufacturing of any structural product is based on the layer-by-layer deposition techniques that follow the path provided by CAD software. In the area of ESDs, studies have rated that the devices such as structural electrodes, separators, charge collectors, and package materials manufactured using 3D printing have shown quick transfer of electrons/

DOI: 10.1201/9781003484325-9

ions between electrode and electrolyte [7]. The large exposed contact area between the electrode and electrolyte was found as the main reason associated with the quick electrons/ions. Another advantage of the 3D printing route has opened the doors of areas of unexplored regions of ESDs in the last decade [8,9].

Although, the performance of the common electrodes, and ESDs manufactured using conventional/traditional techniques, are always questionable because poor consistency of performance properties was found in the prepared ESDs. Moreover, the steps involved in the conventional manufacturing techniques were also too tedious. In this context, the example of manufacturing Lithium-ion batteries using conventional routes consisted of materials mixing, slurry preparations, electrode preparation, electrode cutting, electrolyte interactions, and other different processes [10–12]. These different processes proceeded step by step, therefore the cost and time owned for the preparation of Lithium-ion batteries were high. These too-long processing steps involved in conventional techniques for the fabrication of batteries induced a major effect on the life cycle, accuracy, and performance of the manufactured electrodes. Contrarily, freeze casting, chemical vapor deposition, and intertwining techniques were found more advanced routes for the manufacturing of ESDs. However, these techniques were associated with problems of complex processing steps and high fabrication costs. Therefore, these techniques are not able to find suitable options against traditional processes. Besides, the traditional processes were also connected with the problem of imprecise control over the parameters that played a major role in the context of controlling the performance characteristics of manufactured electrodes/ESDs [11,13].

In contrast, 3D printing has many different advantages and differences over traditional and complex manufacturing routes (the same can be observed in Table 9.1) [14]. The first advantage of 3D printing is the fast and repeatable production of electrodes; packaging materials and current collectors of ESDs can be produced from a single 3D printing machine, which simplifies the production cost, consistency, and accuracy of manufactured products [15]. Second, the internal characteristics such as microstructural properties and dimensional parameters of fabricated electrodes, current collectors, and packaging materials of ESDs are governable by regulating the properties of prepared inks and predefined programming parameters in 3D printing. Further, areal loading characteristics and better electrochemical performance of fabricated electrodes can be controlled using 3D printing technologies. From the previously mentioned advantages, it could be observed that the 3D printing manufacturing route is one of the most demanding processes and promising routes for the fabrication of ESDs [16]. Moreover, 3D printing is also found capable of solving the problems of wastage of raw materials that generally occurred with traditional manufacturing routes. In the present chapter, we are providing a brief introduction regarding the classifications of 3D printing technologies and then emphasis our attention on the progress of 3D printing in the context of preparing the ESDs. Further, in this section, we will discuss the materials adopted for the fabrications of ESD [17–20].

TABLE 9.1

Details of Advantages and Differences of 3D Printing Technologies over Traditional Processes

SL No.	Processes	Characteristics
1.	3D printing	Requires minimum time for the preparation of electrodes/ charge collectors/ packaging materials for ESDs.
		Easy to regulate the microstructural properties and areal load-bearing characteristics of structural ESDs.
		Less wastage of raw materials.
		Easy to produce intricate shapes and objects.
2.	Traditional processes (Like materials mixing, slurry preparations, electrode preparation, electrode cutting, electrolyte interactions, electrode fabrication, electrolyte addition, device assembly, and other different processes)	Too many steps are involved which makes the process too tedious and time-consuming.
		Difficult to control the dimensional accuracy.
		Imprecise control over microstructural and load-bearing characteristics of ESDs.
		Poor control over the waste of raw materials.
		Challenging for the intricate shapes and objects.

9.2   CLASSIFICATIONS OF 3D PRINTING TECHNOLOGIES

The process starts with 3D printing with the development of a virtual computer-aided design (CAD) model. The virtual model was achieved by designing software such as Creo, Solid Works, or AutoCAD. Once the 3D CAD model has been developed the model is saved into .stl (Standard Tessellation Language) format. The .stl format helps to store the details of the model's surface information in the coordinates of triangulated section. The .stl file format has characteristics of a *de-facto* standard (i.e., this is a universally acceptable file format for all 3D printers). Further, the .stl file converts into "G-codes" format after the slicing process. Slicing includes the generation of a 2D cross-section of the entire 3D model. Then, the printer starts the fabrication of the desired model by the deposition of materials in a layer-by-layer fashion [21]. The current 3D printing technologies that are most explored for fabricating the structural ESDs are direct ink writing (DIW), fused deposition modeling (FDM), stereolithography (SLA), inkjet printing (IJP), freeze nano printing (FNP), and selective laser melting (SLM). Among them, DIW, SLM, FDM, IJP, and SLA are widely used for the fabrication of structural ESDs [22]. A brief introduction to widely used 3D printing for the fabrication of structural ESDs is discussed here.

9.2.1   DIRECT INK WRITING

DIW is an extrusion-based 3D printing process wherein semi-liquid phase ink is dispensed through a small size nozzle under a controlled flow rate and deposited on

(a)   (b)

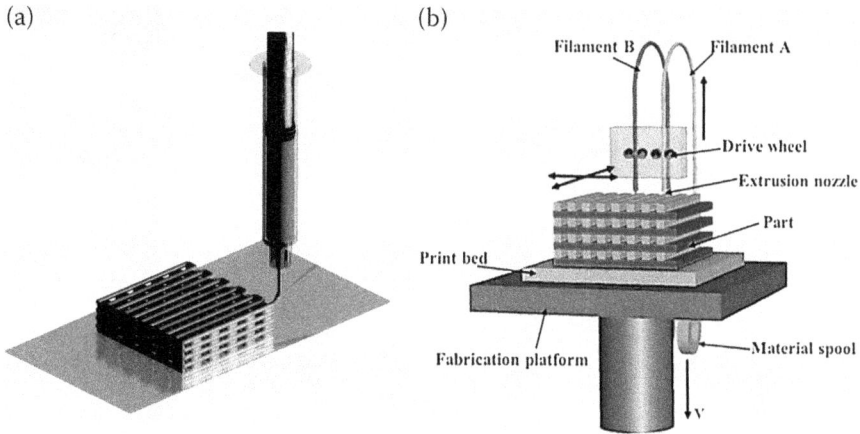

FIGURE 9.1 Schematic diagram of 3D printer (a) direct ink writing [24], and (b) fused deposition modeling.

the substrate as per the path provided by the sliced file of the 3D model. It is the easiest technique among other 3D printing techniques, and it is compatible with all types of materials like plastics, powders, and composites. DIW-based 3D printing consisted of three stages: (1) preparation of semi-liquid phase ink, (2) deposition of ink on the substrate as per the path provided by CAD software, and (3) drying of deposited part; the same can be inferred from Figure 9.1(a) [23].

DIW has recently gained noticeable attention in the field of ESDs because this unique process can print any miniature part by extruding the filament at room temperature. Mainly, solid features having lateral dimensions of less than 1 micrometer can be smoothly printed using a DIW-based 3D printing process [24]. Moreover, the rheological properties of semi-liquid phase ink in DIW have played a vital role in controlling the properties of the prepared structure. In this context, the shear-thinning behavior of prepared ink was found better characteristic to build any intricate structure [25–29]. Printable ink is the main key factor, especially in DIW-based technology. Here, are the two important recommendations that are associated with preparing for 3D-printed ESDs. The first one is the shear-thinning behavior of prepared ink that we mentioned in the earlier paragraph. In brief, shear-thinning behavior resembles that with an increase in shear rate, the viscosity of ink decreases. Further, it will smooth the flow of ink through the nozzle and fabricate intricate structures. Another one is the mechanical strength of prepared ink that aids to prepare robust structures that can be used for mechanical foundation/support. Li et al. [23] have presented the earliest study of the fabrication of an energy storage structural device for LIBs device using prepared ink by DIW technology (from Figure 9.2(a)).

In this study, the authors prepared the ink by mixing LMO powder, PVDF, and CB in NMP solvent with a regulated or optimized ratio of amounts. The results revealed that an advanced structural electrode with better performance of mechanical properties was prepared by using an optimized ratio of mixture

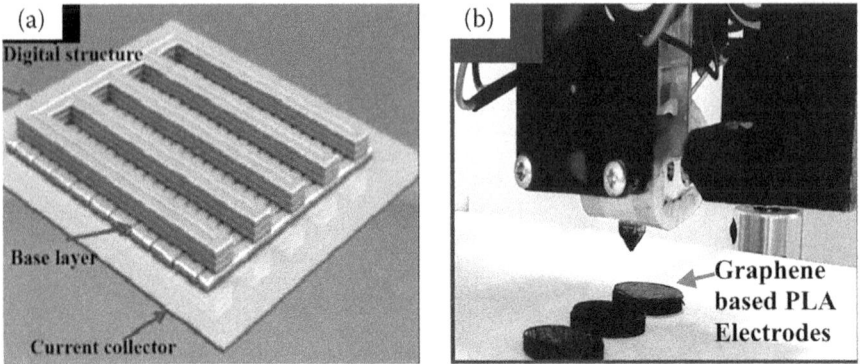

FIGURE 9.2 (a) Schematic diagram of hybrid 3D structure prepared by DIW, and (b) optical image of graphene-based PLA electrodes fabricated using fused deposition modeling (FDM) technique.

amounts. Moreover, a higher utilization rate of Li+ ions was observed in prepared 3D electrodes, which were found much more advanced than traditional electrodes. Besides, Park et al. [30] have stated another printable ink that was used to create the densified 3D structural electrodes. These dense 3D structures were fabricated by curable acrylate-based materials whose rheological properties were found very highly controllable. Thereby, better mechanical property-inhibited electrode 3D structures were successfully prepared. Wang et al. [31] employed DIW to showcase the creation of all-fiber Li-ion batteries, holding potential for wearable electronics applications. In this approach, inks resembling industrial slurries utilized in Li-ion battery production (composed of active material particles – either LTO or LFP –, PVDF-co-HFP, and carbon nanotubes) were developed. This ink formulation was carefully adjusted to possess superior yield stress and storage modulus compared to typical slurries. The intention behind this modification was to improve the printability of the inks. By immersing single fibers into an ethanol coagulation bath after printing, followed by soaking in a gel electrolyte and subsequent twisting to form a 1D battery, the researchers successfully generated these all-fiber batteries. Notably, the presence of the polymeric binder within the fiber electrodes provided both flexibility and mechanical strength, enabling seamless integration into textile materials.

9.2.2 Fused Deposition Modeling

In the FDM technique, a continuous filament of thermoplastic materials like polylactic acid (PLA), acrylonitrile butadiene styrene (ABS), and polyamide are used to fabricate the desired 3D model. The filament is heated in nozzle head above the melting point of materials and further, the materials are extruded on the platform. The thermoplasticity of extruded materials is decided by the adhering property between the deposited layers (can be seen in Figure 9.1(b)). Moreover, the layer thickness, infill density, printing speed, and orientation of filament deposition are the main process parameters that proportionally affect the mechanical properties of printed parts. Moreover, limited thermoplastic materials, poor mechanical strength, as well as poor surface properties, are the main demerits of FDM-based 3D-printed parts [31].

In recent studies, graphene-based polylactic acid (PLA) filaments have been utilized for the fabrication of 3D disc electrode structures for ESDs by using the FDM technique (this can be inferred from Figure 9.2(b)). These 3D-printed electrodes were associated with the characteristics of better electrochemical and physicochemical properties. Further, it was applied for freestanding electrodes of lithium-ion batteries and solid-state supercapacitors. While the electrochemical performance of the prepared electrode was found not very good. Even, this work has shown the possibility to fabricate intricate structure electrodes using 3D printing techniques [32].

The formulation of the negative electrode of Li+ ions battery using FDM technology has been reported by Maurrel et al. [32]. In this study, the authors initially prepared the composite filament by synthesizing Li_2TP (Lithium Terephthalate) and PLA in an extruder. The printability of the FDM was improved by the incorporation of poly dimethyl ether as a plasticizer while the electrical performances were improved via mixing carbon black. Thereby, 3D-printed electrode discs were printed with different infill (30%) patterns (rectilinear, gyroid, and archimedean) (can be observed in Figure 9.3) [32]. The low infill density pattern in fabricated electrodes induced high impregnation tendency of electrolytes. Thereby, better interaction between electrode and electrolyte was established, consequently, the electrical capacity of electrodes has been improved. Foster et al. [33] detailed the production process of a 3D-printed negative electrode disc, measuring 1 mm in thickness, using a commercially available graphene-based polylactic acid filament (graphene/PLA) as the primary material resource.

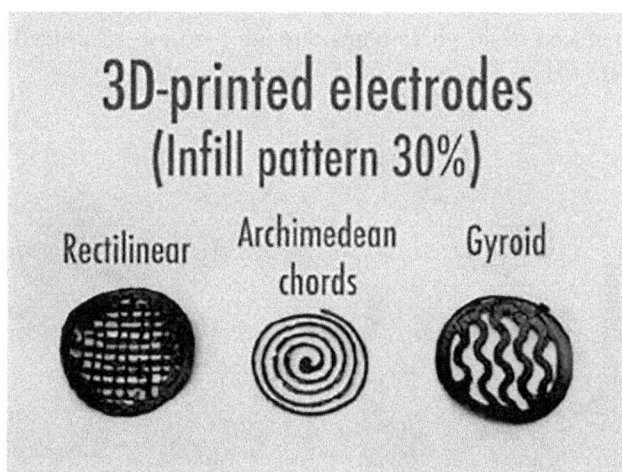

FIGURE 9.3 Different infill patterns-based 3D-printed electrodes prepared using fused deposition modeling [33].

This filament composition consisted of a minor proportion of active material (just 8 wt.% graphene in contrast to 92 wt.% PLA). This translated to a relatively modest quantity of active material, with 103 mg present per cm^3 of the composite. Consequently, when subjected to current densities of 10mAg^{-1} (C/37), the electrode exhibited discharge-specific capacities of 15.8 mAh g^{-1} for the active material (equivalent to 1.26 mAh g^{-1} of the overall composite, considering its electrode volume of 1.63 mAh cm^{-3}).

9.2.3 STEREOLITHOGRAPHY

SLA is one of the oldest 3D printing technologies that use UV (ultraviolet) light or electron beam for the initiation of a chain reaction in a monomer solution. The monomer solutions are mainly epoxy or acrylic-based and instantly convert into polymer chains as UV light is exposed over them. In this polymerization stage, resins inside the desired patterns get solidified to support the next layers (shown in Figure 9.4(a)) [34]. Further, the resins that are unreacted during the printing are precisely removed after completing the process. Moreover, the post-processing stage in SLA plays vital for inducing better mechanical strength in the printed parts. The dispersion of ceramic nanoparticles in the bath of resin can be used to fabricate ceramic polymer composite. In SLA, high resolution around 10-micrometer parts can easily print [34].

Similar to FDM, SLA needs conductive agents for the preparation of ESDs. In this context, silver nitrate and multiwall carbon nanotube (MWCNTs) have been used in the acrylic and polyethylene glycol diacrylate-based resin for the fabrication of structural ESDs. However, the electrical conductivity of fabricated electrodes was found very limited. Moreover, the addition of silver nanowires as conductive fillers in acrylic-based resins was used to construct highly durable and mechanically stable microstructural designs of ESDs. Initially, the conductivity of fabricated structures was found very high around 200 micro-ohms, later it was reduced to 40 micro-ohms through pyrolysis of printed parts (from Figure 9.4(b)) [35].

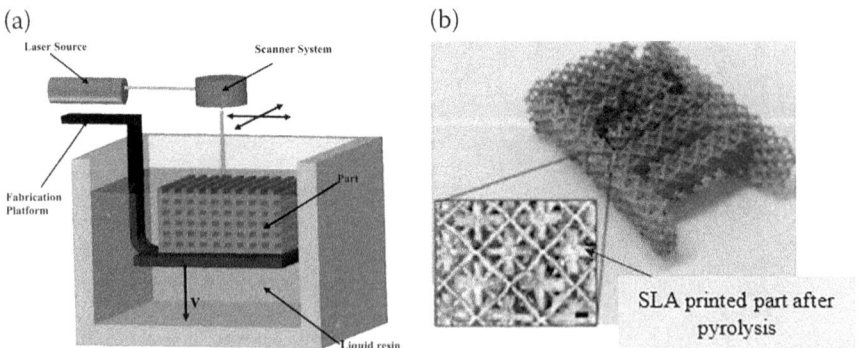

FIGURE 9.4 (a) Schematic diagram of SLA and (b) SLA printed ESDs after pyrolysis [35].

9.2.4 INK JET PRINTING

Ink jet printing (IJP) is one of the main additive manufacturing processes that is widely used to create flexible electromechanical storage devices like capacitors and batteries. IJP is considered as most promising technology over other additive manufacturing techniques because it allows more flexibility in the low-cost manufacturing of ESDs and direct writing liberty [36]. In this technique, a balanced ratio of ceramic solutions is deposited in the form of droplets on the substrate via an injection nozzle. Further, a continuous stream of droplets is deposited to form a continuous pattern that solidifies to induce maximum mechanical strength to hold subsequent layers of printed materials (from Figure 9.5(a)). In this context of materials utilization, mainly two types of inks are used: (1) wax-based inks and (2) liquid-based suspension. The wax-based inks are melted and dropped on the cold substrate (to solidify quickly) where the liquid-based suspension is deposited by evaporation. Moreover, optimum printing speed, nozzle size, uniform particle size distribution, and deposition rate are the main parameters that directly affect the quality of the printed parts in IJP [37]. In a recent study, Zhang et al.[38] presented the preparation of high-performance ink by combining the IJP technique and viscous water-based MXene. In brief, MXene is kind of viscous fluid that associated with several properties such as good mechanical properties, better electrical conductivity, high hydrophilicity and diverse element composition. MXene-based micro-supercapacitors offers high

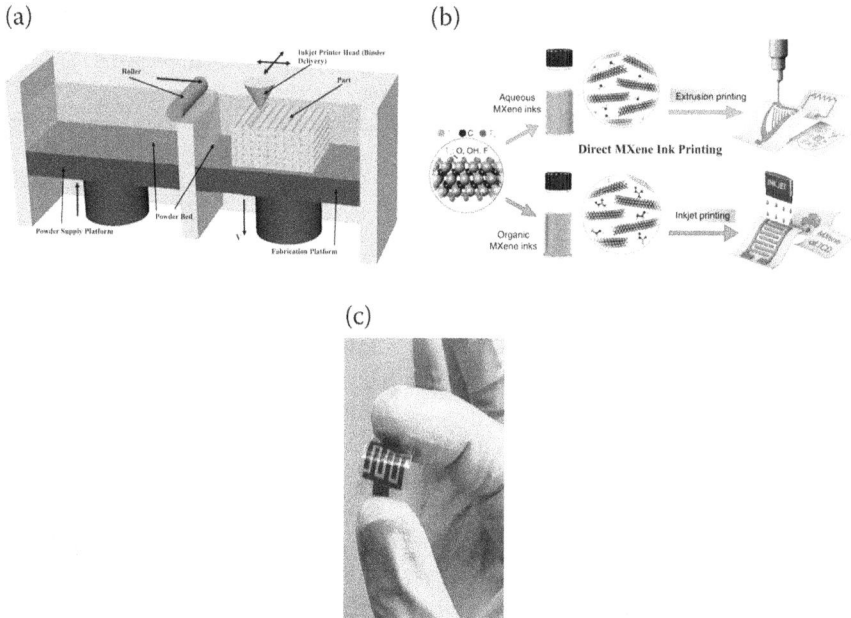

FIGURE 9.5 (a) Schematic diagram of inkjet printing, (b) illustration of MXene-based ink printing, and (c) digital photograph of MXene-based screen-printed micro supercapacitor [38].

electron transfer and provides high response rates in solid-state devices. Therefore, it does not require a metallic current collector. Moreover, there is high capacitance performance of MXene-based micro-supercapacitors, as shown in Figure 9.5(b), wherein the sensitivity of the capacitors has been checked by slightly bending or twisting the capacitors. In addition, the digital photo of a screen-printed micro supercapacitor is shown in Figure 9.5(c) [38].

The fabrication of an alkaline zinc–silver micro 3D pillar battery using inkjet printing has been reported by Ho et al. [39], wherein silver (Ag) was initially mixed with n-tetradecane for preparing the ink for 3D printing of both positive and negative electrodes. In this study, Ag is used as a current collector for depositing the zinc onto the negative electrode, and SiO_2 is used for the positive electrode. Furthermore, the conductivity and porosity of the 3D-printed electrodes significantly impacted the overall performance of the battery [39]. In a study conducted by Y. Gu et al. [38], cathodes for Li-ion batteries were crafted through the employment of the IJP method. They utilized water-soluble LFP inks (consisting of LFP, carbon black, and sodium carboxymethyl cellulose in an 8:1:1 ratio) with an initial pH value of 9.13. These inks were deposited onto a current collector made of carbon nanotube paper. Impressively, the resulting cathodes demonstrated a noteworthy capacity of 151.3mAhg-1 at a rate of 0.1 C, accompanied by a remarkable coulombic efficiency of 94.4%. This outcome underscores the viability of this technique as a promising avenue for the production of rechargeable Li-ion batteries.

9.3   MATERIALS ADOPTED FOR THE FABRICATION OF EESDS

The present section states different materials and composites used for the fabrication of electronic ESDs (EESDs).

9.3.1   CARBON NANOTUBE-BASED MATERIAL

The unique characteristics of single-walled and multi-walled carbon nanotubes (CNT), such as improved carrier mobility, nano-scale size, and long mean free path, have attracted scientists to use the same for the fabrication of ESDs. Single-walled carbon nanotube (SWNT) is used for making thin-filmed electrodes. Functionalized SWNT dispersed in the aqueous solution of sodium dodecyl sulfate is used as the ink material for inkjet 3D printing [40]. The printing is performed over a polymeric layer (polyvinyl alcohol), cloth fabrics, and SiO_2/Si. A gel of polyvinyl alcohol (PVA) and phosphoric acid is used as the electrolyte acting as a separator between two layers of printed SWNT. Next-generation ESD such as Lithium–Sulfur battery (Li–S) is infused with single-walled carbon nanotubes (SWNT) using extrusion 3D Printing. SWNTs are infused with electronically conductive sulfur and are used for direct writing 3D printing [41]. The ink preparation is a key challenge for DIW without clogging the nozzle or extruder. Usually, carrier solvent like cyclohexyl pyrrolidone (CHP) is used to disperse sulfur fused with SWNT using hotplate and ultrasonication. These types of printed electrodes produce a stable charge capacity in the range of 700–800 mAh g^{-1} compared to traditionally fabricated 2D graphene–sulfur Li–S batteries [42].

Multi-walled carbon nanotubes (MWNTs) are also used for electrode fabrication due to enhanced conductivity and porosity. It is used to fabricate electrodes using ink-based 3D printing. Unlike SWNTs ink preparation, isopropyl alcohol and deionized (DI) water are used to disperse MWNTs followed by ultrasonication [43]. Functionalized Multi-walled carbon nanotubes (MWCNTs) are used to fabricate lightweight, flexible supercapacitors with excellent cycling stability using an inkjet 3D printer [44]. Polyethylene-terephthalate (PET) fused with MWCNTs is used for ink preparation for successful printing.

9.3.2 GRAPHENE-BASED MATERIAL

Graphene and its oxides are used in 3D printing, improving mechanical and electrical characteristics of the components. It has potential applications for the development of ESDs due to the increased surface area and enhanced conductivity. FDM is one of the conventional 3D printing methods for the fabrication of 3D disc electrodes [32]. Figure 9.6 shows the developed method to fabricate the electrode using FDM and a variety of disc-shaped electrodes. The filament is commercially available and made of polylactic acid (PLA)/graphene powder (namely, Black magic). The major advantage of 3D-printed electrodes is the application of freestanding anodes in a lithium-ion battery and as a solid-state supercapacitor. Shen et al. [45] attempted to fabricate micro-supercapacitors using solvent-based extrusion 3D printing. Vanadium pentoxide and graphene-vanadium nitride quantum dots (G-VNQDs) with the graphene

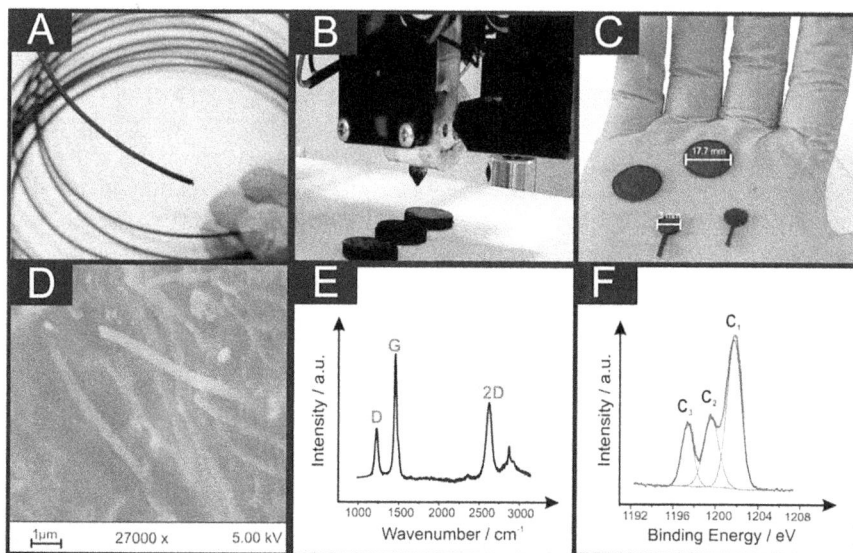

FIGURE 9.6 PLA/graphene filament preparation, 3D printing of disc-shaped electrodes using FDM, variety of 3D-printed discs, and the morphological and physicochemical investigations of 3D-printed electrodes [32].

oxide dispersion are used for the fabrication of the cathode and anode, respectively. Graphene oxide improves the shear-thinning behavior of the solution for DIW. It enhances the viscosity so that the magnitude of the storage modulus is higher as compared to the loss modulus. It helps maintain accurate shape fidelity and dimensional stability. Holey graphene oxide (hGO) dispersed in a solvent like water is used to print Lithium oxide (Li-O_2) cathodes using 3D printing [46]. The hGO is chemically formulated from graphene and is known for its lower contact angle to liquid. Owing to its low particle size (4–25 nm), the loading for the same in the ink is maximized and used for the extrusion 3D printing. The porous electrode helps transfer oxygen, electrolyte, and ions of the electrolyte. Hence, these electrodes deliver enhanced areal capacitance compared to a coil cell with a 2D electrode. Reduced graphene oxide (r-GO) with Ni nanoparticles is used to fabricate the 3D scaffolds for ultra-thick cathodes [47]. The fabrication method follows the steps such as printing 3D scaffolds using r-GO ink, dipping scaffolds in $NiCl_2$, and heating the scaffolds to anchor the Ni nanoparticles over the scaffold surfaces uniformly. The r-GO scaffolds yield improved characteristics such as high specific capacity and a high areal capacity.

Graphene dispersed in aerosol has potential application in the fabrication of ESDs. Graphene nanoplatelets with silica (SiO_2) are used to prepare suitable ink for DIW. The supercapacitors consisting of printed electrodes of these materials are characterized by excellent rate capability and have good retention capacity with enhanced electrical conductivity. Graphene aerosol with mixed-dimensional hybrid aerogel (MDHA) based inks is used for the fabrication of 3D-printed supercapacitors [48]. From the research works it has been found that the 3D-printed energy devices deliver an aerial capacitance of 639.56 mF cm^{-2} at 4 mA cm^{-2} and retained 71.4% at 40 mA cm^{-2}. Graphene hydrogel with polyaniline (GHPANI) are used to fabricate nanohybrid paper electrode using DIW [49]. Ball milling is performed to generate a mixture of graphene and polyaniline. Ultrasonication is provided to the mixture in deionized water (DI) to prepare a concentrated solution of the mixture. The printing method is illustrated in Figure 9.7. The 3D printing of the solution is performed over graphene oxide (GO) paper. Once the reduction reaction is completed, the GHPANI/GO is removed from the substrate. The electrolyte used in these supercapacitors is a mixture of sulfuric acid (H_2SO_4), polyvinyl alcohol (PVA), and DI water. Cellulose is used as the separator between the 3D-printed electrode and the electrolyte. These supercapacitors show enhanced cyclic stability as it retains almost 85% of the initial capacitance after 5000 cycles at a constant current density of 8 A g^{-2}.

9.3.3 METAL-BASED MATERIALS

Compared to Li-ion batteries, metals such as zinc and silver offer enhanced mass energy up to 300 Wh g^{-1} with a time rate of energy transfer of 0.600 W g^{-1}. Moreover, the Zn-Au battery has advantages like being less dangerous and less flammable. Super inkjet printing (SIJP) is performed to fabricate sub-micron

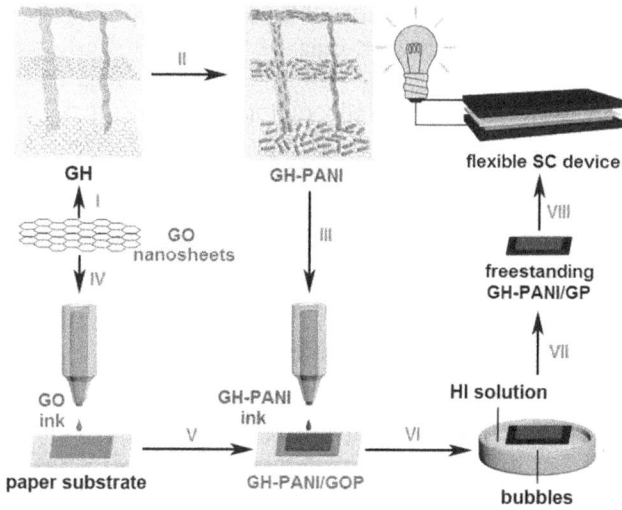

FIGURE 9.7 Schematic illustration of the inkjet printing of GHPANI/GP supercapacitors device [49].

features with an array of pillar-over silver electrodes for micro-batteries [50]. SIJP uses electrohydrodynamic actuation that created droplets of micro-sizes to deposit over the platform. The solution for inkjet printing is usually a paste mixed with silver with an average size of 24 nm and n-tetradecane. The solution is prepared in such a way that it must follow the shear-thinning behavior with appropriate viscosity for the 3D printing process. Potassium hydroxide (KOH) is used for the dispersion of zinc oxide powder with particle size in nanometer to prepare the electrolyte. The 3D-printed pillared electrode yields more energy density compared to the planar electrodes. Stereolithography, with the combination of tape casting method, and dielectric capacitors are manufactured. The resin is customized by dispersing Silver decorated Pb (Zr, TiO$_3$) (PZT) nanoparticles in commercially purchased resin. The manufactured ESD generates an output of about 63 F/g at a current density of 0.5 A/g. Digital light processing (DLP) is used to manufacture 3D electrodes for micro-supercapacitors (MSC) [51]. The photocurable ink mainly consists of commercial resin and silver nanowires. The hierarchical design and complex structure of the electrodes offer reduced electrical resistance and have a potential application in the field of MSC. The 3D-printed zinc-air battery (ZAB) has enormous potential in the field of ESDs because of its higher energy density compared to Lithium-ion batteries. Solvent extrusion direct printing is used to print functional electrodes of both cathode and anode [52]. The printable ink for zinc electrodes mainly comprises zinc powder, carbon black/nanofibers, and polyvinylidene fluoride. The solvent used to disperse these polymers and powders is 2-ethyl pyrrolidone. For air electrodes, ink is prepared from a mixture of graphene oxide, carbon nanotubes, and manganese dioxide nanoparticles. The electrolyte used is a mixture of 6 M potassium hydroxide and 0.2 M zinc acetate. From the charging and discharging cycle test, it is found that the 3D-printed Zinc-air battery has enhanced specific resistance and long-term cyclic performance. 3D-printed Zinc-MnO$_2$ are used for Zinc based micro-batteries. Zinc powder of 50 nm mixed in 5% by weight of poly (vinylidene fluoride-hexafluoropropylene) dissolved in n-methyl-2-pyrrolidone (NMP) solvent was used to print the Zinc electrode. MnO$_2$ powder with conductive filler mixed in the solution of poly (vinylidene fluoride-hexafluoropropylene) solution is used to print the MnO$_2$ electrode. Figure 9.5 shows a schematic of the fabrication of a Zn-MnO$_2$ battery where a gel electrolyte is sandwiched between the electrodes.

9.4 SUMMARY AND FUTURE PERSPECTIVE

In this chapter, we have explained a few innovative layered manufacturing methods to fabricate EESDs. The advantages of using the 3D printing method are complex design to improve the patterning in 2D and 3D as well. Among all the 3D printing techniques, FDM, stereolithography, and solvent extrusion 3D printing are extensively used for the fabrication of electrodes and storage devices. FDM, stereolithography, and solvent extrusion 3D printing use material in the form of filament, resin, and ink, respectively to print the electrodes. Despite substantial progress in the manufacturing of ESDs using additive manufacturing, there are several areas left unexplored. There exist areas such as design aspects, material combinations, and standardization of processes for the fabrication of ESDs. More efforts need to be devoted to investigating the effects of different process parameters of 3D printing methods on the electrochemical performances of 3D-printed devices. There exists an area of investigation to build numerical or empirical relationships between process parameters and storage cycle efficiency of 3D-printed storage devices. Additionally, the effects of generated pores in the structures, printing defects on the energy efficiency, and electrochemical performances need to be investigated accurately to realize further enhancement of the efficiency of 3D-printed storage devices.

REFERENCES

[1] A. S. Aricò, P. Bruce, B. Scrosati, J.-M. Tarascon, and W. van Schalkwijk, "Nanostructured materials for advanced energy conversion and storage devices," *Nat. Mater.*, vol. 4, no. 5, pp. 366–377, 2005, doi: 10.1038/nmat1368.

[2] M.-R. Gao, Y.-F. Xu, J. Jiang, and S.-H. Yu, "Nanostructured metal chalcogenides: Synthesis, modification, and applications in energy conversion and storage devices," *Chem. Soc. Rev.*, vol. 42, no. 7, pp. 2986–3017, 2013, doi: 10.1039/C2CS35310E.

[3] Y.-G. Guo, J.-S. Hu, and L.-J. Wan, "Nanostructured materials for electrochemical energy conversion and storage devices," *Adv. Mater.*, vol. 20, no. 15, pp. 2878–2887, 2008, doi: 10.1002/adma.200800627.

[4] B. Guo, G. Liang, S. Yu, Y. Wang, C. Zhi, and J. Bai, "3D printing of reduced graphene oxide aerogels for energy storage devices: A paradigm from materials and technologies to applications," *Energy Storage Mater.*, vol. 39, no. December 2020, pp. 146–165, 2021, doi: 10.1016/j.ensm.2021.04.021.

[5] C. Fang *et al.*, "Routes to high energy cathodes of sodium-ion batteries," *Adv. Energy Mater.*, vol. 6, no. 5, 2016, doi: 10.1002/aenm.201501727.

[6] V. L. Pushparaj *et al.*, "Flexible energy storage devices based on nanocomposite paper," *Proc. Natl. Acad. Sci. USA*, vol. 104, no. 34, pp. 13574–13577, 2007, doi: 10.1073/pnas.0706508104.

[7] F. Meng, M. Zhang, J. Huang, W. F. Lu, J. M. Xue, and H. Wang, "Additive manufacturing of stable energy storage devices using a multinozzle printing system," *Adv. Funct. Mater.*, vol. 31, no. 9, pp. 1–9, 2021, doi: 10.1002/adfm.202008280.

[8] L. Peng, Y. Zhu, D. Chen, R. S. Ruoff, and G. Yu, "Two-dimensional materials for beyond-lithium-ion batteries," *Adv. Energy Mater.*, vol. 6, no. 11, pp. 1–21, 2016, doi: 10.1002/aenm.201600025.

[9] X. Su, Q. Wu, X. Zhan, J. Wu, S. Wei, and Z. Guo, "Advanced titania nanostructures and composites for lithium ion battery," *J. Mater. Sci.*, vol. 47, no. 6, pp. 2519–2534, 2012, doi: 10.1007/s10853-011-5974-x.

[10] P. Meister, H. Jia, J. Li, R. Kloepsch, M. Winter, and T. Placke, "Best practice: performance and cost evaluation of lithium ion battery active materials with special emphasis on energy efficiency," *Chem. Mater.*, vol. 28, no. 20, pp. 7203–7217, Oct. 2016, doi: 10.1021/acs.chemmater.6b02895.

[11] R. Schmuch, R. Wagner, G. Hörpel, T. Placke, and M. Winter, "Performance and cost of materials for lithium-based rechargeable automotive batteries," *Nat. Energy*, vol. 3, no. 4, pp. 267–278, 2018, doi: 10.1038/s41560-018-0107-2.

[12] S. Zhang *et al.*, "A convenient, low-cost graphene UV-cured additive manufacturing electronic process to achieve flexible sensors," *Chem. Eng. J.*, vol. 451, p. 138521, 2023, doi: 10.1016/j.cej.2022.138521.

[13] L. Xu, M. Ouyang, J. Li, F. Yang, L. Lu, and J. Hua, "Optimal sizing of plug-in fuel cell electric vehicles using models of vehicle performance and system cost," *Appl. Energy*, vol. 103, pp. 477–487, 2013, doi: 10.1016/j.apenergy.2012.10.010.

[14] J. C. Ruiz-Morales *et al.*, "Three dimensional printing of components and functional devices for energy and environmental applications," *Energy Environ. Sci.*, vol. 10, no. 4, pp. 846–859, 2017, doi: 10.1039/c6ee03526d.

[15] J. Wang *et al.*, "Toward high areal energy and power density electrode for Li-ion batteries via optimized 3D printing approach," *ACS Appl. Mater. Interfaces*, vol. 10, no. 46, pp. 39794–39801, Nov. 2018, doi: 10.1021/acsami.8b14797.

[16] X. Tian and K. Zhou, "Additive manufacturing of energy storage devices," in *Additive Manufacturing: Materials, Functionalities and Applications*, K. Zhou, Ed. Cham: Springer International Publishing, 2023, pp. 51–83.

[17] A. E. Jakus, E. B. Secor, A. L. Rutz, S. W. Jordan, M. C. Hersam, and R. N. Shah, "Three-dimensional printing of high-content graphene scaffolds for electronic and biomedical applications," *ACS Nano*, vol. 9, no. 4, pp. 4636–4648, Apr. 2015, doi: 10.1021/acsnano.5b01179.

[18] M. Mao *et al.*, "The emerging frontiers and applications of high-resolution 3D printing," *Micromachines*, vol. 8, no. 4, 2017, doi: 10.3390/mi8040113.

[19] C. Zhu *et al.*, "Supercapacitors based on three-dimensional hierarchical graphene aerogels with periodic macropores," *Nano Lett.*, vol. 16, no. 6, pp. 3448–3456, Jun. 2016, doi: 10.1021/acs.nanolett.5b04965.

[20] C. J. Zhang *et al.*, "Stamping of flexible, coplanar micro-supercapacitors using mxene inks," *Adv. Funct. Mater.*, vol. 28, no. 9, pp. 1–10, 2018, doi: 10.1002/adfm.201705506.

[21] A. Ambrosi and M. Pumera, "3D-printing technologies for electrochemical applications," *Chem. Soc. Rev.*, vol. 45, no. 10, pp. 2740–2755, 2016, doi: 10.1039/c5cs00714c.

[22] J. S. Park, T. Kim, and W. S. Kim, "Conductive cellulose composites with low percolation threshold for 3D printed electronics," *Sci. Rep.*, vol. 7, no. 1, pp. 1–10, 2017, doi: 10.1038/s41598-017-03365-w.

[23] L. Zeng, P. Li, Y. Yao, B. Niu, S. Niu, and B. Xu, "Recent progresses of 3D printing technologies for structural energy storage devices," *Mater. Today Nano*, vol. 12, pp. 1–13, 2020, doi: 10.1016/j.mtnano.2020.100094.

[24] S. Tagliaferri, A. Panagiotopoulos, and C. Mattevi, "Direct ink writing of energy materials," *Mater. Adv.*, vol. 2, no. 2, pp. 540–563, 2021, doi: 10.1039/d0ma00753f.

[25] D. K. Mishra and P. M. Pandey, "Effects of morphological characteristics on the mechanical behavior of 3D printed ordered pore topological Fe scaffold," *Mater. Sci. Eng. A*, p. 140759, 2021, doi: 10.1016/j.msea.2021.140759.

[26] D. K. Mishra and P. M. Pandey, "Mechanical behaviour of 3D printed ordered pore topological iron scaffold," *Mater. Sci. Eng. A*, no. 139293, pp. 1–8, 2020, doi: 10.1016/j.msea.2020.139293.

[27] D. K. Mishra and P. M. Pandey, "Effect of sintering parameters on the microstructure and compressive mechanical response of porous Fe scaffold fabricated by 3D printing and pressure less microwave sintering," *Proc. Inst. Mech. Eng. Part C J. Mech. Eng. Sci.*, pp. 1–16, 2020, doi: 10.1177/0954406220921416journals.sagepub.com/home/pic.

[28] D. K. Mishra and P. M. Pandey, "Experimental investigation into the fabrication of porous biodegradable fe scaffold by microwave sintering of 3D printed green body," 2021, vol. 1, pp. 1–5, doi: 10.1115/MSEC2021-63402.

[29] M. Srivastava, J. Singh, D. K. Mishra, and R. P. Singh, "Review on the various strategies adopted for the polishing of silicon wafer – A chemical perspective," *Mater. Today Proc.*, 2022, doi: 10.1016/j.matpr.2022.02.300.

[30] S. Park et al., "Development of highly energy densified ink for 3D printable batteries," *Energy Technol.*, vol. 6, no. 10, pp. 2058–2064, 2018, doi: 10.1002/ente.201800279.

[31] S. Bose, D. Ke, H. Sahasrabudhe, and A. Bandyopadhyay, "Additive manufacturing of biomaterials," *Prog. Mater. Sci.*, vol. 93. Elsevier Ltd, pp. 45–111, Apr. 2018, doi: 10.1016/j.pmatsci.2017.08.003.

[32] C. W. Foster et al., "3D printed graphene based energy storage devices," *Sci. Rep.*, vol. 7, no. 1, pp. 1–11, 2017.

[33] A. Maurel, R. Russo, S. Grugeon, S. Panier, and L. Dupont, "Environmentally FRiendly Lithium-terephthalate/polylactic Acid Composite Filament Formulation for Lithium-ion Battery 3D-printing via fused deposition modeling," *ECS J. Solid State Sci. Technol.*, vol. 10, no. 3, p. 037004, 2021, doi: 10.1149/2162-8777/abedd4.

[34] T. D. Ngo, A. Kashani, G. Imbalzano, K. T. Q. Nguyen, and D. Hui, "Additive manufacturing (3D printing): A review of materials, methods, applications and challenges," *Compos. Part B Eng.*, vol. 143, pp. 172–196, 2018, doi: 10.1016/j.compositesb.2018.02.012.

[35] U. Gulzar, C. Glynn, and C. O'Dwyer, "Additive manufacturing for energy storage: Methods, designs and material selection for customizable 3D printed batteries and supercapacitors," *Curr. Opin. Electrochem.*, vol. 20, pp. 46–53, 2020, doi: 10.1016/j.coelec.2020.02.009.

[36] C. Li, F. Bu, Q. Wang, and X. Liu, "Recent developments of inkjet-printed flexible energy storage devices," *Adv. Mater. Interfaces*, vol. 9, no. 34, p. 2201051, 2022, doi: 10.1002/admi.202201051.

[37] A. Bhatia and A. K. Sehgal, "Additive manufacturing materials, methods and applications: A review," *Mater. Today Proc.*, May 2021, doi: 10.1016/j.matpr.2021.04.379.

[38] Y. Gao *et al.*, "Printable electrode materials for supercapacitors," *ChemPhysMater*, vol. 1, no. 1, pp. 17–38, 2022, doi: 10.1016/j.chphma.2021.09.002.

[39] T. T. Huang and W. Wu, "Scalable nanomanufacturing of inkjet-printed wearable energy storage devices," *J. Mater. Chem. A*, vol. 7, no. 41, pp. 23280–23300, 2019, doi: 10.1039/c9ta05239a.

[40] P. Chen, H. Chen, J. Qiu, and C. Zhou, "Inkjet printing of single-walled carbon nanotube/RuO 2 nanowire supercapacitors on cloth fabrics and flexible substrates," pp. 594–603, 2010, doi: 10.1007/s12274-010-0020-x.

[41] C. A. Milroy, S. Jang, T. Fujimori, A. Dodabalapur, and A. Manthiram, "Inkjet-printed lithium–sulfur microcathodes for all-printed, integrated nanomanufacturing," *Small*, vol. 13, no. 11, 2017, doi: 10.1002/smll.201603786.

[42] M.-Q. Zhao *et al.*, "Graphene/single-walled carbon nanotube hybrids: One-step catalytic growth and applications for high-rate Li-S batteries," *ACS Nano*, vol. 6, no. 12, pp. 10759–10769, 2012.

[43] C. Milroy and A. Manthiram, "Printed microelectrodes for scalable, high-areal-capacity lithium-sulfur batteries," *Chem. Commun.*, vol. 52, no. 23, pp. 4282–4285, 2016, doi: 10.1039/c5cc10503j.

[44] S. K. Ujjain, P. Ahuja, R. Bhatia, and P. Attri, "Printable multi-walled carbon nanotubes thin film for high performance all solid state flexible supercapacitors," *Mater. Res. Bull.*, vol. 83, pp. 167–171, 2016, doi: 10.1016/j.materresbull.2016.06.006.

[45] K. Shen, H. Mei, B. Li, J. Ding, and S. Yang, "3D printing sulfur copolymer-graphene architectures for Li-S batteries," *Adv. Energy Mater.*, vol. 8, no. 4, p. 1701527, 2018.

[46] X. Tian *et al.*, "Boosting capacitive charge storage of 3D-printed micro-pseudocapacitors via rational holey graphene engineering," *Carbon N. Y.*, vol. 155, pp. 562–569, 2019.

[47] Y. Qiao *et al.*, "3D-printed graphene oxide framework with thermal shock synthesized nanoparticles for Li-CO2 batteries," *Adv. Funct. Mater.*, vol. 28, no. 51, p. 1805899, 2018.

[48] X. Tang *et al.*, "Generalized 3D printing of graphene-based mixed-dimensional hybrid aerogels," *ACS Nano*, vol. 12, no. 4, pp. 3502–3511, 2018.

[49] K. Chi *et al.*, "Freestanding graphene paper supported three-dimensional porous graphene--polyaniline nanocomposite synthesized by inkjet printing and in flexible all-solid-state supercapacitor," *ACS Appl. Mater. & interfaces*, vol. 6, no. 18, pp. 16312–16319, 2014.

[50] K. Murata, J. Matsumoto, A. Tezuka, Y. Matsuba, and H. Yokoyama, "Super-fine ink-jet printing: Toward the minimal manufacturing system," *Microsyst. Technol.*, vol. 12, no. 1, pp. 2–7, 2005.

[51] S. H. Park, M. Kaur, D. Yun, and W. S. Kim, "Hierarchically DEsigned Electron Paths in 3D printed energy storage devices," *Langmuir*, vol. 34, no. 37, pp. 10897–10904, 2018, doi: 10.1021/acs.langmuir.8b02404.

[52] J. Zhang *et al.*, "3D-printed functional electrodes towards Zn-Air batteries," *Mater. Today Energy*, vol. 16, p. 100407, 2020, doi: 10.1016/j.mtener.2020.100407.

10 Applications of 3D Printing in the Biomedical Field
Methods, Biomaterials, and Recent Advancements

Rudranarayan Kandi and Pulak Mohan Pandey

10.1 INTRODUCTION

The first 3D Printing method was patented by C. Hull in 1986 for the invention of stereolithography using photocurable resin [1]. Nowadays, the process is known as rapid prototyping, additive manufacturing, or layered manufacturing. 3D Printing has evolved in various fields like aerospace, biomedical, academics, and robotics. The major advantages of this process are the ease of fabricating complex parts, the least wastage of material, multi-material printing, and fabricating implantable devices. conventional printing methods of polymeric materials include stereolithography (SLA), fused deposition modeling (FDM), selective laser sintering (SLS), and laminated object manufacturing (LOM). The application of 3D Printing can be now seen in various fields of biomedical engineering like fabrication of anatomical prototypes [2], artificial tissue regeneration [3], artificial functional organs [4], prostheses, and implants [5]. Apart from these, bioprinting with biological materials and living cells helps the medical community to face the challenges of lack of donor availability. For the last decades, 3D Printing has been combined with the tissue engineering method to produce scaffolds for regenerative applications. Table 10.1 shows various 3D Printing methods with the biomaterials used and their applications in biomedical engineering. Scaffolds are complex in design and critical to producing with any other conventional manufacturing methods. Porous scaffolds with ordered lattice structures are now produced using 3D Printers, and these scaffolds act as the support structure for the native cells over which they proliferate, attach, and infiltrate to generate and remodel or replace the diseased part. The porous structures of the scaffolds reflect the tissue architecture over which the cells are intended to proliferate. Materials that are used for fabricating biomedical components are called biomaterials. An ideal 3D-printed biomaterial should possess good biocompatibility, antibacterial properties, equivalent mechanical properties as compared to the native tissues, controlled degradation rate, and reduced immunological reaction.

DOI: 10.1201/9781003484325-10

TABLE 10.1
Various 3D Printing Methods with Their Applications in the Biomedical Field [6]

Type of 3D Printing method	3D Printing Technologies	Type of Biomaterials	Pros	Cons	Applications in the Biomedical Field
Filament-based	Fused Deposition Modeling (FDM)	Thermoplastic Polymer and composites of polymers in the form of filament	Low cost, Fast printing	Slow printing, Anisotropy in the part, reduced dimensional accuracy, material constraint	Organ prototypes, Bone scaffolds, Dental applications.
Liquid-based	Stereolithography (SLA), Digital light printing (DLP), Inkjet printing	Photocurable resins, low viscous polymer resin, and hydrogels.	High-resolution printing, improved surface finish	Relatively expensive, difficulty in ink formulation, material constraint.	Patient-specific models, implantable devices, Tissue engineering, Scaffold fabrication.
Powder-based	Selective laser sintering (SLS), Selective laser melting (SLM), Electron beam melting (EBM), Laser-engineered net shaping (LENS)	Polymeric powder, composite polymer powder, metal powders, composite metal powder, ceramic powder.	No support structure required, direct fabrication of functional part, relatively fast, improved material utilization.	Requires high-power laser, costly, residual stress development, and limited biocompatible material availability.	Patient-specific anatomical model, Metal implants, scaffolds for bone regenerative application,
Solvent-based	Solvent-based extrusion printing, Embedded 3D Printing, 3D Bioprinting	Polymeric solution, hydrogel, hydrogel with biological cells.	Multi-material printing, direct functional part printing.	Solvent toxicity, difficult-to-handle rheological properties of the ink	Tissue engineering (Skin, trachea, cornea), vascular stent application, Drug delivery devices.

TABLE 10.2

Classifications of Biomaterials and Their Applications

Biomaterial Type	Applications
Thermoplastic polymer [7] • Synthetic polymers • Natural polymers	Tissue engineering, Anatomical models for training and surgical planning, Bone, and dental applications, cardiovascular stent applications, etc.
Metal and alloys [8]	Orthopedic and dental applications, implants, prostheses, and drug delivery devices.
Ceramic and metal oxides [9]	Orthopedic implants, soft robotics, bionic devices, magnetic bioreactors.
Composites [10]	Tissue engineering, bone regeneration, and microfluidic devices.

Based on the 3D printable biomaterials, the classifications of various biomaterials are shown in Table 10.2. Nowadays, several novel biomaterials and innovative printing methods are used in biomedical applications. However, the fabrication of a fully functional organ is still a challenge to researchers and doctors. This chapter provides detailed information related to various 3D Printing methods, their principles, and biomaterials used for biomedical engineering. Further, it provides details of novel 3D Printing methods with compatible biomaterials for the printing of direct functional parts.

10.2 3D PRINTING OF BIOMATERIALS

The present section focuses on various 3D Printing of biomaterials implemented in the biomedical field. The major types are filament-based, liquid-based, powder-based, and solvent-based 3D Printing.

10.2.1 FILAMENT-BASED 3D PRINTING

This type of 3D Printing method includes FDM, which was developed in 1989 by Crump, Co-founder of Stratasys Inc. FDM is a commonly used, low-cost, and robust 3D Printing method. It uses filaments of thermoplastics (Polylactic acid (PLA), acrylonitrile butadiene (ABS), polyamide, etc.) as the raw material wound over a spool. The filament diameter usually ranges from 1.5 to 3 mm and the nozzle size usually ranges from 0.4 mm to 1 mm. It works on the principle of melting the filament using the heating element present in the print head and depositing the melted filament over the platform in a layer-by-layer process. The rollers are present in the 3D Printing system that push the filament in the forward direction so that melted filament is extruded out and solidified in the normal environment. During the deposition method, the build platform moves along the Z-direction by a distance equal to the layer thickness and the next layer is deposited over the layer. The deposited layers are bonded due to thermal heating and cooled naturally. This process is usually used to fabricate different prototypes of the organs, patient-specific models, and surgical prototypes. Further, this has been used for making customized tissue-engineered scaffolds for both bone and dental applications.

The method has been used to develop deformable, patient-specific anatomical models for orthopedic, cardiovascular, and neurosurgical applications. Ploch et al. [2] used this filament-based 3D Printing method to fabricate realistic models of the human brain using the imaging data obtained from magnetic resonance imaging (MRI). The MRI data were processed and converted to standard tessellated files (STL) ready to be used in 3D Printing. The fabrication process used an FDM 3D Printer to fabricate a human brain model as a negative template of acrylonitrile butadiene styrene (ABS) material. To prepare a mold, silicon paste was used over the negative template, and on solidification, a silicone mold was prepared. Finally, gelatin was cast onto the silicon mold to prepare the replica of the human brain. In the surgical survey, it was found that the model was suitable for training, improving patient education, neurosurgical training, and pre-surgical planning. FDM is used to print cerebral hollow aneurysm models for educational and training purposes. Solid tubes of ABS material are fabricated using the FDM process as per the scanning data from MRI and the tubes are coated with silicon to make an outer layer as flexible as the native tissue. Later, ABS is dissolved in acetone solvent to produce hollow silicon tubes that are exact replications of aneurysms [11]. This method is also used to develop accurate anatomical cerebral aneurysms for clipping simulations [12]. Further, FDM is used for the fabrication of bone scaffolds, dental implants, and orthosis. Polylactic acid (PLA) and PLA-CaCO$_3$ are used to fabricate upper limb orthoses and casts using Fused Filament Fabrication (FFF). The characteristics of 3D-printed orthoses of PLA-CaCO$_3$ such as water resistance, light weight, and improved rigidity help the components more stable and comfortable to the patients [13]. Figure 10.1(a) shows the 3D-printed wrist splints of

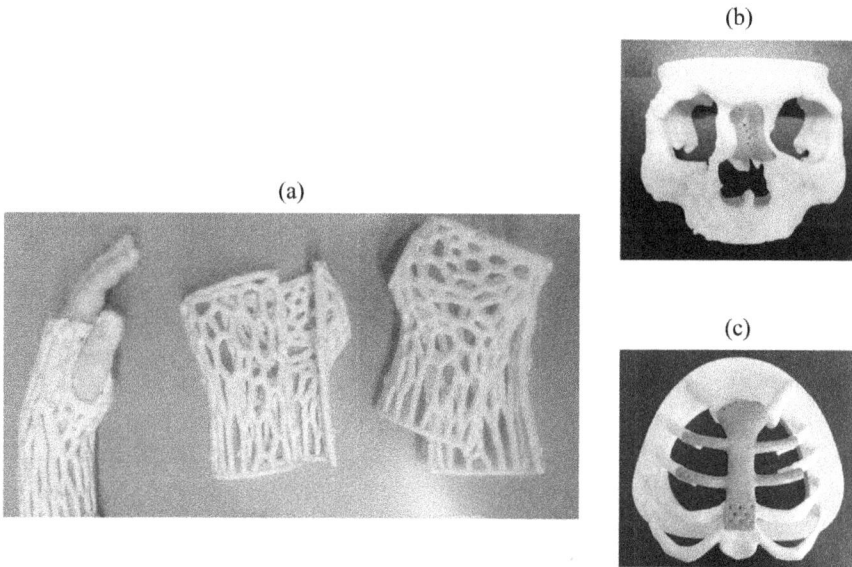

FIGURE 10.1 Application of fused deposition modeling (FDM): (a) wrist splint [13], personalized 3D-printed components for (b) nasal reconstruction, and (c) breastbone made of PEEK [17].

PLA, PLA-CaCO$_3$. Bio-composite filaments made of PLA-magnesium-vitamin E(α-tocopherol) are used to develop anterior cruciate ligament (ACL) screws using low-cost filament-based 3D Printing [14]. Figure 10.1(b) shows the developed ACL screw made of a bio-composite with clear structural integrity and improved strength. It is also commonly used to develop customized scaffolds for bone regenerative applications and dental applications. PLA reinforced with carbonated hydroxyapatite (cHA) [15], hydroxyapatite-modified PLA (PLA-HAp) [16], and Carbon Fiber Reinforced PEEK Composites [17] have been recently used to develop patient-specific customized scaffolds and implants. Figure 10.1(c) shows the clinical application of PEEK using the FDM method to print breastbone and nasal construction.

10.2.2 LIQUID-BASED 3D PRINTING

The first liquid-based 3D Printing, "Stereolithography" (SLA), was developed and commercialized by Chuck Hull in 1986 [1]. The prime components of SLA are the ultraviolet (UV) laser system, scanning system, vat filled with photocurable resin, and movable build platform. The UV ray is used to selectively solidify the photosensitive resin in the vat with each layer equivalent to the sliced layer of the geometrical information fed to the printer. When printing of one layer is complete, the build platform is lowered down by a distance equal to the layer thickness, and again the UV ray selectively cures the second layer. The process continues till the final part is completely built over the platform. Materials like liquid resins (typically acrylate) with modified radicals are used as photocurable resins for this method. The maximum layer thickness for this method ranges from 150 to 250 µm as the laser beam diameter is limited. Since the layer thicknesses are comparatively small and the medium of curing is liquid, the surfaces of the component produced are relatively fine and accurate with improved surface finish. However, the limitations of this process are slow printing, limited material availability, and expensive laser source. Recently, another liquid-based 3D Printing, Digital Light Processing (DLP) is often used for biomedical applications. DLP works on the principle of curing a whole area using a UV light projector for a single layer. The varieties of light sources used in DLP range from classical lamps to modern diodes with varying ranges of intensities of light. The lateral resolution of this method ranges from 10 to 50 µm. Since it solidifies a portion of the area, it is a relatively faster printing process than SLA printing. This method is mostly used in the fabrication of tissue-engineered scaffolds and medical implants.

There are vast applications of liquid-based 3D Printing in pre-surgical applications, training, personalized artificial prostate generation, tissue-engineered scaffolding, and the development of regenerative devices. It is used for the generation of exact models of both hard and soft tissues of cranio-maxillofacial and skin for patient education and pre-surgical planning [18]. The method includes both SLA printing and vacuum casting. The 3D-printed parts are used as the master part to produce silicone molds for vacuum casting. The casting method used various cast materials like polyurethane and silicon to provide a natural representation of soft tissues. In another case study, SLA was used to reconstruct the craniofacial defect [19]. It is also used as the preoperative tool for Total Hip Replacement (THR) [20].

Physical components of the pelvis and femur are built using SLA by selectively curing liquid epoxy resin for better understanding and feasibility of the surgical procedure. Morris et al. [21] demonstrated a preoperative model printed using SLA before the mandibular resection for planning and better preparation for the metallic implantation. In another study, a cranial implant made of carbon fiber reinforce polymer (CFRP) was developed using the SLA technique [22]. The method involved CT scanning of cranial defects, processing, evaluation, and 3D printing of the defect portion using SLA. A wax model of the defect was made manually and the final cranial implant of CFRP was fabricated using the loosen mold. Apart from surgical management, and prostheses fabrication, SLA is extensively used in tissue engineering applications. Baino et al. [23] attempted to fabricate bone scaffolds made of a mixture of photocurable resin and ceramic, hydroxyapatite (HAP). Post-sintering the green scaffold, the sintered printed scaffolds of HAP were obtained for bone tissue regeneration. Shie et al. [3] used light-cured and water-based polyurethane for the fabrication of customized scaffolds for cartilage regeneration. Figure 10.2 shows the fabrication of scaffolds for both bone and cartilage regenerative applications where SLA process has been adopted. Sodian et al. [24]

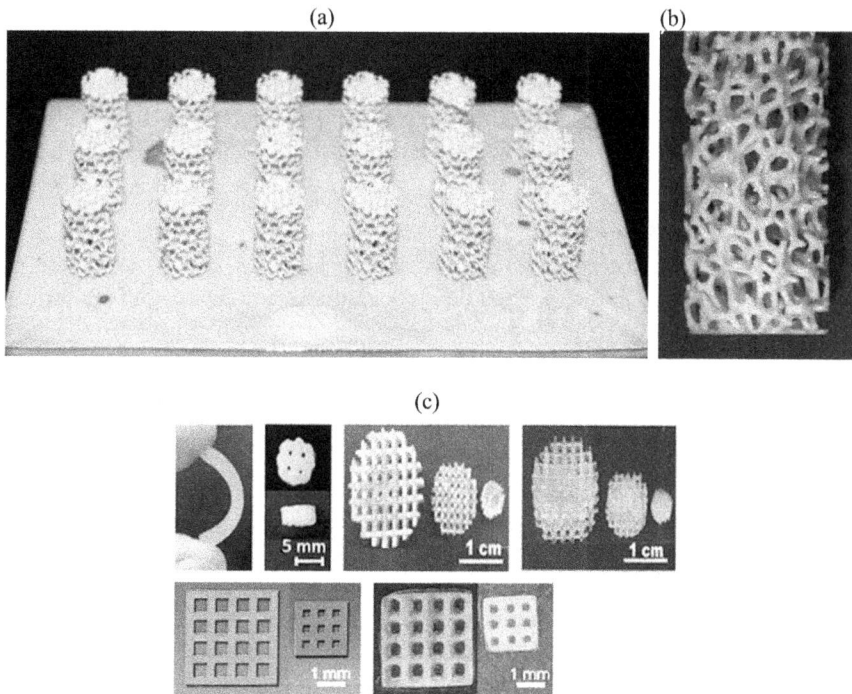

FIGURE 10.2 Application of 3D Printing using a mixture of photocuring resin and hydroxyapatite in bone tissue regeneration (a) fabricated bone scaffolds over work platform, (b) sintered scaffold made of hydroxyapatite after removing photocured resin [23], and (c) printed scaffold of water-based photocurable resin using DLP for cartilage tissue regeneration [3].

investigated the fabrication and its functional testing of aortic valve of thermo-plastic poly-3-hydroxyalkanoate-co-3-hydroxyhexanoate (PHOH) and poly-4-hydroxybutyrate (P4HB) printed via SLA.

Recently, Mott et al. [25] developed a novel non-toxic resin for bone tissue engineering applications using a digital micromirror device (DMD). The resin was a mixture of poly (propylene fumarate) (PPF), TiO_2, and oxybenzone to control the cure depth and avoid dark cure. Cardiac patches with microvascular channels were fabricated using SLA to study the therapeutic potential of the bone marrow-derived mesenchymal stem cells (BM-MSCs) reducing the degradation of myocardial tissue postinfarction [26]. Poly (ethylene glycol) di-methacrylate (PEGDMA, photoini-tiators, and pre-gel PEGDMA with and without MSCs were used for preparing the photosensitive resin for the fabrication of the cardiac patch. In another application of tissue engineering, a tabletop SLA was used to print a biomimetic bone matrix model that contained both breast cancer cells and bone stromal cells [27]. The in-vitro investigations depicted the interaction of different types of cells among each other showing an artificial bone microenvironment suitable to study the breast cancer progression in bone. A 3D triculture model made of hydrogels and gelatin methacrylate (GelMA) was developed using a custom-built DMD for investigating the performances of various biological cells for biomimetic liver model [28]. In a similar research, artificial liver made of photosensitive liver decellularized extra-cellular matrix (dECM), GelMA, and photoinitiators were printed using the DMD method [29].

10.2.3 POWDER-BASED 3D PRINTING

10.2.3.1 Selective Laser Sintering

A typical SLS printer contains major parts such as a powder bed, high-power CO_2 laser, scanner, powder roller, powder delivery platform, and fabrication platform. It works on the principle of sintering the powder layer over another layer using a laser source until the final part is built. The roller spreads the power over the work platform from the powder delivery platform. Sintering takes place just below the melting temperature of the powder material when the laser beam selectively traces over the surface. The powder bed is heated just below its melting point to avoid thermal distortion helping the fusion of consecutive layers. Once a single layer is sintered, the work platform moves down by the layer thickness, and power spreading is done again with the help of the roller. Again, the laser beam selectively sinters the new layer over the previous layers. The un-sintered powders will act as the support materials for the component to be printed. The cycle is repeated until the final part is built completely. A wide range of plastic powders, metals, and ceramics are used in SLS to achieve a variety of applications in biomedical engineering such as surgical training, medical prototype development, device development, tissue engineering scaffolds, and drug delivery systems.

SLS has been used in several surgical preoperative planning by fabricating highly accurate anatomical models. Polymer and its composite powders such as nylon, polycarbonate, and acrylic have been used in SLS for the fabrication of various surgical models and elements. Silva et al. [30] studied the dimensional accuracy of the prototype developed for craniomaxillary anatomy using SLS and the 3DP process. The result suggested the prototypes printed using SLS were more accurate and useful for maxillofacial operations. Schrank et al. [31] investigated the dimensional accuracy of ankle-foot prostheses fabricated by SLS. The dimensional deviations of various components were measured and compared to the CAD model. It was found that the accuracy of the printed parts was well within the ranges of the tolerance. Jardini et al. [32] fabricated the bio-model for a patient's cranial defect so that patient-specific biocompatible implant can be developed. The bio-model was made of polyamide (PA 2200) and the SLS (EOS) was used for the fabrication of the same from the commuted scanning data. SLS has been used for the fabrication of various scaffolds for tissue regenerative applications. Shuai et al. [33] developed composite scaffolds of poly (3-hydroxybutyrate-co-3-hydroxyvalerate) (PHBV) and calcium silicate (CS) for bone tissue engineering using the SLS method. The researchers found that the incorporation of CS in PHBV-enhanced osteogenic differentiation can be used for bone regenerative applications. Figure 10.3 shows the scaffold printed using SLS with scanning electron microscopy morphology of the constructs of the scaffolds. In another investigation, the researchers attempted to fabricate porous composite scaffolds of polyether ether ketone/polyglycolic acid (PEEK/PGA) incorporated with hydroxyapatite (HAP) for bone tissue regeneration using the SLS method [33]. The sintered scaffolds of PEEK/PGA/HAP were found to be improved in terms of cell attachment proliferation compared to PEEK/PGA scaffolds. Gayer et al. [34] investigated the process for the development of

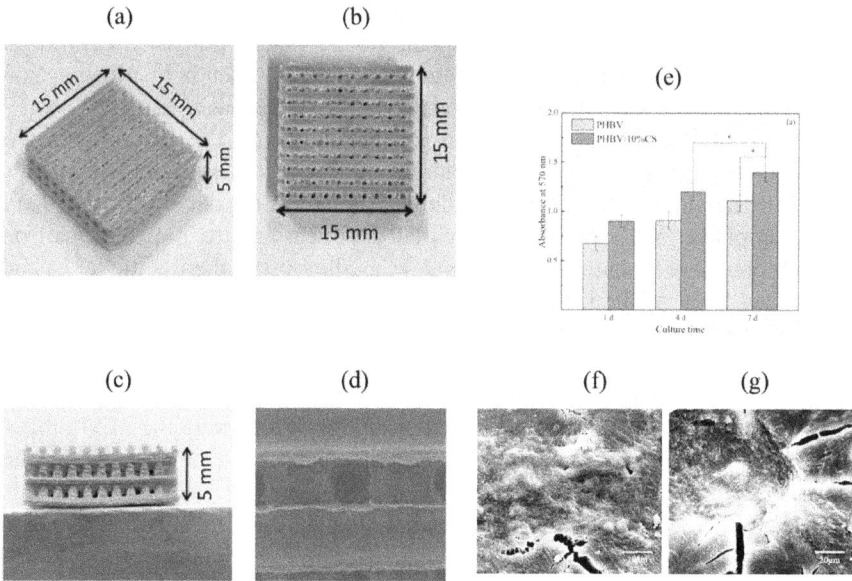

FIGURE 10.3 Application of SLS in bone tissue regeneration (a-c) Macroscopic view of a sintered scaffold, (d) SEM image of each struct in the scaffold, (e) cell viability assessment of scaffolds for days 1, 4 and 7, SEM morphology of MG3cells attached to the surface of (f) PHBV and (g) PHBV/ 10% CS [33].

patient-specific cranial implants of polylactide/calcium carbonate composite powder using SLS. The sintered material surface was found to be a suitable environment for the growth of osteoblast cells suitable for bone regenerative application.

10.2.3.2 Selective Laser Melting (SLM)

SLM is powder-based 3D Printing where the powder material is melted through a high-power laser beam and solidified to print a layer. This process is like the SLS process except for high-power laser, melting of powder material, and generation of comparatively stronger components. The initial layers are fused to the base plate and the next layers are melted over the layers. The unused powders are used as the support materials for the overhanging structures. There are large varieties of polymeric and metal powders used in this process such as polyamide, stainless steel, titanium and its alloys, ceramic, and Cobalt–Chromium alloy powders. This process is used to fabricate customized scaffolds for bone tissue regeneration, artificial hip joints, and knee implants in the field of biomedical engineering.

SLM is used to develop metallic anatomic micro and macro structures of dense steel and titanium metal. Wehmoller et al. [35] used the SLM method to fabricate a detailed structure of the cortical lower jaw, spongiosa and mandibular canal, and tubular bone with screw connection for support structures of high-alloyed steel and titanium powder. Fukuda et al. [36] used SLM to fabricate metallic implants of titanium with pore sizes varying from 500 to 1200 μm. The investigation suggested that SLM is a very suitable manufacturing method to fabricate accurate and detailed structures without sacrificing the complexity of the component. The interconnected pore in an implant affects the osteo-induction and bone growth in the site. The implant with a macro pore size of 500 μm yielded the highest osteo-induction value compared to other implants with pore sizes ranging from 500–1200 μm. Recently, SLM has been used to develop functionally graded porous metallic components mimicking the bone structure for load-bearing applications [37]. Structures built with honeycomb lattice units yielded the maximum strength and modulus suitable for the development of load-bearing implants. Xin et al. [38] compared the release of toxic metal ions from the components of Cobalt–Chromium (Co–Cr) alloy using the traditional casting method and SLM method. The results showed that the ion release was less from the 3D-printed part and the cell proliferation was higher over the surface of SLM-printed parts. Hence, SLM-fabricated parts for biomedical implants showed better biocompatibility compared to the parts fabricated using the traditional casting method. Recently, titanium implants have been developed using the SLM process and the surfaces of the components are modified with micro particles and nanoscale nanospheres [39]. The overall osteointegration, cell attachment, and cell differentiation of modified components yielded better results compared to the SLS-printed implants.

10.2.3.3 Electron Beam Melting

This process works on the principle of melting the metal powder using a high-power focused electron beam. The beam of electrons always accelerated to the powder materials; hence, the material must be electrically conductive. To avoid contamination with the atmospheric air, the process is performed inside a vacuum chamber. This process is comparatively faster than SLS and SLM due to the high power of the electron beam and higher scanning rate. Materials such as Ti-6Al-4V, Cobalt–Chromium alloy, and non-ferrous alloys are used for making parts. This process has been used in medical industries by fabricating artificial implants, artificial maxillofacial plates, scaffolds, acetabular cups, and knee prostheses.

The electron beam melting (EBM) process has been used in developing open-cell titanium components for bone implants. Heinl et al. [40] used the EBM process to fabricate cellular structures of titanium to adapt the possible mechanical properties of the native bone. The main advantage of the EMB process is that it prevents the degeneration of the material properties due to processing inside a vacuum chamber. In another work, researchers fabricated interconnected cellular structures of titanium alloy (Ti-6Al-4V) for bone implants. The prime hypothesis of the investigation was to minimize the shielding effect by matching the mechanical properties of the structures with the native bone [41]. Harrysson et al. [42] fabricated the hip step implant of Ti-6Al-4 V using the EBM method with tailored mechanical properties of bone mimicking stiffness. Finite element analysis was performed to verify the printed structure and the stress shielding effect was checked for successful implantation. Koike et al. [43] investigated the mechanical, grindability, and corrosion properties of components fabricated using EBM and the investigations suggested that the as-fabricated component can have some application in dentistry. Yang et al. [44] made a comparative study to check the biocompatibility and stability of titanium surgical screws fabricated using EMB and conventional commercially available screws. Both the screws were implanted in different sheep animals' bodies at the anterior cervical vertebral site. Histopathological results suggested that EBM fabricated screws were suitable for the native environment with a higher degree of osseointegration and anti-corrosion properties. Further, the study suggested that no surface modification is required for bone implants fabricated using the EBM process. Figure 10.4 illustrates the fabricated surgical screw using EBM and conventional smooth surgery screw and their SEM morphologies.

10.2.3.4   Laser-Engineered Net Shaping

It is a powder-based technology developed in the 1990s and commercialized by Optomec in 1998. It uses a high-power laser source connected to an integrating part and powder-feeding nozzles. The laser beam creates a melt pool at the selected site and the feeding nozzles feed the metallic power through the feeding nozzle. The metal powder is melted in a melt pool developed over the substrate or any layer surface. The melted powder is quickly solidified over the substrate, or the previous layer and the platform is lowered down by a distance equal to the layer thickness of the component. This process uses metal powders such as titanium and titanium alloys, stainless steel, and a mixture of powders for the fabrication of titanium implants, hip implants, craniofacial implants, and dental applications.

Bala et al. [45] fabricated porous scaffolds of tantalum (Ta) to check their suitability in bone regenerative applications. The as-fabricated Ta scaffolds

(a)   (c)

(b)   (d)

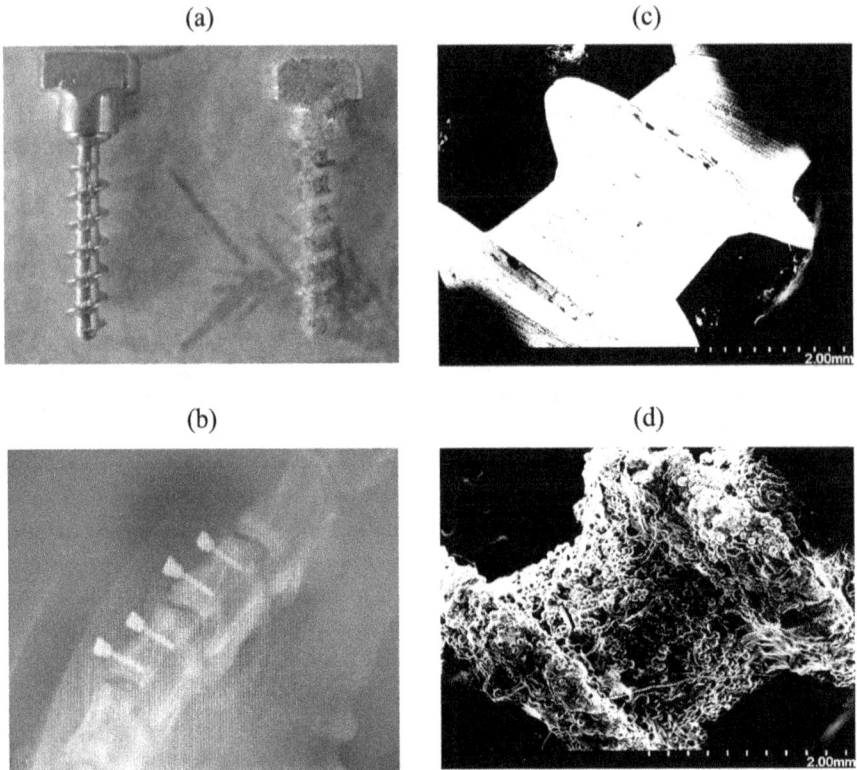

FIGURE 10.4   Illustration of (a) surgical screw pin fabricated using EBM and commercially available screw pin, (b) implant site of surgical screws, SEM images of (c) commercially available screw pin, and (d) EBM printed screw pin [44].

yielded mechanical properties close to the human cortical bone and strong cellular attachment and differentiation. Laser-engineered net shaping (LENS) was used to fabricate components of titanium-silver for biomedical applications in terms of anti-microbial properties and biocompatibility [46]. It was found that the as-printed part showed improved anti-microbial properties, mechanical properties, and biocompatibility suitable for orthopedic implants. LENS was used to fabricate the load-bearing implants of CoCrMo alloy for bone implants [47]. The porosity of the fabricated structure was controlled by the LENS method so that the relative density of the implant could match the native bone strength to eradicate the stress shielding effect. The in-vitro biological assessment showed that laser-processed CoCrMo alloy was non-toxic and could be a potential alloy for the fabrication of load-bearing implants. In another study, LENS was used to fabricate functional porous structures of Ti-6Al-4V alloy for bone implants [5]. The relative densities were varied as the porosity ranges were varied to match with the strength of the human cortical bone. In-vivo investigations depicted that there was a large deposition of calcium over the implant and across the pores, accelerating the tissue regeneration at the implant site.

10.2.4 SOLVENT-BASED EXTRUSION 3D PRINTING

Solvent extrusion-based 3D Printing works on the principle of extruding solution through a nozzle over the build platform. The print head usually consists of a disposable syringe that moves horizontally and vertically as per the G-Code given to the printer. The print heads commonly used are plunger-based (driven by a mechanical plunger), pneumatic-based (driven by compressed air), and screw based (driven by the rotation of the screw). The stepper motors usually control the movement of the print head in X- and Z-direction. One stepper motor controls the build platform in the Y-direction. It is noted that the stored solution is dispensed either by compressed air or liner movement of the plunger or rotation of the screw. Solution-based extrusion printing, nowadays, is associated with the extrusion of bioinks that include both biomaterials and biological cells or living tissue or decellularized tissue material to improve the biocompatibility of the printed components. The process is bioprinting and the term was first introduced in an international conference on bioprinting in 2004 [48]. Various 3D Printing methods are now using biological materials for bioprinting, and among these, most of the bioprinting is done using solution-based extrusion printing. For successful printing, the bioink must have shear thinning behavior and must not contain any toxic solvent. A wide range of biomaterials such as alginate, gelatin, hyaluronic acid, Polyethylene glycol (PEG), polycaprolactone (PCL), polylactic acid (PLA) with and without biological materials can be printed using solution-based extrusion printing. It has a wide range of applications in the field of biomedical such as skin, bone, heart, dental, vasculature, drug delivery devices, and artificial tissue generation.

Dispensing-based 3D Printing was used for the fabrication of scaffolds for bone regenerative applications. Synthetic polymer, polylactic acid (PLA), and tricalcium phosphate (β-TCP) with various hydrogels were used for the preparation of bioink for the extrusion printing [49]. The bioprinter (Hyrel 3D, SDS-5 Extruder, GA, USA) used was a plunger-based type, and the flow rate was controlled by a syringe pump. The printed scaffolds of PLA/β-TCP/chitosan were found to have enhanced cell viability, antibacterial properties, and improved mechanical properties for bone regenerative applications. Dong et al. [50] used 3D Bioprinter (Regenovo, China) to fabricate a novel bioactive scaffold of hydroxyapatite-chitosan-silica-silica-silica hybrid (HA-CSH) for bone tissue engineering. The equipment used was a pneumatic-based bioprinter with a nozzle size of 400 μm. The incorporation of HA in the scaffold material not only improved the mechanical properties of the scaffolds but also improved the cell viability and osteoinductivity. Extrusion 3D Printing was used to fabricate full-thickness artificial skin of silk and gelatin [51]. A pneumatic-based extrusion printer (Fiber Align, Aerotech Inc., Pittsburgh, USA) with a nozzle size of 210 μm was used to print the dermal construct. The multilayered construct consisted of various cells (fibroblast and keratinocyte) proliferated at different layers. In another investigation, a bilayer membranous skin construct of alginate and gelatin was fabricated using an open-source pneumatic 3D bioprinter with heated printheads [52]. The artificial construct yielded excellent biocompatibility (cell viability > 95%) when seeded with human cells and 3D Bioprinting can be a promising method to develop artificial skin for future skin engineering. Another application of dispensing-based 3D printing is the fabrication of cardiovascular stents. Singh et al. [53] developed a novel fabrication technique using the extrusion method where the print head deposited material over a rotating mandrel to fabricate bioresorbable cardiovascular stents. In another investigation, Kandi et al. [54–56] developed a novel approach to fabricate customized tubular scaffolds for tracheal regenerative application. Extrusion printing was used for coating natural polymer (alginate/gelatin) over the synthetic tubular scaffolds to improve the biocompatibility of the printed scaffolds [57]. Apart from these applications, there are several applications of extrusion-based bioprinting in the biomedical field like artificial cartilage printing [58], nerve tissue printing [59], and vascular construction.

10.3   RECENT DEVELOPMENTS IN NOVEL 3D PRINTING OF BIOMATERIALS

There are several novel 3D printing methods adopted to print biomaterials in the fields of tissue engineering, artificial organ development, and drug delivery. In a novel approach, 3D writing inside a granular gel medium was performed [60]. Water-based polymer structures were used for writing inside a granular gel medium. Using this novel technique, materials like silicone, colloidal, hydrogels, and living cells were printed inside a granular gel medium to create complex patient-specific parts, closed shells, and tubular networks generating potential applications in the medical field. In another investigation, microvascular networks were produced using omnidirectional printing (ODP) [61]. The ODP works on the printing of fugitive ink inside a photocurable gel reservoir filled with a Pluronic F127-diacrylate matrix. Highly et al. [62] investigated alternative bioprinting of various components for biomedical applications with ink composed of jammed microgels. Figure 10.5 (B) shows the

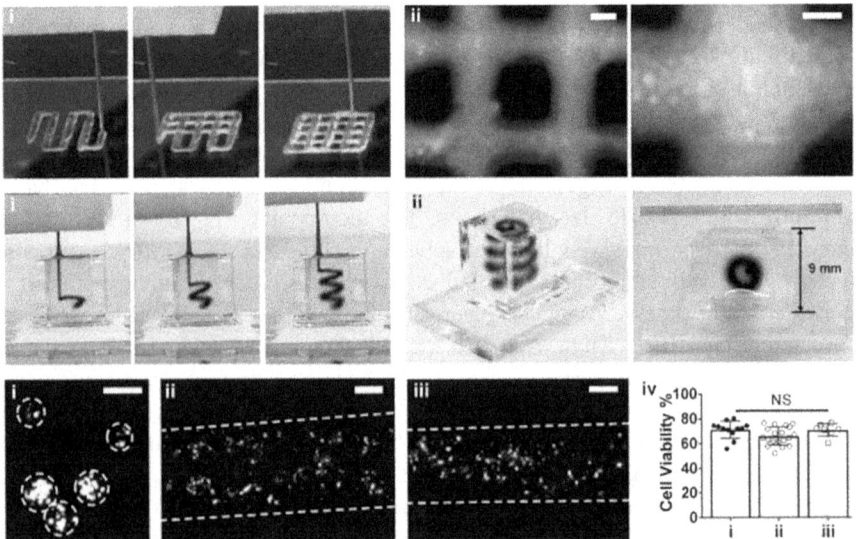

FIGURE 10.5   Illustration of jammed microgel ink printing over a glass slide and inside a gel reservoir [62].

3D Printing of jammed microgels over the surface and inside the support matrix with various characterizations. The jammed microgels showed shear thinning behavior and were printable on the surface and inside the support matrix. The jammed inks yielded excellent cell viability and were made using various materials like thiol-ene cross-linked hyaluronic acid (HA), photo-cross-linked poly (ethylene glycol), and thermo-sensitive agarose. A novel 3D Printing has been developed to fabricate customized components by using Computed Tomography (CT) images, called computed axial lithography (CAL), and the mechanism is like a stereolithography process. The advantage of CAL is the fabrication of support-free parts, flexible and smooth parts without staircase effect. Kelly et al. ([63] demonstrated a novel printing where concurrent printing was performed by curing photosensitive liquid using a dynamically evolving light pattern inside a rotating platform. In recent years, a new dimension, i.e., time, has been included in the 3D Printing technology to change the shape of the 3D-printed components in presence of an external stimuli, called 4D Printing. Pandey et al. [64] demonstrated a thermally activated tubular 3D-printed part of PCL and PLA for tracheal scaffolding application. Tracheal stents of polylactic acid and magnetic iron oxide (Fe_3O_4) were manufactured to check the effect of magnetic field on the stent expansion [65]. The 4D Printed tracheal stent was able to expand within 40 s under magnetic field showing a vast scope of research in the field of 4D Printing of composite shape memory polymers. Apart from these, several research works investigated 4D Printing with different external stimuli like light [66], water [67], and pH [68].

10.4 SUMMARY AND FUTURE PROSPECTIVES

In this chapter, various layered manufacturing methods used in the biomedical field to fabricate organ prototypes, artificial organs, personalized implants, and drug delivery devices using biomaterials are covered. Applications of different biomaterials in various fields such as implant development, tissue engineering, cardiovascular, and regenerative medicine are addressed in detail. Apart from conventional 3D Printing methods, a few recently developed novel methods such as volumetric additive manufacturing, embedded printing, and 4D Printing are highlighted with their working principles. Despite progress in the 3D Printing of biomaterials, there are many challenges present, such as the impossibility of printing large volumes and the limited availability of biocompatibility materials. Very few biomaterials are available with mechanical properties close to human tissues. So, there are several prospects in this area such as the development of novel biomaterials with equivalent properties of human tissue and enhanced biocompatibility, creation, and strengthening of the market to optimize the clinical use of artificial implants. More focus is required in developing multi-material 3D Printing with cost-effectiveness for smaller volume production.

REFERENCES

[1] Hull CW (1984) Apparatus for production of three-dimensional objects by stereolithography. United States Patent, Appl, No 638905, Filed

[2] Ploch CC, Mansi CSSA, Jayamohan J, Kuhl E (2016) Using 3D printing to create personalized brain models for neurosurgical training and preoperative planning. World Neurosurg 90:668–674. 10.1016/j.wneu.2016.02.081

[3] Shie MY, Chang WC, Wei LJ, et al (2017) 3D printing of cytocompatible water-based light-cured polyurethane with hyaluronic acid for cartilage tissue engineering applications. Materials 10:. 10.3390/ma10020136

[4] Liu F, Wang X (2020) Synthetic polymers for organ 3D printing. Polymers 12:1765

[5] Bandyopadhyay A, Espana F, Balla VK, et al (2010) Influence of porosity on mechanical properties and in vivo response of Ti6Al4V implants. Acta Biomater 6:1640–1648

[6] Ahangar P, Cooke ME, Weber MH, Rosenzweig DH (2019) Current biomedical applications of 3D printing and additive manufacturing. Appl Sci 9:1713

[7] Tetsuka H, Shin SR (2020) Materials and technical innovations in 3D printing in biomedical applications. J Mater Chem B 8:2930–2950

[8] Ni J, Ling H, Zhang S, et al (2019) Three-dimensional printing of metals for biomedical applications. Mater Today Bio 3:100024

[9] Mishra DK, Kandi R, Pathak DK, Sharma P (2023) 3D-Printed Metal Oxides for Biomedical Applications. In: 3D Printing. CRC Press, pp 369–382

[10] Ghilan A, Chiriac AP, Nita LE, et al (2020) Trends in 3D printing processes for biomedical field: Opportunities and challenges. J Polym Environ 28: 1345–1367

[11] Frölich AMJ, Spallek J, Brehmer L, et al (2016) 3D printing of intracranial aneurysms using fused deposition modeling offers highly accurate replications. Am J Neuroradiol 37:120–124. 10.3174/ajnr.A4486

[12] Mashiko T, Otani K, Kawano R, et al (2015) Development of three-dimensional hollow elastic model for cerebral aneurysm clipping simulation enabling rapid and low cost prototyping. World Neurosurg 83:351–361

[13] Varga P, Lorinczy D, Toth L, et al (2019) Novel PLA-CaCO3 composites in additive manufacturing of upper limb casts and orthotics—A feasibility study. Mater Res Express 6:45317

[14] Antoniac I, Popescu D, Zapciu A, et al (2019) Magnesium filled polylactic acid (PLA) material for filament based 3D printing. Materials 12:719

[15] Mondal S, Nguyen TP, Hoang G, et al (2020) Hydroxyapatite nano bioceramics optimized 3D printed poly lactic acid scaffold for bone tissue engineering application. Ceram Int 46:3443–3455

[16] Oladapo BI, Zahedi SA, Adeoye AOM (2019) 3D printing of bone scaffolds with hybrid biomaterials. Compos Part B Eng 158:428–436

[17] Han X, Yang D, Yang C, et al (2019) Carbon fiber reinforced PEEK composites based on 3D-printing technology for orthopedic and dental applications. J Clin Med 8:240

[18] Seitz H, Tille C, Irsen S, et al (2004) Rapid prototyping models for surgical planning with hard and soft tissue representation. In: International Congress Series. pp 567–572

[19] Mankovich NJ, Cheeseman AM, Stoker NG (1990) The display of three-dimensional anatomy with stereolithographic models. J Digit Imaging 3:200–203. 10.1007/BF03167610

[20] De Momi E, Pavan E, Motyl B, et al (2005) Hip joint anatomy virtual and stereolithographic reconstruction for preoperative planning of total hip replacement. In: International Congress Series. pp 708–712

[21] Morris CL, Barber RF, Day R (2000) Orofacial prosthesis design and fabrication using stereolithography. Aust Dent J 45:250–253. 10.1111/j.1834-7819.2000. tb00259.x

[22] Wurm G, Tomancok B, Holl K, Trenkler J (2004) Prospective study on cranioplasty with individual carbon fiber reinforced polymere (CFRP) implants produced by means of stereolithography. Surg Neurol 62:510–521. 10.1016/j.surneu.2004.01.025

[23] Baino F, Magnaterra G, Fiume E, et al (2022) Digital light processing stereolithography of hydroxyapatite scaffolds with bone-like architecture, permeability, and mechanical properties. J Am Ceram Soc 105:1648–1657

[24] Sodian R, Loebe M, Hein A, et al (2002) Application of stereolithography for scaffold fabrication for tissue engineered heart valves. ASAIO J 48:12–16. 10.1097/00002480-200201000-00004

[25] Mott EJ, Busso M, Luo X, et al (2016) Digital micromirror device (DMD)-based 3D printing of poly(propylene fumarate) scaffolds. Mater Sci Eng C 61:301–311. 10.1016/j.msec.2015.11.071

[26] Melhem MR, Park J, Knapp L, et al (2017) 3D printed stem-cell-laden, microchanneled hydrogel patch for the enhanced release of cell-secreting factors and treatment of myocardial infarctions. ACS Biomater Sci Eng 3:1980–1987. 10.1021/acsbiomaterials.6b00176

[27] Zhou X, Zhu W, Nowicki M, et al (2016) 3D Bioprinting a cell-laden bone matrix for breast cancer metastasis study. ACS Appl Mater Interfaces 8:30017–30026. 10.1021/acsami.6b10673

[28] Ma X, Qu X, Zhu W, et al (2016) Deterministically patterned biomimetic human iPSC-derived hepatic model via rapid 3D bioprinting. Proc Natl Acad Sci USA 113:2206–2211. 10.1073/pnas.1524510113

[29] Ma X, Yu C, Wang P, et al (2018) Rapid 3D bioprinting of decellularized extracellular matrix with regionally varied mechanical properties and biomimetic microarchitecture. Biomaterials 185:310–321. 10.1016/j.biomaterials.2018.09.026

[30] Silva DN, De Oliveira MG, Meurer E, et al (2008) Dimensional error in selective laser sintering and 3D-printing of models for craniomaxillary anatomy reconstruction. J Cranio-Maxillofacial Surg 36:443–449

[31] Schrank ES, Stanhope SJ (2011) Dimensional accuracy of ankle-foot orthoses constructed by rapid customization and manufacturing framework. J Rehabil Res Dev 48:31–42. 10.1682/JRRD.2009.12.0195

[32] Jardini AL, Larosa MA, Macedo MF, et al (2016) Improvement in Cranioplasty: Advanced prosthesis biomanufacturing. Procedia CIRP 49:203–208. 10.1016/j.procir.2015.11.017

[33] Shuai C, Guo W, Gao C, et al (2017) Calcium silicate improved bioactivity and mechanical properties of poly (3-hydroxybutyrate-co-3-hydroxyvalerate) scaffolds. Polymers 9:175

[34] Gayer C, Ritter J, Bullemer M, et al (2019) Development of a solvent-free polylactide/calcium carbonate composite for selective laser sintering of bone tissue engineering scaffolds. Mater Sci Eng C 101:660–673

[35] Wehmöller M, Warnke PH, Zilian C, Eufinger H (2005) Implant design and production—a new approach by selective laser melting. In: International Congress Series. pp 690–695

[36] Fukuda A, Takemoto M, Saito T, et al (2011) Osteoinduction of porous Ti implants with a channel structure fabricated by selective laser melting. Acta Biomater 7:2327–2336

[37] Xiong Y-Z, Gao R-N, Zhang H, et al (2020) Rationally designed functionally graded porous Ti6Al4V scaffolds with high strength and toughness built via selective laser melting for load-bearing orthopedic applications. J Mech Behav Biomed Mater 104:103673

[38] Xin XZ, Xiang N, Chen J, Wei B (2012) In vitro biocompatibility of Co--Cr alloy fabricated by selective laser melting or traditional casting techniques. Mater Lett 88:101–103

[39] Sun X, Lin H, Zhang C, et al (2022) Improved osseointegration of selective laser melting titanium implants with unique dual micro/nano-scale surface topography. Materials 15:7811

[40] Heinl P, Rottmair A, Körner C, Singer RF (2007) Cellular titanium by selective electron beam melting. Adv Eng Mater 9:360–364

[41] Heinl P, Müller L, Körner C, et al (2008) Cellular Ti--6Al--4V structures with interconnected macro porosity for bone implants fabricated by selective electron beam melting. Acta Biomater 4:1536–1544

[42] Harrysson OLA, Cansizoglu O, Marcellin-Little DJ, et al (2008) Direct metal fabrication of titanium implants with tailored materials and mechanical properties using electron beam melting technology. Mater Sci Eng C 28:366–373

[43] Koike M, Martinez K, Guo L, et al (2011) Evaluation of titanium alloy fabricated using electron beam melting system for dental applications. J Mater Process Technol 211:1400–1408

[44] Yang J, Cai H, Lv J, et al (2014) Biomechanical and histological evaluation of roughened surface titanium screws fabricated by electron beam melting. PLoS One 9:e96179

[45] Balla VK, Bodhak S, Bose S, Bandyopadhyay A (2010) Porous tantalum structures for bone implants: fabrication, mechanical and in vitro biological properties. Acta Biomater 6:3349–3359

[46] Maharubin S, Hu Y, Sooriyaarachchi D, et al (2019) Laser engineered net shaping of antimicrobial and biocompatible titanium-silver alloys. Mater Sci Eng C 105: 110059

[47] España FA, Balla VK, Bose S, Bandyopadhyay A (2010) Design and fabrication of CoCrMo alloy based novel structures for load bearing implants using laser engineered net shaping. Mater Sci Eng C 30:50–57

[48] Mironov V, Reis N, Derby B (2006) Bioprinting: A beginning. Tissue Eng 12: 631–634

[49] Aydogdu MO, Oner ET, Ekren N, et al (2019) Comparative characterization of the hydrogel added PLA/β-TCP scaffolds produced by 3D bioprinting. Bioprinting 13:e00046

[50] Dong Y, Liang J, Cui Y, et al (2018) Fabrication of novel bioactive hydroxyapatite-chitosan-silica hybrid scaffolds: Combined the sol-gel method with 3D plotting technique. Carbohydr Polym 197:183–193

[51] Admane P, Gupta AC, Jois P, et al (2019) Direct 3D bioprinted full-thickness skin constructs recapitulate regulatory signaling pathways and physiology of human skin. Bioprinting 15:e00051

[52] Liu P, Shen H, Zhi Y, et al (2019) 3D bioprinting and in vitro study of bilayered membranous construct with human cells-laden alginate/gelatin composite hydro-gels. Colloids Surf B Biointerfaces 181:1026–1034

[53] Singh J, Pandey PM, Kaur T, Singh N (2021) A comparative analysis of solvent cast 3D printed carbonyl iron powder reinforced polycaprolactone polymeric stents for intravascular applications. J Biomed Mater Res Part B Appl Biomater 109:1344–1359

[54] Kandi R, Pandey PM, Majood M, Mohanty S (2021) Fabrication and characterization of customized tubular scaffolds for tracheal tissue engineering by using solvent based 3D printing on predefined template. Rapid Prototyp J 27:421–428. 10.1108/RPJ-08-2020-0186

[55] Kandi R, Pandey PM (2021) Statistical modelling and optimization of print quality and mechanical properties of customized tubular scaffolds fabricated using solvent-based extrusion 3D printing process. Proc Inst Mech Eng Part H J Eng Med 235:1421–1438. 10.1177%2F09544119211032012

[56] Kandi R, Sachdeva K, Dey S, et al (2022) A facile 3D bio-fabrication of customized tubular scaffolds using solvent-based extrusion printing for tissue-engineered tracheal grafts. J Biomed Mater Res - Part A 1–16. 10.1002/jbm.a.37458

[57] Kandi R, Sachdeva K, Pandey PM, Mohanty S (2023) Fabrication of hybrid tubular scaffolds using direct ink writing for tracheal regenerative application. J Mater Sci 58:4937–4953. 10.1007/s10853-023-08313-w

[58] Chawla D, Kaur T, Joshi A, Singh N (2020) 3D bioprinted alginate-gelatin based scaffolds for soft tissue engineering. Int J Biol Macromol 144:560–567. 10.1016/j.ijbiomac.2019.12.127

[59] Kucukgul C, Ozler SB, Inci I, et al (2015) 3D bioprinting of biomimetic aortic vascular constructs with self-supporting cells. Biotechnol Bioeng 112:811–821

[60] Bhattacharjee T, Zehnder SM, Rowe KG, et al (2015) Writing in the granular gel medium. Sci Adv 1:e1500655

[61] Wu W, DeConinck A, Lewis JA (2011) Omnidirectional printing of 3D microvascular networks. Adv Mater 23:H178--H183

[62] Highley CB, Song KH, Daly AC, Burdick JA (2019) Jammed microgel inks for 3D printing applications. Adv Sci 6:1801076

[63] Kelly BE, Bhattacharya I, Heidari H, et al (2019) Volumetric additive manufacturing via tomographic reconstruction. Science (80-) 363:1075–1079

[64] Pandey H, Mohol SS, Kandi R (2022) 4D printing of tracheal scaffold using shape-memory polymer composite. Mater Lett 329:133238. 10.1016/j.matlet.2022.133238

[65] Zhang F, Wen N, Wang L, et al (2021) Design of 4D printed shape-changing tracheal stent and remote controlling actuation. Int J Smart Nano Mater 12:375–389. 10.1080/19475411.2021.1974972

[66] Yang H, Leow WR, Wang T, et al (2017) 3D printed photoresponsive devices based on shape memory composites. Adv Mater 29:1701627

[67] Sydney Gladman A, Matsumoto EA, Nuzzo RG, et al (2016) Biomimetic 4D printing. Nat Mater 15:413–418

[68] Nadgorny M, Xiao Z, Chen C, Connal LA (2016) Three-dimensional printing of pH-responsive and functional polymers on an affordable desktop printer. ACS Appl Mater Interfaces 8:28946–28954

Index

Note: Locators in *italics* represent figures and **bold** indicate tables in the text.

For Product Safety Concerns and Information please contact our EU
representative GPSR@taylorandfrancis.com
Taylor & Francis Verlag GmbH, Kaufingerstraße 24, 80331 München, Germany

www.ingramcontent.com/pod-product-compliance
Lightning Source LLC
Chambersburg PA
CBHW060354220326
41598CB00023B/2916

9 7 8 1 0 3 2 7 7 6 8 3 5